Stamatios Papadakis and Georgios Lampropoulos (Eds.)
Artificial Intelligence in Higher Education

Also of Interest

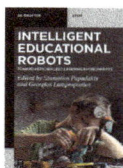

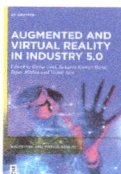

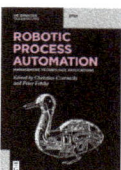

Artificial Intelligence in Higher Education

—

Generative AI, Personalized Learning, Digital Transformation

Edited by
Stamatios Papadakis and Georgios Lampropoulos

DE GRUYTER

Editors

Georgios Lampropoulos
Visiting Lecturer
Department of Education
School of Education
University of Nicosia
Nicosia 2417
Cyprus
and
Postdoctoral Researcher
Department of Preschool Education
School of Education
University of Crete
74100 Rethymno
Greece
and
Postdoctoral Researcher
Department of Applied Informatics
School of Information Sciences
University of Macedonia
54636 Thessaloniki
Greece
lamprop.geo@gmail.com
https://orcid.org/0000-0002-5719-2125

Stamatios Papadakis
Assistant Professor
Department of Preschool Education
School of Education
University of Crete
Rethymnon Campus
Gallos, 74100
Rethymnon, Crete
Greece
stpapadakis@uoc.gr
https://orcid.org/0000-0003-3184-1147

ISBN 978-3-11-914829-0
e-ISBN (PDF) 978-3-11-220639-3
e-ISBN (EPUB) 978-3-11-220640-9

Library of Congress Control Number: 2026930376

Bibliographic information published by the Deutsche Nationalbibliothek
The Deutsche Nationalbibliothek lists this publication in the Deutsche Nationalbibliografie;
detailed bibliographic data are available on the Internet at http://dnb.dnb.de.

© 2026 Walter de Gruyter GmbH, Berlin/Boston, Genthiner Straße 13, 10785 Berlin
Cover image: Khanchit Khirisutchalual/ iStock/Getty Images Plus
Typesetting: Integra Software Services Pvt. Ltd.

www.degruyterbrill.com
Questions about General Product Safety Regulation:
productsafety@degruyterbrill.com

Preface

Artificial intelligence is becoming a core part of the educational domain. Due to its capabilities, it has shown great potential to bring about a digital transformation of teaching and learning in which stakeholders in education have to rethink and reform their practices.

What makes the integration of artificial intelligence in education distinctive is not simply the many new tools that it brings, but the pace and scale of its adoption. For example, generative artificial intelligence has become widespread in a matter of a couple of months. Their use has resulted in changes to the way classrooms, laboratories, libraries, and other aspects of education function. However, several ethical considerations have arisen.

Therefore, the adoption of artificial intelligence in education carries both promises and risks. On the one hand, there lies the potential for more flexible, responsive, and inclusive learning environments. Adaptive systems can respond to different student needs, reduce repetitive tasks for faculty, and support new forms of collaboration. On the other hand, there are challenges, such as uneven access to resources, uncertainties about data use, and the possibility of narrowing intellectual work to what machines can easily process. Higher education institutes have to weigh these issues in real time, often without clear guidance or precedent.

The purpose of this volume is to take look into these rapid changes that the use of artificial intelligence brings to teaching and learning practices in higher education. It gathers contributions from different contexts and disciplines, each examining how artificial intelligence is being used in higher education, what opportunities it offers, and what challenges it creates. Therefore, this volume presents diverse perspectives into the use of artificial intelligence, which, although might differ, agree that higher education must be open to and engage actively with upcoming technological changes. Finally, another common aspect among the chapters is that artificial intelligence will not replace the human factor of teaching and learning but it will reshape the conditions in which these activities take place.

https://doi.org/10.1515/9783112206393-202

Contents

List of Contributing Authors

Georgios Lampropoulos
Department of Education
School of Education
University of Nicosia
Nicosia, Cyprus
and
Department of Applied Informatics
School of Information Sciences
University of Macedonia
Thessaloniki, Greece
and
Department of Preschool Education
School of Education
University of Crete
Crete, Greece

Stamatios Papadakis
Department of Preschool Education
School of Education
University of Crete
Crete, Greece

Daniela Conti
Department of Humanities
University of Catania
Catania, Italy

Giuseppe Romeo
Department of Humanities
University of Catania
Catania, Italy

Santo F. Di Nuovo
Department of Educational Sciences
University of Catania
Catania, Italy

Iván Claudio Suazo Galdames
Universidad Autónoma de Chile
Vicerrectoría de Investigación y Doctorados
Santiago de Chile, Chile

Alain Manuel Chaple-Gil
Universidad Autónoma de Chile
Facultad de Ciencias de la Salud
Santiago de Chile, Chile

Talip Gönülal
Department of Foreign Languages Education
Erzincan Binali Yıldırım University
Erzincan, Turkey

Rama Yusvana
Faculty of Engineering Technology
Tun Hussein Onn University of Malaysia (UTHM)
Pagoh Higher Education Hub
Panchor 84600
Johor, Malaysia

Azli Nawawi
Faculty of Engineering Technology
Tun Hussein Onn University of Malaysia (UTHM)
Pagoh Higher Education Hub
Panchor 84600
Johor, Malaysia

Ivy Shen
Southeast Missouri State University
Cape Girardeau
MO, USA

Luis Manuel Cerdá-Suárez
Faculty of Economics and Business
Universidad Internacional de La Rioja (UNIR)
Logroño
La Rioja, Spain

Rita Cersosimo
Department of Education
University of Genoa
Genoa, Italy

Gonzalo Lorenzo
Department of Developmental Psychology and
Didactics
University of Alicante
Alicante, Spain

Andrea Cerdán-Chacón
Department of Developmental Psychology and
Didactics
University of Alicante
Alicante, Spain

https://doi.org/10.1515/9783112206393-204

Alejandro Lorenzo-Lledó
Didactics and School Organization Department
University of Granada
Granada, Spain

Milan Lazic
Department of Applied Psychology and Human
Development
OISE
Toronto
ON, Canada

Jenny Jun
Department of Applied Psychology and Human
Development
OISE
Toronto
ON, Canada

Earl Woodruff
Department of Applied Psychology and Human
Development
OISE
Toronto
ON, Canada

Serkan Savaş
Department of Computer Engineering
Kırıkkale University
Kırıkkale, Turkey

Zacharias Andreadakis
Department of Pedagogy, Religion and Social
Studies
Western Norway University of Applied Sciences
Bergen, Norway

Georgios Lampropoulos*, Stamatios Papadakis

Transforming Teaching and Learning Through Generative Artificial Intelligence

Abstract: Artificial intelligence (AI) is increasingly being used in educational settings. The rapid advancement and spread of generative AI (GenAI) have led stakeholders in education to examine how to effectively integrate it into their practices. GenAI has shown great potential in transforming teaching and learning practices. This is particularly evident in higher education where GenAI is being more widely used. This chapter examines the use of AI in education and focuses on the adoption of GenAI in teaching and learning activities and introduces the studies contained within this volume. It goes over the specifications and main outcomes of the related studies and provides inputs regarding the applicability of GenAI in education. Finally, the chapter highlights the important role of educators and instructional designers, emphasizes the need to develop education stakeholders' AI literacy and digital competencies, and calls for more longitudinal and experimental studies across diverse educational contexts to be conducted.

Keywords: Artificial intelligence, AI, generative artificial intelligence, GenAI, higher education, teaching, learning, digital transformation, review

1 Introduction

When applied in educational settings, artificial intelligence can bring about changes that influence both educators and students [1, 2]. To better understand the impact that artificial intelligence can have on educational practices, the research regarding its adoption and integration into classrooms is increasing [3, 4].

Moreover, artificial intelligence systems mimic the way humans think, reason, act, learn, and communicate [2, 5], perceive their surroundings, and act in an intelligent and autonomous manner [6, 7]. Hence, they are increasingly being used in education since they can enrich teaching and learning practices [8–10]. When designed and developed in a student-centered manner while having specific learning goals to achieve, artificial intelligent systems can provide personalized learning experiences

*Corresponding author: Georgios Lampropoulos, Department of Education, School of Education, University of Nicosia, Nicosia, Cyprus; Department of Applied Informatics, School of Information Sciences, University of Macedonia, Thessaloniki, Greece; Department of Preschool Education, School of Education, University of Crete, Crete, Greece, e-mail: lamprop.geo@gmail.com
Stamatios Papadakis, Department of Preschool Education, School of Education, University of Crete, Greece, e-mail: stpapadakis@uoc.gr

https://doi.org/10.1515/9783112206393-001

that can positively affect learning outcomes [11, 12]. On the same note, it is becoming increasingly important for both teachers and students to develop their artificial intelligence literacy skills [13].

Artificial intelligence can be used in various ways in education, including virtual tutors, chatbots, and recommender systems, and offer various benefits [14, 15]. Of particular interest is the adoption of generative artificial intelligence (GenAI) in education since it can support both teachers and students [16]. Specifically, studies have highlighted how GenAI can be used by teachers from the initial design of their course and educational material to the execution [17, 18]. When using GenAI, students can engage in collaborative learning experiences which promote personalized learning, tailored feedback, and improved formative assessment [19–21]. Using a variety of prompts, GenAI can create multimedia and human-like content [22]. Hence, when used as an educational means, it has the potential to transform teaching and learning approaches [23, 24]. Despite the benefits it can yield, there are several challenges that need to be addressed, such as technical limitations, data security, and privacy issues, as well as ethical, social, and moral concerns [25–27].

To explore this emerging field of study, this book aims to examine the transformative role of artificial intelligence in higher education, focusing on its impact on teaching, learning, and research. Specifically, looking into the potential of artificial intelligent to enhance student experiences and academic productivity, it offers insights into the challenges and opportunities of integrating artificial intelligence and GenAI into educational settings. The remaining chapters within this edited volume focus on artificial intelligence interventions across different educational settings.

2 Overview of the Chapters

This section provides an overview of the remaining chapters contained within this edited volume. The related chapters have focused on the use of artificial intelligence in educational settings with an increased emphasis on the use of GenAI.

Chapter 2 focuses on exploring the cognitive, pedagogical, and ethical implications of GenAI in higher education. Specifically, it presents a novel model in which GenAI acts as a knowledgeable cognitive partner. The chapter also goes over key challenges in the field, including artificial intelligence literacy, assessment redesign, cognitive engagement, and digital upskilling. Finally, it highlights that artificial intelligence can influence human knowledge and inquiry and emphasizes that a balanced integration that sustains cognitive autonomy and cultivates a culture of reflective, critical, and sustainable learning needs to be promoted.

Chapter 3 looks into how GenAI can be used to enrich personalized learning within higher education. Specifically, the study adopts a systematic review approach following the PRISMA (Preferred Reporting Items for Systematic Reviews and Meta-

Analyses) guidelines and using the Mixed Methods Appraisal Tool to examine 57 related studies through a thematic synthesis and strengths, weaknesses, opportunities, and threats analysis to identify key benefits, challenges, and strategic considerations. Additionally, the study suggests that GenAI can improve academic performance, enhance engagement, and support faculty in designing adaptive and inclusive curricula. However, it also highlights critical concerns that need to be addressed, which involve data privacy risks, algorithmic bias, academic dishonesty, unequal access, and limited institutional preparedness. Finally, it points out that if used responsibly, ethically, and in alignment with pedagogical goals, GenAI can positively affect higher education.

Chapter 4 examines higher education students' viewpoints regarding the ethical dimensions and boundaries of artificial intelligence in education. The related outcomes highlight that students have a clear mindset regarding what is considered as an acceptable level of artificial intelligence assistance and what is regarded as unethical artificial intelligence use and can distinguish between substitutive use and assistive use cases. However, it is pointed out that several factors, including the nature of the tasks, level of personal contributions, transparency, and intentions of using artificial intelligence, can influence their perspectives.

Chapter 5 explores how artificial intelligence tools and gamified pedagogies can be used in higher education to enhance students' knowledge retention and engagement. The chapter goes over emerging artificial intelligence-driven tools and connects empirical case studies with theoretical frameworks, such as self-determination theory. It highlights that using artificial intelligence tools to provide personalized learning and adopting gamification elements can positively affect students' collaboration, self-directed learning, and intrinsic motivation. Finally, it goes over existing challenges and issues, including ethical concerns, content homogenization, over-reliance on automated systems' academic integrity, lack of training, and digital divide.

Chapter 6 focuses on providing authentic learning experiences in higher education through the use of GenAI. It goes over the recent advancements of artificial intelligence and its educational applicability and emphasizes the importance of authentic learning experiences for both students and teachers. Additionally, it highlights the potential of GenAI to aid in providing effective project-based activities and personalized learning as well as in supporting authentic assessment.

Chapter 7 looks into the integration of artificial intelligence tools in Latin America higher education and how it shapes student-teacher interactions. Specifically, their comparative study involves 385 participants across universities in Chile and Spain and analyzes how students' learning experiences can be influenced by providing them with personalized feedback and by their interactions with teachers. It is pointed out that artificial intelligence tools can both aid and complicate the engagement, support, and communication between teachers and students, and as a result, the chapter highlights the need for a balanced integration of artificial intelligence in education and provides strategies to mitigate these challenges.

Chapter 8 encourages rethinking inclusive design in higher education and reforming Universal Design for Learning (UDL) practices through the adoption of GenAI. Specifically, drawing from training experiences of preservice teachers, the study presents a novel model that focuses on UDL principles, artificial intelligence literacy, and the Technological Pedagogical Content Knowledge framework and seeks to foster reflective and accessible course design among university faculty. Finally, it highlights the role of GenAI as a design companion that prompts inclusive thinking and adaptive planning and provides strategies for integrating artificial intelligence tools to scaffold UDL-oriented faculty training.

Chapter 9 examines the role of artificial intelligence in supporting autistic students based on the UDL framework. It points out the ability of artificial intelligence to adapt the tasks and learning material in real time based on students' performance and characteristics. The nonexperimental quantitative study, which involved higher education students, reveals that artificial intelligence could create effective communicative environments for autistic students and that it could offer meaningful feedback in real time. Finally, it comments upon the need to provide teachers with additional training opportunities on how to effectively integrate artificial intelligence into educational settings and to consider students' characteristics (age, prior knowledge of artificial intelligence, etc.) when introducing artificial intelligence in classrooms.

Chapter 10 focuses on increasing students' agency and active involvement in the educational settings through the use of artificial intelligence. Specifically, drawing from cognitive science, philosophy, and education, the study presents a novel framework for designing and developing artificial intelligence tutors that can self-regulate across three key areas, namely survival, emotional awareness, and ethical reflection. It points out that to promote students' deeper understanding, development, and agency, artificial intelligence should act as a conscious collaborator in learning. Finally, it emphasizes the need for artificial intelligence to support meaningful engagement and growth and discusses the related ethical and philosophical imperatives.

Chapter 11 explores the potentials and challenges of utilizing artificial intelligence in educational environments. Specifically, the study goes over the history and advancements of artificial intelligence, presents various artificial intelligence applications, systems, and related technologies, and analyzes their advantages and limitations. It reveals the ability of artificial intelligence to support educational material and activities generation, to offer effective evaluation and assessment, to provide guidance and feedback, as well as to track and monitor students' progress and performance as key benefits. Finally, it highlights ethics as a key challenge in the adoption of artificial intelligence in education.

Finally, Chapter 12 puts emphasis on improving critical feedback and academic productivity through the use of artificial intelligence agents and large language models. It highlights the importance of effective assessment and feedback since it shapes arguments, refines methods, and fosters intellectual growth. The study presents a framework in which artificial intelligence is treated as a supplement to scholarly attention and comments upon how artificial intelligence agents can be used to provide critical feedback.

3 Conclusion

Artificial intelligence and GenAI have showcased great promise in enriching the educational domain and transforming teaching and learning practices. The outcomes of the chapters contained in this volume highlight the benefits that can be yielded when introducing such technologies and tools in classrooms but also point out key challenges that need to be addressed. Common aspects within the chapters are the potentials of artificial intelligence to aid both teachers and students and the need to adopt co-design approaches that involve education stakeholders throughout the design and development process. Collectively, the chapters advocate that the adoption and integration of artificial intelligence into education should be carefully planned, that emphasis should be put on adopting human-centered and inclusive approaches, and that although there might be changes to the role of education stakeholders, human relationships should remain at the core of education. All in all, artificial intelligence has the potential to constitute an effective educational tool that can support and enrich teaching and learning practices across educational levels.

References

[1]   Brynjolfsson E and Mcafee A. Artificial intelligence, for real. *Harvard Business Review*. 2017;1:1–31.

[2]   Ertel W. *Introduction to Artificial Intelligence*. Springer; 2018.

[3]   Song P and Wang X. A bibliometric analysis of worldwide educational artificial intelligence research development in recent twenty years. *Asia Pacific Education Review*. 2020;21:473–486. https://doi.org/10.1007/s12564-020-09640.

[4]   Hinojo-Lucena F-J, Aznar-Díaz I, Cáceres-Reche M-P and Romero-Rodríguez J-M. Artificial intelligence in higher education: A bibliometric study on its impact in the scientific literature. *Education Sciences*. 2019;9:51. https://doi.org/10.3390/educsci9010051.

[5]   Li D and Du Y. *Artificial Intelligence with Uncertainty*. CRC press; 2017. https://doi.org/10.1201/9781315366951.

[6]   Stone P, Brooks R, Brynjolfsson E, Calo R, Etzioni O, Hager G, et al. Artificial intelligence and life in 2030: The one hundred year study on artificial intelligence. 2016. https://doi.org/10.48550/ARXIV.2211.06318.

[7]   Haenlein M and Kaplan A. A brief history of artificial intelligence: On the past, present, and future of artificial intelligence. *California Management Review*. 2019;61:5–14. https://doi.org/10.1177/0008125619864925.

[8]   Chen L, Chen P and Lin Z. Artificial intelligence in education: A review. *IEEE Access*. 2020;8:75264–75278. https://doi.org/10.1109/access.2020.2988510.

[9]   Holmes W, Bialik M and Fadel C. *Artificial Intelligence in Education: Promises and Implications for Teaching and Learning*. Center for Curriculum Redesign; 2020.

[10]   Chiu TKF, Xia Q, Zhou X, Chai CS and Cheng M. Systematic literature review on opportunities, challenges, and future research recommendations of artificial intelligence in education. *Computers and Education: Artificial Intelligence*. 2023;4:100118. https://doi.org/10.1016/j.caeai.2022.100118.

[11]   Topali P, Ortega-Arranz A, Rodríguez-Triana MJ, Er E, Khalil M and Akçapınar G. Designing human-centered learning analytics and artificial intelligence in education solutions: A systematic literature review. *Behaviour & Information Technology*. 2025;44:1071–1098. https://doi.org/10.1080/0144929x.2024.2345295.

[12]   Wang S, Wang F, Zhu Z, Wang J, Tran T and Du Z. Artificial intelligence in education: A systematic literature review. *Expert Systems with Applications*. 2024;252:124167. https://doi.org/10.1016/j.eswa.2024.124167.

[13]   Mustafa MY, Tlili A, Lampropoulos G, Huang R, Jandrić P, Zhao J, et al. A systematic review of literature reviews on artificial intelligence in education (AIED): A roadmap to a future research agenda. *Smart Learning Environments*. 2024;11. https://doi.org/10.1186/s40561-024-00350-5.

[14]   Zhai X, Chu X, Chai CS, Jong MSY, Istenic A, Spector M, et al. A review of artificial intelligence (AI) in education from 2010 to 2020. *Complexity*. 2021:1–18. https://doi.org/10.1155/2021/8812542.

[15]   Lampropoulos G and Papadakis S. The educational value of artificial intelligence and social robots. In: *Social Robots in Education*. Cham: Springer; 2025. pp. 3–15. https://doi.org/10.1007/978-3-031-82915-4_1.

[16]   Bozkurt A, Junhong X, Lambert S, Pazurek A, Crompton H, Koseoglu S, et al. Speculative futures on ChatGPT and generative artificial intelligence (AI): A collective reflection from the educational landscape. *Asian Journal of Distance Education*. 2023;18:53–130.

[17]   Samala AD, Rawas S, Wang T, Reed JM, Kim J, Howard N-J, et al. Unveiling the landscape of generative artificial intelligence in education: A comprehensive taxonomy of applications, challenges, and future prospects. *Education and Information Technologies*. 2025;30:3239–3278. https://doi.org/10.1007/s10639-024-12936-0.

[18]   Alkhasawneh SN, Lampropoulos G and Hernández-Leo D. Generative AI in learning design: A systematic review. *Lecture Notes in Computer Science*. 2025;17–30. https://doi.org/10.1007/978-3-031-98459-4_2.

[19]   Su J and Yang W. Unlocking the power of ChatGPT: A framework for applying generative AI in education. *ECNU Review of Education*. 2023;6:355–366. https://doi.org/10.1177/20965311231168423.

[20]   Mao J, Chen B and Liu JC. Generative artificial intelligence in education and its implications for assessment. *TechTrends*. 2024;68:58–66. https://doi.org/10.1007/s11528-023-00911-4.

[21]   Baîdoo-anu D and Ansah LO. Education in the era of generative artificial intelligence (AI): Understanding the potential benefits of ChatGPT in promoting teaching and learning. *Journal of AI*. 2023;7:52–62. https://doi.org/10.61969/jai.1337500.

[22]   Sharples M. Towards social generative AI for education: Theory, practices and ethics. *Learning: Research and Practice*. 2023;9:159–167. https://doi.org/10.1080/23735082.2023.2261131.

[23]   Giannakos M, Azevedo R, Brusilovsky P, Cukurova M, Dimitriadis Y, Hernandez-Leo D, et al. The promise and challenges of generative AI in education. *Behaviour & Information Technology*. 2024:1–27. https://doi.org/10.1080/0144929x.2024.2394886.

[24]   Peres R, Schreier M, Schweidel D and Sorescu A. On ChatGPT and beyond: How generative artificial intelligence may affect research, teaching, and practice. *International Journal of Research in Marketing*. 2023;40:269–275. https://doi.org/10.1016/j.ijresmar.2023.03.001.

[25]   Mittal U, Sai S, Chamola V and Sangwan D. A comprehensive review on generative AI for education. *IEEE Access*. 2024;12:142733–142759. https://doi.org/10.1109/access.2024.3468368.

[26]   Al-kfairy M, Mustafa D, Kshetri N, Insiew M and Alfandi O. Ethical challenges and solutions of generative AI: An interdisciplinary perspective. *Informatics*. 2024;11:58. https://doi.org/10.3390/informatics11030058.

[27]   Wang N, Wang X and Su Y-S. Critical analysis of the technological affordances, challenges and future directions of generative AI in education: A systematic review. *Asia Pacific Journal of Education*. 2024;44:139–155. https://doi.org/10.1080/02188791.2024.2305156.

Daniela Conti*, Giuseppe Romeo, Santo F. Di Nuovo

Generative Artificial Intelligence in Higher Education: Opportunities, Risks, and the Future of Critical Thinking and Human Reasoning

Abstract: This chapter examines the cognitive, pedagogical, and ethical implications of generative artificial intelligence (GenAI) in higher education. While GenAI offers unprecedented opportunities to personalize learning, support metacognition, and foster reflective inquiry, it also poses significant risks, such as cognitive offloading, reduced critical thinking, and over-reliance on algorithmic outputs. Drawing on John Dewey's philosophy, the chapter proposes an educational model in which artificial intelligence (AI) acts not as a substitute for human thought but as a knowledgeable cognitive partner. Neuroscientific research indicates that passive use of AI may impair memory, reasoning, and executive control, especially among younger users. However, intentionally designed learning environments for promoting critical engagement with AI, metacognitive scaffolding, and delayed AI integration can mitigate these effects. The chapter addresses four key challenges: AI literacy, assessment redesign, cognitive engagement, and digital upskilling. In conclusion, AI is profoundly reshaping human inquiry and knowledge. To prevent algorithmic convenience from leading to intellectual decline, we must promote a balanced integration that sustains cognitive autonomy and cultivates a culture of reflective, critical, and sustainable learning.

Keywords: Generative AI, critical thinking, neuroeducational implications

1 Introduction

Generative artificial intelligence (GenAI) is profoundly transforming higher education by offering tools for personalized learning, automating assessment processes, and enabling the digital transformation of educational practices. However, integrating such technologies requires a critical reflection on their cognitive, linguistic, and pedagogi-

---

***Corresponding author: Daniela Conti**, Department of Humanities, University of Catania, Catania, Italy, e-mail: daniela.conti@unict.it

Giuseppe Romeo, Department of Humanities, University of Catania, Catania, Italy, e-mail: g.romeo.uni@gmail.com

Santo F. Di Nuovo, Department of Educational Sciences, University of Catania, Catania, Italy, e-mail: s.dinuovo@unict.it

https://doi.org/10.1515/9783112206393-002

cal consequences, especially regarding the central role of reasoning in human development.

Our digitalized world presents significant challenges to contemporary education. A review by Dutton et al. [1] indicates that linguistic impoverishment and reduced vocabulary can contribute to the decline of average IQ in developed countries, highlighting the link between language, complex thought, and brain plasticity. Thus, Gen Z's reliance on typed, fast, and fragmented social media communication may impair deep linguistic processing and limit the ability to structure complex reasoning [2]. Additionally, social media creates "filter bubbles" reinforcing preexisting beliefs and limiting exposure to diverse perspectives, which may also hinder critical thinking [3].

Further studies have shown that passive and uncritical use of GenAI can foster reliance on automated responses, reducing the need for students to elaborate, compare, and argue. This decline in active thinking can directly impact neural plasticity, as the lack of cognitive engagement hinders the strengthening of synaptic connections involved in learning, memory, problem-solving, and executive functions [4]. If neglected, this may lead to a weakening of metacognitive skills and a gradual loss of critical and autonomous thinking.

However, a recent survey [5] revealed differing perceptions about artificial intelligence (AI) use between students and faculty, with students appreciating its support for learning, while faculty express concerns about its impact on soft skills and intellectual autonomy. Both groups agree on the urgent need to enhance digital literacy and adopt an ethical, regulated approach to AI in education.

On the other hand, tools such as AI-enabled Intelligent Assistants show that adaptive learning environments can be designed to support students while respecting their cognitive needs. When used consciously and integrated into a balanced educational context, AI can support learning by facilitating access to information and personalizing educational pathways [6].

In a context marked by cognitive erosion and anxiety over AI diffusion, promoting human agency and active participation emerges as a central pillar for sustainable AI development [7]. In this regard, Dewey's educational philosophy offers valuable guidance, emphasizing the learner's active role and the importance of intelligent and reflective engagement in the learning process.

The remainder of this chapter is organized as follows: Section 2 explores the active role of learning, focusing on reflective thought and the purposeful use of technology. Section 3 examines AI as an "external memory" and its impact on cognitive processes, addressing cognitive offloading, the erosion of critical thinking, and the neuroeducational implications of AI use. It also discusses neuroplasticity and cognitive training as potential responses to these challenges. Section 4 outlines contemporary educational challenges and provides practical recommendations, including redesigning assessment metrics to emphasize process over product, scaffolding GenAI as a reflective partner, upskilling learners in self-regulated learning, human-AI teaming, and digital literacy, as well as active learning-based prompt engineering. Finally, Section 5 concludes the chapter by summarizing key insights and suggesting directions for future research and practice.

2 The Active Role of Learning

The rise of GenAI in higher education prompts renewed scrutiny of foundational educational aims, especially the cultivation of critical thinking. Dewey's seminal work *How We Think* [8] offers a compelling philosophical framework for reassessing these aims. Dewey views reflective thinking not merely as a method but as an educational ideal centered on cultivating habits of inquiry, judgment, and problem-solving. Within this tradition, critical thinking is inherently experiential and evidence-based, grounded in the continuity of thought and action, and aimed at fostering responsible and autonomous learners.

Dewey's educational thought can still guide teaching and learning in the twenty-first century, given its ever-changing and exploratory perspective. In a modern society that values inquiry, innovation, and adaptability, Dewey's emphasis on creativity, experiential learning, and reflective thinking offers a powerful framework [9].

Dewey's entire activist pedagogy is grounded in the philosophical premise that experience is the result of an interaction between the organism and its natural and social environment, a transactional relationship marked by mutual adaptation. Within this process, intelligence may function passively, adjusting to existing conditions, or actively, by modifying the environment to fulfill emerging needs and aims [10]. Activist pedagogy advances what Dewey calls the *method of intelligence* [10, p. 180], which involves engaging learners in the resolution of problematic situations arising from their lived experiences, an approach he also refers to as the "*experimental* attitude" [10, p. 318, p. 400].

Within this dynamic field of lived experience, Dewey's notion of technology reaffirms the experimental and active role individuals play in shaping and being shaped by their environment. In Dewey's view, technology is the invention, development, and reflective use of tools, techniques, and artifacts to address perceived problems. This broad conception highlights technology as a form of intelligent action, the application of natural and human energies in ways that serve human purposes. Crucially, this understanding underscores the inseparable link between intelligence, problem-solving, and technological activity. For Dewey, intelligence is intrinsic to technology. It is a form of inquiry rather than a mechanical routine. However, when a certain technology stops being guided by reflection and no longer responds to new problems, it degenerates into mere *habit*, losing its genuinely intelligent character [11]. Instead, according to Dewey, the central aim of education is the cultivation of the habit of reflective thinking. In this case, "habit" is conceived as a stable, permanent attitude. Reflective thinking, for Dewey, stands in contrast to impulsive or routine action; it makes action intentional, deliberate, and guided by a conscious aim [8]. Rather than relying on one's own habitual responses, traditional customs, or inherited beliefs, reflective thought is rooted in *personal inquiry*. It is proactive, inventive, and continuously evolving. Reflective thinking consists of this foundational activity of inquiry, sustained by the personal traits and moral commitment that nourish it [8]. Thus, it is not a single faculty but a coordinated set of mental processes and personal dispositions. It originates in the natural curiosity of the human mind and is typically triggered by a state of doubt or uncertainty. Its function, as Dewey describes, is to move from a doubtful to a settled situation. This reflective process unfolds in distinct stages: it begins with the recognition of a problem through observation, which evokes imaginative suggestions or hypotheses. These are then subjected to logical reasoning and experimental testing, ultimately leading to a conclusion that is warranted. In essence, it mirrors the scientific method [8].

Dewey asserts that education should aim to cultivate students' capacities for critical reflection, problem-solving, and the practical application of their knowledge. Therefore, learning is fundamentally learning "how to learn" [8, p. 78]. Good learning experiences foster growth. As we grow, our range of skills expands, and we gain confidence in facing new challenges. The aim of school education, according to Dewey, is to ensure the continuity of education, that is, to structure experiences in ways that sustain the conditions for ongoing growth and make future learning possible [10]. In view of this, critical thinking enables such knowledge *transfer* [8, 12]. Specifically, the cultivation of reflective thinking serves as a safeguard against the formation of false or harmful beliefs, insofar as it requires that beliefs be grounded in evidence. Accordingly, the pedagogical task is to refine and guide the mind's native abilities, especially *curiosity*, so they progress from casual inference to attitudes of alert, cautious, and thorough inquiry. To achieve this, educators need to personalize learning to effectively guide such native dispositions [8].

Dewey also emphasizes that the educational value does not lie solely in mastering static bodies of knowledge, the ready-made classifications found in textbooks, but in acquiring the methods by which knowledge is discovered and constructed. Drawing an analogy with geographical explorations, Dewey argues that education should orient students toward the act of exploring the mental *process* itself rather than merely handing them the map, that is, the finished *product* [8, p. 73]. Such a reflective habit not only enables the continuous growth of knowledge but also draws upon prior experience and linguistic competence; hence, Dewey underscores the importance of language proficiency as essential for expressing, sharing, and refining thought [8].

2.1 Reflective Thought and Purposeful Use of Technology

In line with Dewey's emphasis on curiosity, reflective thought, and autonomous learning, modern research shows that the four levels of reflection align with stages of cognitive learning [13]. When learners are sufficiently motivated, exposure to new information activates stimulated and descriptive reflection, which corresponds to the initial stages of Bloom's taxonomy, that is, remembering and understanding, as learners recall and interpret prior knowledge. Dialogic reflection follows, supporting the synthesis of new information into the learner's personal knowledge base, aligning with reconceptualization and knowledge construction. Finally, critical reflection involves the evaluation and application of knowledge, corresponding to the higher-order stage of learning cognitive processes [13].

Within this framework, technology-enhanced learning environments can motivate students and support reflection by providing access to diverse information sources that build essential background knowledge. Simultaneously, challenging tasks that foster higher-order thinking, reflection, and conceptual change are still needed to ensure profound learning [13]. Indeed, ChatGPT demonstrates a strong positive effect on student's learning performance (effect size $g = 0.867$) and a moderately positive impact on enhancing learning perception and fostering higher-order thinking [14]. However, while these tools can improve short-term task performance (e.g., essay scores), they may not enhance intrinsic motivation, knowledge gain, or knowledge transfer, potentially leading to stagnation in long-term skill development [15].

These are precisely the educational conditions Dewey emphasized as vital for meaningful learning. Without thoughtful pedagogical guidance, learners risk developing uncritical and passive habits in their use of GenAI [16], echoing Dewey's early warnings about the unreflective use of technology [11]. As Dewey wrote, automatic practices lose their educative effect when they no longer lead to new perceptions and connections; they restrict rather than expand the horizon of meaning [10, p. 102]. This tendency to prioritize convenience over cognitive engagement has long been reinforced by digital technologies, fostering "copy and paste" behaviors from early search engines and Wikipedia to today's AI-based platforms. Such habits, while quick and efficient, pose a serious challenge to learner autonomy and adaptability skills, which are increasingly being demanded by the contemporary job market. As we will explore in the following section, these behavioral patterns also have demonstrable effects on cognitive processing and neural development.

3 Artificial Intelligence as "External Memory" and the Transformation of Cognitive Processes

AI is rapidly reshaping every facet of human life from work and communication to education and deeper cognitive processes, such as thought and memory. Since the time of Socrates, the value of self-knowledge has occupied a central place in philosophical discourse. However, as Plato observed, this self-awareness often proves vague, distorted, and unreliable. Psychological research supports this insight, showing that self-knowledge can be both confabulated and disconnected from reality [17–19].

This premise underscores humanity's enduring fascination with understanding ourselves and our cognitive capacities. An intriguing study indicates that, even among healthy individuals, metacognitive judgments tend to correlate, although imperfectly, with actual cognitive performance [20]. Neuroscientific studies explain this discrepancy by identifying distinct brain regions responsible for metacognitive monitoring, such as the anterior prefrontal cortex, which function independently from those involved in task execution [21]. In this light, technology becomes more than a tool. Digital environments that prompt reflection, self-assessment, and feedback can improve both self-regulation and metacognitive awareness. A recent study found that students using a metacognitive support system in online learning showed marked gains in critical thinking and communication, thanks to enhanced monitoring of their own cognitive processes [22]. Thus, AI has the potential not only to enhance learning but also to deepen our understanding of ourselves as thinking agents. With its capacity to process vast datasets, generate personalized content, and support complex decision-making, AI opens new frontiers in education and cognition.

However, this rapid proliferation of technology raises fundamental questions about its impact on the human mind: what are the cognitive and neurofunctional consequences of uncritical or excessive AI use?

A key phenomenon in this discussion is the "Google effect" [23], which describes how individuals tend to forget factual details while remembering where to find the information digitally. This behavior reflects a profound transformation in human memory from internal storage to reliance on external digital supports.

Today's AI is far more sophisticated than simple search engines and functions as a form of *transactive memory* [24]. It acts not only as a repository of information but also as an active partner in cognition, capable of generating content, proposing hypotheses, offering solutions, and guiding decisions.

From a neuroscientific perspective, neuroimaging studies have found that frequent and intensive use of digital tools and AI can induce structural brain changes. These include variations in gray matter density, particularly in frontal and temporal regions, and altered functional connectivity within neural networks responsible for attention, memory, and executive control [25, 26].

Moreover, instant and unlimited access to information may reduce the need for deep cognitive processing, such as critical evaluation, source assessment, and personal integration of knowledge. This can encourage a form of mental laziness, where reflective thought is replaced by passive trust in AI-generated answers.

3.1 Cognitive Offloading and the Erosion of Critical Thinking

The concept of cognitive offloading [27] refers to the strategy by which individuals deliberately delegate complex cognitive functions, such as calculation, memorization, or planning, to external tools in order to reduce mental effort and increase efficiency. While this practice can free up cognitive resources to tackle more complex tasks, excessive and passive use of AI, especially GenAI applications, risks impairing the exercise of executive functions essential for critical thinking, independent reasoning, and cognitive self-regulation.

Evidence from our study and recent literature [28, 29] shows that intensive use of AI tools is linked to reduced reasoning abilities and a diminished tendency to question received information. This phenomenon is particularly pronounced among younger individuals, who tend to rely more heavily on AI and exhibit lower critical evaluation skills compared to participants with higher education.

Sparrow et al. [23] highlighted how reliance on search engines reduces episodic memory, which refers to the ability to recall contextual and personal details of experiences, leading to superficial information processing with poor integration and reflection, thus compromising the depth of critical thinking. Similarly, automated writing tools, such as Grammarly or E-rater, while improving grammatical correctness, tend to standardize texts and stifle argumentative creativity, threatening the development of autonomous analytical skills [30].

Qualitative data from interviews confirm these effects: many participants with lower educational attainment recognize a passive dependence on AI, often without critically analyzing AI-generated content. One participant with a high school diploma admitted, "I use AI because it simplifies everything, but sometimes I feel like I'm losing my own problem-solving skills" (P221). Another said, "I don't really think critically when using AI; I just follow what it suggests" (P607, some college). Conversely, participants with higher education levels demonstrated more skeptical and critical attitudes, as illustrated by a doctoral student: "While I use AI tools regularly, I always make sure to critically evaluate the information I receive. My education has taught me the importance of not accepting things at face value, especially with AI, which can sometimes offer biased or incomplete information" (P601).

This gap highlights the crucial role education plays in mitigating the negative effects of cognitive offloading on critical thinking skills. The growing trust placed in AI technologies, driven by their perceived reliability and convenience, fosters a dependence that reduces active cognitive engagement. As Gerlich [28] showed, this phenomenon of delegating cognitive tasks to external tools results in diminished participation in deep cognitive processes, leading to an erosion of critical thinking. The reliance on content generated by virtual influencers, which amplifies trust in AI-provided information, further reinforces this dynamic of reduced intellectual autonomy.

In summary, while cognitive offloading can be a valuable support in managing mental load, excessive and passive dependence on AI poses a real risk to critical thinking abilities. The educational challenge is therefore to promote a conscious and balanced use of technologies that encourage critical reflection, autonomous verification of information, and sustained activation of higher cognitive functions.

3.2 Neuroeducational Implications of AI Use

In educational settings, the introduction of AI-powered adaptive learning platforms has improved personalization and teaching effectiveness by tailoring content to individual learner profiles and pacing. However, these benefits are jeopardized if AI use supplants students' intellectual autonomy and personal reflection. Halpern [12] emphasizes that critical thinking is essential not only for meaningful learning, but also for professional success and active societal engagement.

Studies by Freeman et al. [31] and Deslauriers et al. [32] consistently show that active learning, which is characterized by participation, problem-solving, debate, and self-regulation, produces significantly better outcomes than passive or purely technological methods.

The 2025 study by Gerlich [28] contributes significantly to our understanding of AI's impact on critical thinking. It reveals that frequent AI use, when unaccompanied by active cognitive engagement, can weaken users' critical faculties. This effect is mediated by cognitive offloading by delegating mental tasks to machines and reducing the need for self-reflection. The effect is especially pronounced in younger users, who tend to rely heavily on AI and perform worse in critical-thinking assessments compared to older users. Conversely, higher levels of education correlate with stronger critical reasoning, suggesting that education can buffer against cognitive automation.

Importantly, this does not imply that AI should be avoided. Gerlich [28] highlights that AI can enhance learning, if used actively and thoughtfully. The study shows that reducing cognitive offloading through targeted educational interventions, such as critical-thinking exercises and reflective learning environments, can attenuate AI's negative impact on cognitive function. Thus, technological integration in education must be complemented by efforts that cultivate reflective, critical, and conscious interaction with digital tools.

These conclusions align with growing interest in media literacy, which can be defined as the ability to critically assess digital content, including AI-generated material. Fostering an active rather than passive orientation toward technology not only protects cognitive abilities but also cultivates individual responsibility and autonomy in knowledge construction.

3.3 Neuroplasticity and Cognitive Training: A Potential Response

Neuroplasticity, which refers to the brain's capacity to remodel its structure and function in response to experience and learning, underpins the potential for cognitive recovery and improvement well into adulthood. Cognitive training interventions, which focus on specific functions or strategies [33], have been shown to enhance neural activation and connectivity within networks responsible for memory, attention, and executive control [26, 34].

The concept of "neural scaffolding" [35] describes how the brain can recruit compensatory circuits to maintain or improve performance in the face of age-related decline or functional deficits. Social and cognitive programs like Synapse or Experience Corps demonstrate that stimulating mental and social activities can significantly boost executive processes and prefrontal function which relate to critical regions for critical thought and self-regulation. These findings, once again, suggest that the antidote to the risks of passive AI use is not technological rejection, but the creation of learning environments that encourage sustained cognitive engagement by integrating technology with experiences that meaningfully exercise the mind.

Further neuroscientific evidence supports this. In the first brain scan study of AI users, 83.3% of students who used ChatGPT could not recall a single sentence from essays they had written just minutes earlier. In contrast, students who wrote without AI showed strong recall. Even when ChatGPT users were later asked to write without AI, their brains remained under-engaged, suggesting residual cognitive atrophy. The most successful group began without AI and used it later: they showed stronger memory, better neural connectivity, and the highest overall performance [36].

These findings suggest that how and when AI is used in learning deeply affects cognitive outcomes. AI should be introduced strategically, as a complement and not as a substitute to human effort and understanding.

New technologies, when applied thoughtfully, can increase engagement and curiosity in learners accustomed to digital environments. However, their integration requires an ethical and pedagogically grounded approach that supports, rather than replaces, the core functions of both teachers and students. This could allow a "collaborative intelligence" shared among humans and AI [37]. Doing so demands a clear identification of the most pressing educational challenges today, which we will outline in the next section.

4 Contemporary Educational Challenges

New technologies hold the promise of enriching education, particularly through the integration of GenAI into curriculum development. Tools like ChatGPT can support educators in creating teaching materials, including rubrics, lesson plans, interactive exercises, reflective questions, and personalized case studies [38].

However, their responsible, inclusive, and pedagogically sound implementation is essential to truly support learning and creativity in the digital era. As AI-generated content and immersive environments (e.g., extended reality) become increasingly embedded in education, educators face the dual challenge of adapting to diverse learner needs while upholding creativity, ethical standards, and integrity. GenAI enables personalized content creation and also introduces significant risks, including misinformation, plagiarism, and intellectual property concerns. Moreover, today's learners (Gen Z or neomillennials) are often referred to as digital natives; yet many still lack critical digital literacy.

This gap can lead to the uncritical use of unreliable sources, increasing the risk of academic misconduct and the spread of misinformation [39]. If not critically integrated, GenAI may further hinder original thinking and promote passive learning, as users may not question the validity of AI-generated content and seek quick solutions. This over-reliance risks undermining the development of independent judgment and thoughtful inquiry. Additionally, questions around intellectual property challenge traditional notions of originality and assessment in education [16].

As such, issues related to information literacy and a tendency toward "copy and paste" practices need to be addressed [39]. We propose Dewey's prerequisites for reflective thinking as a useful framework to: (I) identify passive uses of AI based on unmet conditions for reflective thought in human-AI interactions, and (II) design interventions to promote critical thinking [40]. The following section outlines practical applications of this framework.

4.1 Recommendations

4.1.1 Redesign Assessment Metrics: Process over Product

In response to these challenges, Dewey's emphasis on learning as process over product can guide a revision of traditional assessment models. Product-oriented assessments often encourage students to rely on GenAI just to complete assignments [41]. To counter this, assessment must prioritize higher-order thinking and authentic evaluations [38]. Project-based assessments, real-world problem-solving tasks, capstone experiences, and case studies promote original thought and knowledge application. These approaches require students to work overtime and involve continuous instructor feedback, reducing the temptation to misuse GenAI [41]. In view of this, ChatGPT has proven especially effective in problem-based learning environments, where it can assist with scenario generation, feedback, and scaffolding [14].

To foster originality and personal inquiry, additional assessment metrics may evaluate coherence of student work, cross-referencing with reliable sources, and integration of personal insights to refine outputs. These practices reflect Bloom's taxonomy stages of analysis and synthesis [16]. Assessing originality may also involve requiring students to gather and interpret evidence or develop empirical findings to support their arguments [42].

To further promote higher-order thinking, students should critically evaluate AI-generated content as well. Thus, assessment metrics could include critiquing and revising AI-generated essays to analytical and creative skills [16]. In addition, AI literacy is essential for critical thinking development. A lack of understanding often leads students to accept AI-generated outputs as accurate and unbiased, which can result in misinformation and poor academic judgment [43].

4.1.2 Scaffolding: GenAI as a Reflective Partner

To support higher-order thinking with ChatGPT, scaffolding is essential. Embedding ChatGPT into educational frameworks, such as Bloom's taxonomy, would help address its lack of creativity and critical reasoning, making it a more effective tool for advanced learning [14].

In this regard, Dewey's theory of reflective thinking also offers valuable guidance. According to Dewey, the starting condition for reflection is curiosity, followed by a sense of doubt or perplexity. As such, these prerequisites need to be encouraged [8]. However, when AI delivers instant, polished, and overly confident answers, learners may miss this crucial moment of cognitive dissonance that sparks reflection and inquiry [40]. To support this, students could be encouraged to highlight parts of GenAI responses that confuse or interest them. They might also provide feedback to the AI system by formulating prompts to generate responses better suited to their thinking process [40].

Another effective strategy would be to introduce "friction" in the learning process, small steps that prompt students to pause and reflect before using AI outputs [40]. Romeo and Conti's [44] review highlights that increased active engagement with AI responses is key to preventing over-reliance and the risk of progressive deskilling. In strategies such as human-first decision protocols, users make their own judgments before consulting AI, and frictional interface design can function as debiasing tools.

In educational contexts, these strategies can be translated into assignments that require initial independent work, followed by optional AI feedback or comparison with AI-generated responses. For instance, students could submit their answers before accessing AI output or be presented with multiple, and sometimes conflicting, AI-generated viewpoints that they must critically evaluate.

Thus, making AI support accessible only after reaching specific milestones by creating "AI-free zones" or implementing guardrails may encourage students to engage with problems through their own reasoning first [40]. This is especially important for digital natives, who may be more inclined to default to automated solutions. Therefore, ChatGPT and similar tools should be integrated flexibly into teaching, not as a shortcut, but as an intelligent tutor, collaborative partner, and reflective aid in the learning process [14].

Crucially, before engaging with GenAI, students also need to develop a solid foundation of disciplinary knowledge, coupled with AI literacy, to assess the accuracy and relevance of AI-generated content [42]. Domain expertise and AI literacy serve as a protective factor against inaccuracies in AI outputs [44].

4.1.3 Upskilling: Self-Regulated Learning, Human-AI Teaming, and Digital Literacy

To cope with deskilling concerns, timely upskilling is necessary. Capabilities such as creativity, autonomous, and *self-regulated learning* (SRL) are increasingly vital as AI reshapes traditional learning processes [45].

At the start of learning new information, cognitive effort and persistence are essential [40]. According to Dewey, teachers should not shield students from these initial moments of struggle but should frame them as opportunities to test and develop their intelligence [8]. Educational goals, as outlined in Bloom's taxonomy, involve acquiring different types of knowledge, engaging in cognitive processes, and employing metacognitive strategies to regulate one's learning [40]. Yet when learners offload these tasks to *large language models* (LLMs), they risk missing chances to develop foundational cognitive skills, especially remembering, applying, analyzing, and evaluating. Furthermore, some general critical issues regarding the reliance on LLM agents have been highlighted by recent studies: at the semantic level, LLMs can infer from human language even partial, misleading, or inaccurate modalities, "toxic" in some cases, and these would in turn become objects of learning [45]. Some researchers have recently described LLMs (e.g., ChatGPT) as "weapons of mass disinformation" [46].

Over-reliance on AI-generated responses may bypass such essential cognitive and cultural challenges and diminish metacognitive engagement. Delegating learning regulation to adaptive technologies can compromise students' ability to monitor and guide their own thinking [15]. In line with Dewey's emphasis on autonomous learning as a condition for knowledge transfer [8], research confirms that learners achieve better outcomes and transfer knowledge more effectively when they engage in SRL. Thus, cultivating SRL skills and sustaining metacognitive activity are essential to avoiding what Fan et al. [15] call "metacognitive laziness." Scaffolding is needed to help learners ethically divide tasks with AI and actively strengthen their cognitive capacities [15].

To support this, learners should still engage actively with AI during routine knowledge work. Developing *coordination* skills (e.g., by outsourcing only summarizing and information searching tasks) is key to effective human-AI collaboration [47].

Another essential component is *AI literacy*, which equips students with the ability to assess the quality, relevance, and ethical implications of AI-generated content. This is not only necessary for navigating an AI-driven world but also empowers students to question and influence AI's societal role [16]. Understanding how AI systems work and how they shape human reasoning and values is central to using them responsibly. Encouraging humanistic thinking, including ethical, philosophical, and historical perspectives, is essential to ensure AI use aligns with human needs and values [47].

Digital competence, the ability to effectively use digital technologies to accomplish tasks, is also critical. Both educators and students must learn to use GenAI tools not just to consume content, but to create it. This includes the ability to craft clear, purposeful prompts to guide GenAI in producing relevant and meaningful outputs [41]. Training in prompting helps both students and educators shift from passive recipients to active co-creators of their learning experiences [41]. This process demands clarity and precision in thinking, as learners must carefully formulate prompts and evaluate AI responses. By reflecting on the quality of their prompts and the outputs they receive, students enhance their metacognitive awareness, learning not just to solve problems, but how they solve them [48].

Developing skills in prompt engineering strengthens students' ability to transform GenAI from a static information source into an interactive partner that fosters deeper understanding. To support this, prompt engineering and hands-on practice with AI tools should be embedded in active learning curricula [49]. Thus, incorporating both AI competencies and AI literacy into education requires tailored curricula and deliberate instructional strategies [38]. Without this, there is a real risk that AI integration will reinforce convenience and automation over critical inquiry and intellectual independence.

5 Conclusions

AI is profoundly transforming the ways in which humans inquire, acquire, and develop knowledge. The potentials offered by these technologies, which manifest in unprecedented efficiency, a high degree of personalization, and immediate, virtually limitless access to information, represent an epochal shift in educational, professional, and cognitive processes. However, these benefits must not obscure the cognitive and pedagogical risks inherent in the passive and uncritical use of AI. Phenomena, such as cognitive offloading, the progressive reduction of episodic memory, declines in critical thinking abilities, and the standardization of responses constitute real and well-documented threats that demand an integrated intervention, combining neuroscientific insights with innovative educational methodologies.

In this context, promoting a culture of critical awareness and active cognitive engagement is essential, particularly when utilizing instruments based on LLMs, such as ChatGPT. The three qualities that Dewey considers fundamental to critical thinking, that is openness, heartedness, and responsibility [8], must be placed at the core of learning and knowledge construction processes, so that interaction with digital technologies does not devolve into passive delegation but instead becomes an opportunity for reflection, critical evaluation, and personal inquiry. Although the adage "use it or lose it" suggests that sustained mental activity helps maintain and preserve cognitive functions, this premise, while intuitively plausible, requires support and integration within more rigorous theoretical and empirical frameworks that consider the complexity of neuroplastic and cognitive dynamics involved.

In this regard, a close collaboration among educators, neuroscientific researchers, and policymakers is necessary to design learning environments and instructional strategies capable of fully harnessing the potential of AI, without compromising the irreplaceable role of the human mind. It is through such a balanced and conscious integration that we can prevent algorithmic convenience from evolving into intellectual decline, thereby ensuring sustainable, harmonious cognitive development oriented toward fostering critical autonomy in the digital era.

In conclusion, the educational challenge of our time lies in promoting the use of AI not merely as a means to facilitate information access, but as a tool that encourages the cultivation of responsible digital citizenship which refers to the application of these technologies ethically, reflectively, and productively. Only in this way can technological innovation become a lever for cognitive empowerment, preserving and strengthening the intellectual, creative, and critical capacities that constitute the core of human identity.

References

[1] Dutton E, van der Linden D and Lynn R. The negative Flynn effect: A systematic literature review. *Intelligence*. 2016;59:163–169.

[2] Ezeudo CO. Exploring the impact of social media on language use and literacy skills. *Crowther Journal of Arts And Humanities*. 2024 Dec 11;1(6). pp. 86–99.

[3] Barberá P. *Social Media, Echo Chambers, and Political Polarization. Social Media and Democracy: The State of the Field, Prospects for Reform*. 2020 Aug 31. pp. 34–55.

[4] Green CS and Bavelier D. Exercising your brain: A review of human brain plasticity and training-induced learning. *Psychology and Aging*. 2008;23(4):692–701. doi: 10.1037/a0014345.

[5] Haroud S and Saqri N. Generative AI in higher education: Teachers' and students' perspectives on support, replacement, and digital literacy. *Education Sciences*. 2025;15(4):396. doi: 10.3390/educsci15040396.

[6] Sajja R, Sermet Y, Cikmaz M, Cwiertny D and Demir I. Artificial intelligence-enabled intelligent assistant for personalized and adaptive learning in higher education. *Information*. 2024;15(10):596. doi: 10.3390/info15100596.

[7] Romeo G and Conti D. Beyond automation: Reshaping human–artificial intelligence interaction. *Sistemi Intell*. 2024;36(3):641–648. doi: 10.1422/115336.

[8] Dewey J. *How We Think: A Restatement of the Relation of Reflective Thinking to the Educative Process*. Boston: D.C. Heath and Company; 1933.

[9] Chen Y. The relevance of Dewey's educational theory to 'teaching and learning in the 21st century'. *Studies in Social Science & Humanities*. 2023 Mar 31;2(4):65–68.

[10] Dewey J. *Democracy and Education: An Introduction to the Philosophy of Education*. New York: Macmillan; 1930.

[11] Hickman LA. *John Dewey's Pragmatic Technology*. Indiana University Press; 1990.

[12] Halpern DF. Teaching critical thinking for transfer across domains. *American Psychologist*. 1998;53 (4):449–455. doi: 10.1037/0003-066X.53.4.449.

[13] Strampel K and Oliver R. Using technology to foster reflection in higher education. In: *ICT: Providing Choices for Learners and Learning. Proceedings Ascilite Singapore*. 2007 Dec.

[14] Wang J and Fan W. The effect of ChatGPT on students' learning performance, learning perception, and higher-order thinking: Insights from a meta-analysis. *Humanities and Social Sciences Communications*. 2025 May 6;12(1):1–21. doi: 10.1057/s41599-025-01326-0.

[15] Fan Y, Tang L, Le H, Shen K, Tan S, Zhao Y, Shen Y, Li X and Gašević D. Beware of metacognitive laziness: Effects of generative artificial intelligence on learning motivation, processes, and performance. *British Journal of Educational Technology*. 2025 Mar;56(2):489–530. doi: 10.1111/ bjet.13544.

[16] Nguyen KV. The use of generative AI tools in higher education: Ethical and pedagogical principles. *Journal of Academic Ethics*. 2025;1–21. doi: 10.1007/s10805-025-09434-9.

[17] Carruthers P. *The Opacity of Mind: An Integrative Theory of Self-knowledge*. New York (NY): Oxford University Press; 2011.

[18] Nisbett RE and Wilson TD. Telling more than we can know: Verbal reports on mental processes. *Psychological Review*. 1977;84(3):231–259. doi: 10.1037/0033-295X.84.3.231.

[19] Wilson TD and Dunn EW. Self-knowledge: Its limits, value, and potential for improvement. *Annual Review of Psychology*. 2004;55:493–518. doi: 10.1146/annurev.psych.55.090902.141954.

[20] Schwartz B and Metcalfe J Methodological problems and pitfalls in the study of human metacognition. In: Metcalfe J and Shimamura A, editors *Metacognition: Knowing about Knowing*. Cambridge (MA): MIT Press; 1996. pp. 227–251.

[21] Vaccaro AG and Fleming SM. Thinking about thinking: A meta-analysis of neuroimaging studies. *Brain and Neuroscience Advances*. 2018;2:2398212818810591. doi: 10.1177/2398212818810591.

[22] Pereles A, Ortega-Ruipérez B and Lázaro M. The power of metacognitive strategies to enhance critical thinking in online learning. *Journal of Technology and Science Education*. 2024;14(3):831. doi: 10.3926/jotse.2721.

[23] Sparrow B, Liu J and Wegner DM. Google effects on memory: Cognitive consequences of having information at our fingertips. *Science*. 2011;333(6043):776–778. doi: 10.1126/science.1207745.

[24] Wegner DM. Transactive memory: A contemporary analysis of the group mind. In: Mullen B and Goethals GR (eds.). *Theories of Group Behavior*. Springer; 1987. pp. 185–208.

[25] Olesen PJ, Westerberg H and Klingberg T. Increased prefrontal and parietal activity after training of working memory. *Nature Neuroscience*. 2004;7(1):75–79. doi: 10.1038/nn1165.

[26] Astle DE, Barnes JJ, Baker K, Colclough GL and Woolrich MW. Cognitive training enhances intrinsic brain connectivity in childhood. *Journal of Neuroscience*. 2015;35(16):6277–6285. doi: 10.1523/ JNEUROSCI.4399-14.2015.

[27] Risko EF and Gilbert SJ. Cognitive offloading. *Trends in Cognitive Sciences*. 2016;20(9):676–688. doi: 10.1016/j.tics.2016.07.002.

[28] Gerlich A. AI and the erosion of critical thinking: Cognitive offloading and youth vulnerability. *Journal of Cognitive Technology*. 2025;29(2):101–120. doi: 10.1080/15398285.2025.1012368.

[29] Zhang Y, Li H and Tan W. Artificial intelligence and critical reasoning: A longitudinal study. *Educational Technology Research and Development*. 2024;72(1):45–65. doi: 10.1007/s11423-023-10326-4.

[30] Perelman L. Critique of automated essay scoring. *Journal of Writing Assessment*. 2012;9(1):1–7.

[31] Freeman S, Eddy SL, McDonough M, Smith MK, Okoroafor N, Jordt H, et al.. Active learning increases student performance in science, engineering, and mathematics. *Proceedings of the National Academy of Sciences of the United States of America*. 2014;111(23):8410–8415. doi: 10.1073/pnas.1319030111.

[32] Deslauriers L, McCarty LS, Miller K, Callaghan K and Kestin G. Measuring actual learning versus feeling of learning in response to being actively engaged in the classroom. *Proceedings of the*

National Academy of Sciences of the United States of America. 2019;116(39):19251–19257. doi: 10.1073/pnas.1821936116.

[33]   Jolles DD and Crone EA. Training the developing brain: A neurocognitive perspective. *Frontiers in Human Neuroscience.* 2012;6:76. doi: 10.3389/fnhum.2012.00076.

[34]   Park DC and Bischof GN. The aging mind: Neuroplasticity in response to cognitive training. *Dialogues in Clinical Neuroscience.* 2013;15(1):109–119.

[35]   Reuter-Lorenz PA and Park DC. How does it STAC up? Revisiting the scaffolding theory of aging and cognition. *Neuropsychology Review.* 2014;24(3):355–370. doi: 10.1007/s11065-014-9270-9.

[36]   Kosmyna N, Hauptmann E, Yuan YT, Situ J, Liao XH, Beresnitzky AV, Braunstein I and Maes P. Your brain on ChatGPT: Accumulation of cognitive debt when using an AI assistant for essay writing task. *arXiv Preprint arXiv:2506.08872.* 2025 Jun 10.

[37]   Di Nuovo S. Could (and should) we build "collaborative intelligence" with artificial agents? A social psychological perspective. *Qeios.* 2023(6):1–4. doi: 10.32388/ZZ2ZRM.

[38]   Bektik D, Ullmann TD, Edwards C, Herodotou C and Whitelock D. AI-powered curricula: Unpacking the potential and progress of generative technologies in education. *Ubiquity Proceedings.* 2024;4 (1):38. doi: 10.1145/3576764.

[39]   Stephanidis C, Salvendy G, Antona M, Duffy VG, Gao Q, Karwowski W, Konomi SI, Nah F, Ntoa S, Rau PL and Siau KSHCI. Grand challenges revisited: Five-year progress. *International Journal of Human–Computer Interaction.* 2025 Feb 4;1–49. doi: 10.1080/10447318.2025.1002505.

[40]   Singh A, Taneja K, Guan Z and Ghosh A. Protecting human cognition in the age of AI. *arXiv:2502.12447.* 2025 Feb 18.

[41]   Francis NJ, Jones S and Smith DP. Generative AI in higher education: Balancing innovation and integrity. *British Journal of Biomedical Science.* 2025;81:14048. doi: 10.1080/09674845.2025.14048.

[42]   Chiu TK. Future research recommendations for transforming higher education with generative AI. *Computers and Education: Artificial Intelligence.* 2024 Jun 1;6:100197. doi: 10.1016/j.caeai.2024.100197.

[43]   Bhuman C and Nkala M. Cultivating critical thinking in the age of AI: Educational strategies for a data-driven world [report]. 2024 Sep; doi: 10.13140/RG.2.2.34210.03526.

[44]   Romeo G and Conti D. Exploring automation bias in human-AI collaboration: A review and implications for explainable AI. *AI & Society.* 2025. doi: 10.1007/s00146-025-02422-7.

[45]   Kasirzadeh A and Gabriel I. In conversation with artificial intelligence: Aligning language models with human values. *Philosophy & Technology.* 2023;36(2):27–41. doi: 10.1007/s13347-022-00550-x.

[46]   Sison AJG, Daza MT, Gozalo-Brizuela R and Garrido-Merchán EC. ChatGPT: More than a "weapon of mass deception" – Ethical challenges and responses from the human-centered artificial intelligence (HCAI) perspective. *International Journal of Human–Computer Interaction.* 2023;40(17):4853–4872. doi: 10.1080/10447318.2023.2205570.

[47]   Markauskaite L, Marrone R, Poquet O, Knight S, Martinez-Maldonado R, Howard S, Tondeur J, De Laat M, Shum SB, Gašević D and Siemens G. Rethinking the entwinement between artificial intelligence and human learning: What capabilities do learners need for a world with AI?. *Computers and Education: Artificial Intelligence.* 2022 Jan 1;3:100056. doi: 10.1016/j.caeai.2022.100056.

[48]   Federiakin D, Molerov D, Zlatkin-Troitschanskaia O and Maur A. Prompt engineering as a new 21st century skill. *Frontiers in Education.* 2024 Nov 29;9:1366434.Frontiers Media SA. doi: 10.3389/feduc.2024.1366434.

[49]   Lee D and Palmer E. Prompt engineering in higher education: A systematic review to help inform curricula. *International Journal of Educational Technology in Higher Education.* 2025 Feb 10;22(1):7. doi: 10.1186/s41239-025-00306-3.

Iván Claudio Suazo Galdames and Alain Manuel Chaple-Gil

From Algorithms to Adaptation: How Generative AI Is Reshaping Personalized Learning in Higher Education

Abstract: Generative artificial intelligence (GenAI) is redefining personalized learning in higher education by enabling adaptive content delivery, real-time feedback, and learner-specific instructional pathways. These technologies, including large language models, intelligent tutoring systems, and artificial intelligence (AI)-driven assessment tools, support student engagement, critical thinking, and individualized instruction. As GenAI becomes increasingly integrated into educational contexts, its implementation raises important pedagogical, ethical, and institutional questions. This chapter presents a systematic review of current empirical research on the use of GenAI in personalized learning environments across higher education. Drawing on 57 peer-reviewed studies selected through a comprehensive search in Scopus, Web of Science, and PubMed, the analysis follows PRISMA (Preferred Reporting Items for Systematic Reviews and Meta-Analyses) 2020 guidelines and applies the Mixed Methods Appraisal Tool to ensure methodological consistency. The review includes a thematic synthesis and SWOT (strengths, weaknesses, opportunities, and threats) analysis to identify key benefits, challenges, and strategic considerations. Findings indicate that GenAI can improve academic performance, enhance engagement, and support faculty in designing adaptive and inclusive curricula. Its most frequently cited benefits include time efficiency, tailored content delivery, and inclusive feedback mechanisms. Nevertheless, the literature also reveals critical concerns: data privacy risks, algorithmic bias, academic dishonesty, unequal access, and limited institutional preparedness. Notably, regional disparities in adoption patterns raise questions about equity and global implementation. This chapter argues that GenAI has transformative potential in higher education when used responsibly, ethically, and in alignment with pedagogical goals. For effective integration, universities must invest in digital infrastructure, develop robust policy frameworks, and foster human-AI collaboration that preserves the relational dimension of teaching and learning.

Keywords: Generative AI, personalized learning, higher education, adaptive learning technologies, educational innovation, artificial intelligence in education, ethical implications

Iván Claudio Suazo Galdames, Vicerrectoría de Investigación y Doctorados, Universidad Autónoma de Chile, Santiago de Chile, Chile, e-mail: ivan.suazo@uautonoma.cl
Alain Manuel Chaple-Gil, Facultad de Ciencias de la Salud, Universidad Autónoma de Chile, Santiago de Chile, Chile, e-mail: alain.chaple@uautonoma.cl

https://doi.org/10.1515/9783112206393-003

1 Introduction

The exploration of new forms of personalized learning in higher education is crucial due to the increasingly diverse student populations characterized by variations in prior knowledge, learning styles, and sociocultural backgrounds. Personalized learning approaches aim to tailor educational experiences to individual student needs, thereby enhancing engagement and improving learning outcomes.

Personalized learning emphasizes student-centered strategies that prioritize engagement and active participation in the learning process. These approaches recognize the uniqueness of each learner and encourage autonomy, thereby fostering critical thinking and deeper comprehension [1]. Traditional educational models often struggle to accommodate this level of personalization; therefore, innovative pedagogical practices are essential for addressing the needs of diverse learners. Research shows that when educators implement student-centered strategies, students demonstrate improved performance and retention rates [2], underlining the effectiveness of personalized approaches.

Heutagogy, or self-determined learning, reinforces the relevance of personalization by allowing students to construct their own educational pathways based on their unique experiences and contexts [3]. This learner-driven flexibility proves especially beneficial for individuals from diverse cultural and academic backgrounds, as it enhances social learning and fosters inclusive engagement with peers and educators [4].

Acknowledging the role of cultural diversity in educational environments is equally important, as it enriches learning processes and supports the development of personal and professional skills particularly for international students [5, 6]. Personalized learning methodologies that address and adapt to this diversity contribute to an inclusive academic atmosphere, fostering cooperation and mutual respect among students from different backgrounds [7].

At the same time, technological advances have introduced new opportunities for implementing personalized learning strategies. Digital platforms can now support instructional adaptation by aligning content and assessment with students' preferences, learning styles, and academic progress [8]. Studies show that integrating technology into personalized learning significantly improves student engagement and achievement, especially when designed with inclusivity and usability in mind [9, 10]. These developments underscore the potential of technology to transform education, while also highlighting the importance of equitable access and thoughtful implementation.

Within this landscape of pedagogical and technological change, generative artificial intelligence (GenAI) has emerged as a powerful innovation capable of redefining personalized learning in higher education. Tools such as large language models, intelligent tutoring systems, and adaptive learning platforms enable dynamic and context-aware instructional support. GenAI can customize content delivery, provide immediate feedback, and generate learner-specific trajectories, thereby facilitating more effective, engaging, and student-centered learning environments [11–13].

In parallel, GenAI supports educators by enhancing assessment systems and offering insights into student performance through artificial intelligence (AI)-driven analytics [14, 15]. These capabilities help to optimize curriculum design and pedagogical decisions, contributing to a more efficient and responsive educational ecosystem. The adoption of GenAI is also supported by students, who recognize its value for future professional preparation and academic achievement [16, 17].

However, despite its potential, the integration of GenAI also presents challenges. Ethical concerns such as data privacy, academic integrity, and equity of access must be addressed to avoid unintended consequences [18–20]. There are also technical and pedagogical limitations, including the need for faculty training, resistance to technological change, and anxiety about AI's impact on human-centered education [21, 22]. The absence of consistent institutional policies and regulatory frameworks further complicates the responsible and effective use of these technologies.

Given the transformative possibilities and the emerging tensions around GenAI in higher education, a comprehensive and systematic analysis is needed. This review aims to critically examine the benefits and challenges associated with the implementation of GenAI in personalized learning contexts within higher education. By synthesizing empirical evidence from peer-reviewed literature, the review identifies key applications, evaluates pedagogical and ethical implications, and proposes evidence-based recommendations. In doing so, it contributes to a deeper understanding of how GenAI can be effectively and equitably integrated into the evolving landscape of higher education.

To structure the exploration of this topic, the chapter is organized into five main sections. Following this introduction, the methodology section outlines the systematic review process, including the search strategy, eligibility criteria, data extraction, and quality appraisal methods used to ensure rigor and transparency. The results section presents the key findings of the review, highlighting both the benefits and challenges of implementing GenAI in personalized learning environments, along with a detailed SWOT (strengths, weaknesses, opportunities, and threats) analysis. The discussion section contextualizes these findings by examining their pedagogical, ethical, and institutional implications, as well as disparities in global adoption patterns. It also proposes strategic considerations for effective and equitable integration of GenAI in higher education. Finally, the conclusion offers a synthesis of the main insights, articulates practical recommendations for policymakers and educators, and identifies limitations and directions for future research.

2 Methodology

This systematic review followed the guidelines of the PRISMA (Preferred Reporting Items for Systematic Reviews and Meta-Analyses) statement to ensure that the research process was transparent, comprehensive, and reproducible. By adhering to these standards, the review provided an accurate synthesis of current evidence regarding the benefits and challenges of using GenAI in personalized learning within higher education. Each methodological element was designed to strengthen the rigor and replicability of the study.

The PICO model (population, intervention, comparison, and outcome) is widely used in systematic reviews within the health sciences to frame research questions and guide the review process. However, this model can be adapted to nonclinical disciplines, such as education, by using the PICo framework (population, interest, and context). The modified version reflects the unique characteristics of educational research, particularly in exploring interventions, learning outcomes, and contextual influences.

In this systematic review, the PICo framework has been applied to ensure the research question is specific, focused, and aligned with the review's objectives. Below is an explanation of how each component of PICo has been defined and applied within this study:

Population (P): The population component in this review refers to groups directly affected using GenAI in personalized learning. Specifically, this includes students and educators in higher education institutions, such as universities and colleges. The focus on this population ensures that the review captures both the experiences of learners and the pedagogical implications for instructors.

Interest (I): The interest (or intervention) in this review is the use of GenAI in personalized learning environments. GenAI encompasses technologies such as large language models, AI-powered tutoring systems, and adaptive learning platforms. By focusing on this specific technological innovation, the review aims to evaluate its potential to customize learning experiences, enhance student engagement, and support tailored teaching strategies.

Context (Co): The context component addresses the setting or environment in which the intervention occurs. For this review, the context is higher education, which includes universities, colleges, and other post-secondary institutions. The emphasis on this context allows the review to assess how GenAI operates within formal academic settings, considering factors such as institutional resources, curriculum design, and faculty-student interactions.

2.1 Search Strategy

A comprehensive search strategy was implemented to maximize the identification of relevant literature. The search was conducted across three databases (PubMed, Scopus, and Web of Science (WoS)), which provided broad coverage of all disciplines. The search terms combined controlled vocabulary (such as MeSH terms) and free-text keywords, ensuring that both formally indexed studies and those using emerging terminology were included. Boolean operators (AND, OR) and search modifiers were applied to refine the search and capture articles discussing GenAI, personalized learning, and higher education. No restrictions were placed on publication year, language, or document type to guarantee that all potentially relevant studies were considered. Databases, formulations, and filters applied were described in Table 1.

Table 1: Formulation employed in each database.

Database	Formulation	Filter
PubMed	(((("Generative AI"[Title/Abstract] OR "Artificial Intelligence"[Title/Abstract] OR "Machine Learning"[Title/Abstract] OR "ChatGPT"[Title/Abstract] OR "Large Language Models"[Title/Abstract]) AND ("Personalized Learning"[Title/Abstract] OR "Adaptive Learning"[Title/Abstract] OR "Intelligent Tutoring Systems"[Title/Abstract] OR "AI-driven education"[Title/Abstract] OR "Student-centered learning"[Title/Abstract]) AND ("Higher Education"[Title/Abstract] OR "University"[Title/Abstract] OR "College"[Title/Abstract] OR "Postsecondary Education"[Title/Abstract] OR "Tertiary Education"[Title/Abstract]) AND ("Benefits"[Title/Abstract] OR "Advantages"[Title/Abstract] OR "Positive outcomes"[Title/Abstract] OR "Challenges"[Title/Abstract] OR "Limitations"[Title/Abstract] OR "Ethical Issues"[Title/Abstract])) OR (("Artificial Intelligence"[MeSH Terms] OR "Machine Learning"[MeSH Terms] OR "Computational Intelligence"[MeSH Terms]) AND ("Education"[MeSH Terms] OR "Learning"[MeSH Terms] OR "Educational Technology"[MeSH Terms]) AND ("Universities"[MeSH Terms] OR "Education, Higher"[MeSH Terms]) AND ("Benefits"[Title/Abstract] OR "Advantages"[Title/Abstract] OR "Positive outcomes"[Title/Abstract] OR "Challenges"[Title/Abstract] OR "Limitations"[Title/Abstract] OR "Ethical Issues"[Title/Abstract])))	N/A

Table 1 (continued)

Database	Formulation	Filter
Scopus	(TITLE-ABS-KEY("Generative AI" OR "Artificial Intelligence" OR "ChatGPT" OR "Machine Learning" OR "Large Language Models") AND TITLE-ABS-KEY("Personalized Learning" OR "Adaptive Learning" OR "Intelligent Tutoring Systems" OR "AI-driven education" OR "Student-centered learning") ANDTITLE-ABS-KEY("Higher Education" OR "University" OR "College" OR "Postsecondary Education" OR "Tertiary Education") AND TITLE-ABS-KEY("Benefits" OR "Advantages" OR "Positive outcomes" OR "Challenges" OR "Limitations" OR "Ethical Issues"))	AND (LIMIT-TO (DOCTYPE, "ar"))
WoS	TS=("Generative AI" OR "Artificial Intelligence" OR "Machine Learning" OR "Large Language Models") AND TS=("Personalized Learning" OR "Adaptive Learning" OR "Intelligent Tutoring Systems" OR "AI-driven education" OR "Student-centered learning") AND TS= ("Higher Education" OR "University" OR "College" OR "Postsecondary Education" OR "Tertiary Education") AND TS=("Benefits" OR "Advantages" OR "Positive outcomes" OR "Challenges" OR "Limitations" OR "Ethical Issues")	Refined by document types: article

The selection of the databases PubMed, Scopus, and WoS was based on their extensive multidisciplinary coverage and their recognized relevance for indexing high-impact, peer-reviewed literature in the fields of health sciences, educational research, and technological innovation. These databases were chosen to ensure a comprehensive retrieval of studies addressing the intersection of GenAI and personalized learning in higher education.

Furthermore, the decision to include exclusively peer-reviewed journal articles was made to ensure the methodological rigor and empirical robustness of the synthesized evidence. By focusing on this type of publication, the review aimed to capture studies with clearly defined research questions, validated methodologies, and scholarly relevance, while excluding non-peer-reviewed sources that may lack sufficient academic scrutiny.

Final database search was conducted on January 30, 2025. To stay abreast of new research, we have established search alerts in relevant databases. These alerts will notify us of any newly published articles that could inform for a future umbrella review.

2.2 Eligibility Criteria

The eligibility criteria for study inclusion were peer-reviewed studies focusing on the application of GenAI in personalized learning within higher education settings, involving either students, educators, or institutions. Studies were excluded if they focused on general AI applications unrelated to personalized learning or on education levels outside higher education. By clearly defining these criteria, the review maintained its focus on research directly aligned with the stated objectives.

2.3 Study Selection Process

The study selection process involved a two-stage screening to minimize bias. First, titles and abstracts were screened by two independent reviewers to exclude irrelevant studies. Next, the full texts of selected studies were assessed against the eligibility criteria. Any disagreements between the reviewers were resolved through discussion or consultation with a third reviewer if necessary. This rigorous screening process ensured consistency and prevented the inclusion of irrelevant or low-quality studies. The PRISMA flow diagram was used to document each stage of the screening process, providing a clear visual representation of how studies were identified, screened, and included or excluded.

A calibration exercise was conducted among the authors to ensure consistency in the evaluation of articles considered for inclusion. Inter-rater agreement was assessed using the kappa statistic to quantify the level of concordance in inclusion/exclusion decisions. Kappa values ranging from 0.40 to 0.59 indicate acceptable agreement, values between 0.60 and 0.74 represent adequate agreement, and values of 0.75 or higher reflect excellent agreement.

The study selection followed the PRISMA flowchart methodology, ensuring a systematic and transparent review process. Initially, 566 records were identified from 3 key databases: PubMed ($n = 65$), Scopus ($n = 389$), and WoS ($n = 112$). Prior to the screening phase, duplicate records ($n = 63$) were removed, alongside 251 records automatically flagged as ineligible using automation tools.

Following this step, 252 records were screened based on titles and abstracts, resulting in the exclusion of 171 records that did not meet the inclusion criteria. Consequently, 81 reports were sought for full-text retrieval, all of which were successfully obtained. These 81 full-text reports underwent a detailed eligibility assessment, leading to the exclusion of 24 studies. The reasons for exclusion included studies with incorrect relevance to the research topic ($n = 15$), methodological concerns based on the Mixed Methods Appraisal Tool (MMAT) assessment ($n = 5$), and retracted articles ($n = 4$).

Ultimately, 57 new studies were deemed eligible and included in the final review. These selected studies provide a robust foundation for addressing the research objectives concerning the benefits and challenges of using GenAI in personalized learning within higher education (Figure 1).

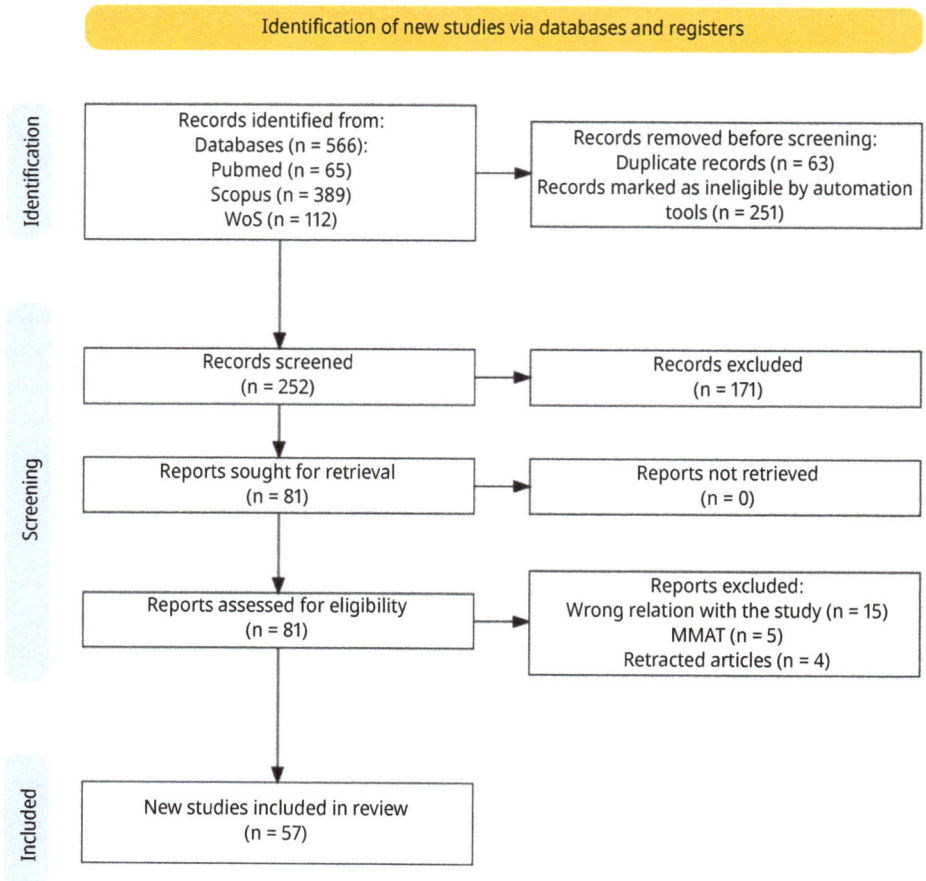

Figure 1: PRISMA flowchart.

2.4 Risk of Bias Assessment

In this systematic review, a formal risk of bias assessment was not conducted due to the anticipated inclusion of nonexperimental articles, such as observational studies and case studies. Unlike randomized controlled trials or experimental research, many of these study types do not follow a standardized intervention-outcome structure, making the application of conventional risk of bias tools, such as the Cochrane Risk of Bias tool or the Newcastle-Ottawa Scale, unsuitable.

Given the diversity of methodologies expected among the articles included, assessing the quality of the evidence requires a more flexible approach. As an alternative, we propose the use of general quality appraisal frameworks, such as the MMAT [23], both of which can accommodate various study designs. These tools assess factors such as the clarity of the research question, appropriateness of the methodology, validity of the findings, and relevance of the study to the review objectives.

To mitigate potential sources of bias during the study selection process, a two-stage screening procedure was implemented. Titles and abstracts were first independently reviewed by two researchers to determine initial eligibility. Full-text screening was subsequently conducted using predefined inclusion and exclusion criteria. Discrepancies in article selection were resolved through discussion, and when necessary, adjudicated by a third reviewer. To assess the level of agreement between reviewers, Cohen's kappa statistic was calculated. This allowed for a quantitative evaluation of inter-rater reliability, with values interpreted according to the established thresholds: values between 0.40 and 0.59 indicated moderate agreement, between 0.60 and 0.74 substantial agreement, and values of 0.75 or higher indicated excellent agreement. This procedure enhanced the transparency and reliability of the selection process.

The following five criteria were applied to each qualitative study:

(D1) whether the qualitative data source was appropriate to address the research question;

(D2) whether the data collection methods were adequate for qualitative research;

(D3) whether the findings were adequately derived from the data;

(D4) whether the interpretation of the results was sufficiently substantiated by the data; and

(D5) whether there was coherence between the data, the analysis, and the interpretation.

Each criterion was rated as "Yes," "No," or "Can't tell," depending on the level of methodological clarity and transparency provided in the published report. The assessments were visually summarized in a risk of bias figure using color coding: green (Yes), yellow (Can't tell), and red (No).

This appraisal allowed for a structured and transparent evaluation of study quality, highlighting both the strengths and potential limitations in methodological rigor across the included qualitative studies.

Furthermore, we plan to evaluate the quality of theoretical and conceptual articles by examining the robustness of their argumentation, the comprehensiveness of the literature cited, and the originality and coherence of their contributions to the topic. For literature reviews included in this study, we will assess their methodology for bias in article selection and synthesis.

2.5 Data Extraction

Data extraction was conducted using a predefined extraction form to maintain consistency and ensure that all relevant information was collected systematically. The extracted data included study details such as the author, year of publication, university, and country, as well as specifics regarding the study design, sample size, discipline, and data collection methods. The form also captured key information on the applications of GenAI, the benefits and challenges identified, and any ethical, technical, or educational considerations.

Moreover, the extraction process incorporated a detailed analysis of SWOT matrix. Strengths focused on identifying internal factors or applications where GenAI demonstrated significant benefits, such as its ability to enhance personalized learning or improve student engagement. Weaknesses, on the other hand, highlighted internal limitations, including issues such as technical deficiencies, inadequate faculty training, or insufficient infrastructure within universities. Opportunities were identified as external prospects for development, such as the potential for new research initiatives, interdisciplinary collaborations, or the integration of AI into emerging educational trends. Threats referred to external risks, including ethical concerns, biases in AI models, or resistance from faculty and students.

Incorporating the SWOT analysis within the data extraction framework proved highly valuable. It provided a more comprehensive understanding of how GenAI could be effectively implemented in higher education. By identifying strengths and opportunities, institutions can maximize the impact of AI applications, while addressing weaknesses and mitigating threats ensures that potential challenges do not become obstacles to progress. This strategic approach enriched the review by offering practical and targeted recommendations for policymakers, educators, and AI developers.

To further ensure accuracy, two reviewers independently extracted data, including the SWOT analysis, and cross-checked their results to resolve any discrepancies.

The synthesis of the data was conducted qualitatively, with the goal of summarizing the benefits, challenges, and key applications of GenAI in personalized learning. The synthesis highlighted patterns, differences, and gaps in literature, providing a comprehensive understanding of the current state of research. If enough comparable studies were identified, a meta-analysis was performed to quantitatively assess the overall effects of GenAI on educational outcomes. This dual approach ensured that both qualitative insights and quantitative evidence were fully leveraged.

The retrieved articles were organized in an Excel spreadsheet and processed using RStudio 2024.12.0 Build 467.

The final step in the methodology involved reporting the findings according to the PRISMA checklist. The review presented the results in a structured and transparent manner, discussing the key findings, limitations, and implications for future research and practice. This approach ensured that the review not only synthesized existing knowledge but also provided actionable recommendations for educators, policymakers, and researchers interested in leveraging GenAI for personalized learning. By following this comprehensive and methodologically sound process, the review contributed to a better understanding of how GenAI could be effectively and ethically integrated into higher education.

All data extracted from the included articles are available in the data repository of Mendeley: https://doi.org/10.17632/2ywvxkgwtj.2

3 Results

3.1 Study Selection Process

The evaluation of articles included in this systematic review was conducted, and most of the articles evaluated scored between 80% and 100%, demonstrating robust methodological design and direct relevance to the review objectives.

For instance, Al-Abdullatif et al. [24] was excluded because it provided a general overview of AI use in education without offering specific insights into personalized learning contexts. Similarly, Ray et al. [25] focused on the application of AI in medical training and surgical simulations, which, while valuable, did not align with the scope of this review addressing general academic settings.

Issa et al. [22] was another study that did not meet the inclusion criteria, as it primarily explored AI literacy in medical education, without addressing adaptive or personalized learning strategies beyond healthcare contexts. Likewise, Oluwad et al. [26] concentrated on clinical medical training, making its findings unsuitable for broader educational applications. Lastly, Wei [27] examined AI applications in piano instruction, which, despite its contribution to music education, did not provide transferable insights for personalized learning across multiple disciplines in higher education.

These exclusions were necessary to ensure the review maintained a focused and rigorous approach, drawing only from studies that directly addressed the application of AI in personalized learning environments within general higher education settings (Figure 1).

3.2 Evaluation and Risk of Bias Assessment

The methodological quality of the qualitative studies included was assessed using MMAT, which evaluates five key dimensions of qualitative research. Specifically, the criteria assessed were as follows: (D1) whether the qualitative data source was appropriate to address the research question; (D2) whether the data collection methods were adequate for qualitative research; (D3) whether the findings were adequately derived from the data; (D4) whether the interpretation of results was sufficiently substantiated by the data; and (D5) whether there was coherence between the data, analysis, and interpretation.

The assessment revealed that most studies met all five criteria, as indicated by the predominance of green cells across the matrix, corresponding to a "Yes" rating. This suggests a generally high level of methodological rigor among the included studies. A smaller number of studies received a "Can't tell" rating (yellow) in one or more dimensions, particularly in D3 and D4, reflecting limitations in reporting clarity or insufficient detail to support a definitive evaluation. Only a few studies were assessed as "No" (red), with most of these ratings concentrated in D4, indicating weaknesses in grounding the interpretation of findings in the empirical data. Overall, the MMAT-based appraisal indicated that the qualitative studies included in this review were largely methodologically sound, although some variability was noted in the robustness and transparency of data interpretation (Figure 2).

3.3 Description of Articles by the Year of Publication

The analysis of the temporal distribution of included articles revealed a marked concentration of publications in recent years. As depicted in Figure 3, the majority of studies (71.9%) were published in 2024, indicating a sharp increase in scholarly attention to the topic during that year. This was followed by 2023, which accounted for 21.1% of the publications. In contrast, only a small fraction of the studies were published in 2025 (5.3%) and 2022 (1.8%). These findings suggest an accelerating research interest in the field, likely driven by the growing relevance and integration of GenAI in higher education contexts.

3.4 Geographical Contributions

The geographical representation of countries in this dataset reveals notable disparities, with China emerging as the most represented country, contributing seven articles, followed by Jordan and Mexico, each with three articles. The dominance of China underscores its strong academic engagement and research output in the field of GenAI and personalized learning. This suggests that Chinese institutions may be at the forefront of technological adaptation and higher education innovation, potentially driven by robust policy initiatives and a thriving academic ecosystem.

In contrast, the least represented regions, including Uganda, the MENA region, and the United Kingdom, each contributed only a single article. This underrepresentation could indicate limited research infrastructure, fewer ongoing projects, or a lack of emphasis on AI in higher education within these regions. For instance, Uganda's singular contribution highlights emerging interest but suggests the need for more targeted academic investment and collaboration opportunities (Figure 4).

3.5 Benefits and Challenges

The integration of GenAI in personalized learning environments has led to numerous positive outcomes across academic settings. Chief among these benefits is enhanced time efficiency, as AI tools optimize student support and reduce administrative burdens on faculty. Personalized learning pathways, adaptive feedback mechanisms, and improved accessibility to knowledge foster student engagement and critical thinking

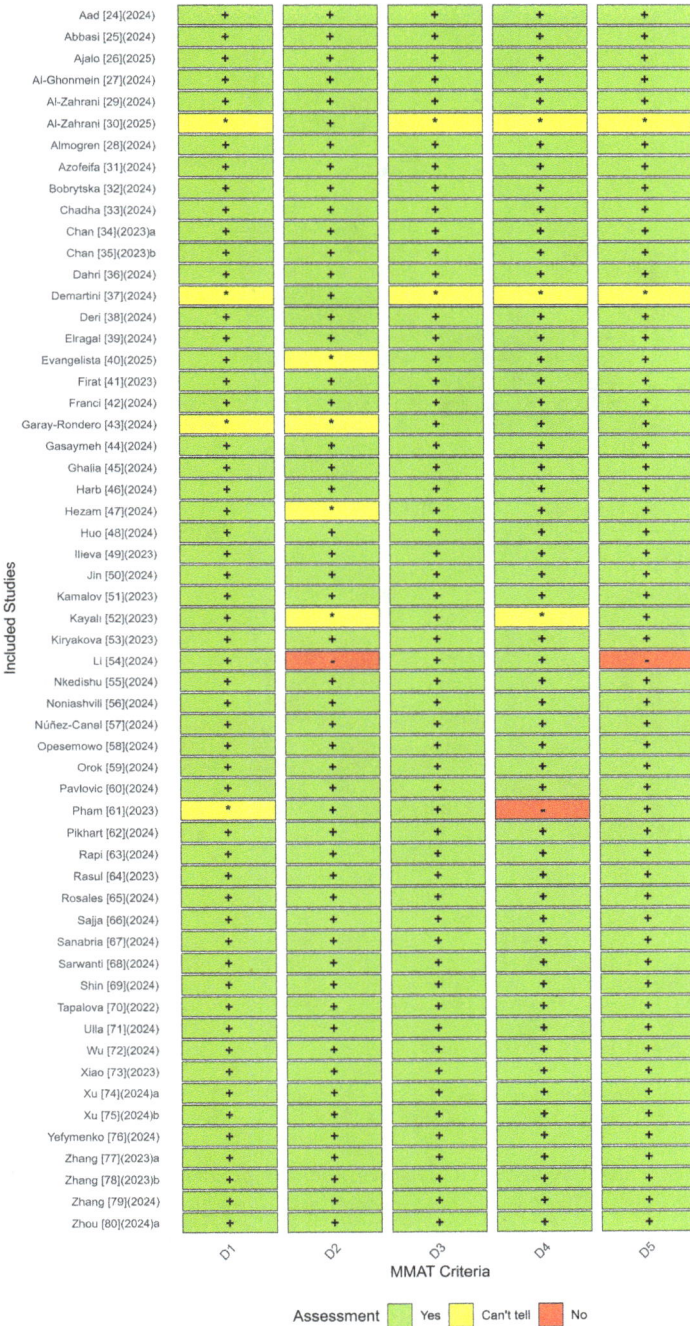

Figure 2: Risk of bias assessment according to the MMAT tool.

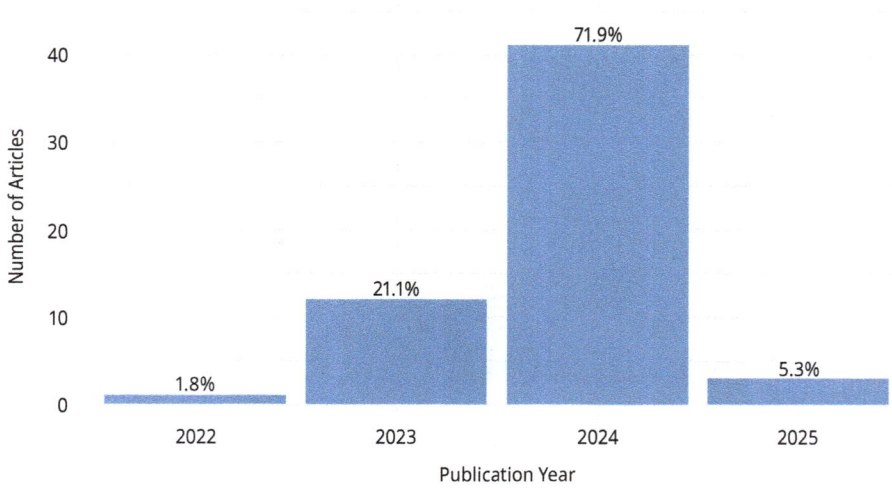

Figure 3: Distribution of articles by the year of publication.

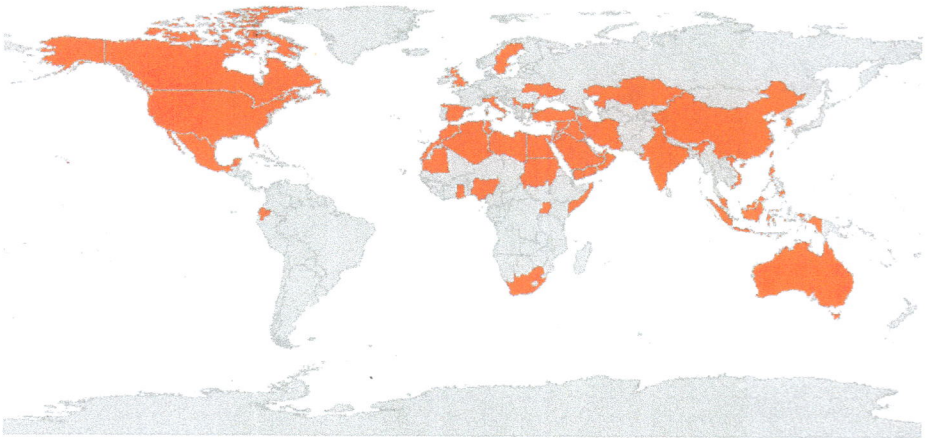

Figure 4: Global distribution of research contributions.

development. Additionally, AI-driven curricula have been associated with improved academic performance, personalized skill development, and adaptability, particularly in technical disciplines like engineering. These tools also facilitate resource optimization and enhance faculty research efforts, creating a symbiotic relationship between teaching and academic growth (Table 2).

Despite these advantages, significant challenges arise with the widespread implementation of GenAI in education. Ethical dilemmas, including the risk of over-reliance on AI, academic dishonesty, and privacy concerns, are at the forefront of these issues. AI systems often exhibit biases that may conflict with cultural and contextual expectations, posing risks to equitable learning. Furthermore, cybersecurity vulnerabilities and limited regulatory frameworks hinder the safe use of these technologies. Practical challenges include scalability issues, insufficient faculty training, and infrastructure gaps, which exacerbate the digital divide and limit AI adoption in certain regions. Without proper policy alignment and bias mitigation strategies, GenAI can inadvertently promote technological dependence, detracting from critical thinking and independent problem-solving (Table 2).

3.6 Critical Synthesis of Main Findings on the Benefits and Challenges of AI for Personalized Learning

A cross-study analysis of the 57 included studies reveals a nuanced and multifaceted landscape regarding the integration of AI in the personalization of learning within higher education (Table 2). The most frequently reported benefits coalesced around three core domains: personalization of learning experiences, enhancement of student engagement and motivation, and efficiency gains for both learners and educators. Personalized feedback, adaptive content delivery, and individualized learning trajectories were cited as central mechanisms driving improved academic performance and learner satisfaction. Additionally, AI was reported to contribute to enhanced inclusivity, particularly for students with disabilities, and to support higher-order cognitive processes such as critical thinking, problem-solving, and creativity.

Despite these promising affordances, a number of recurrent and substantive challenges were documented. Ethical concerns such as the risks of academic dishonesty, bias in algorithmic outputs, and threats to student privacy emerged as a dominant theme. Over-reliance on AI systems was another critical issue, frequently associated with reduced learner autonomy and cognitive dependency. Furthermore, several studies highlighted structural and contextual barriers, including inadequate faculty training, infrastructural limitations, and disparities in access to digital resources, which collectively pose obstacles to equitable implementation. A notable concern across studies was the lack of robust institutional policies and evaluation frameworks to govern AI integration, contributing to inconsistent practices and outcomes.

Interestingly, many studies emphasized the dual nature of AI, portraying it simultaneously as an enabler of pedagogical innovation and as a source of risk and uncertainty. This ambivalence was particularly evident in discussions surrounding academic integrity, where AI tools were acknowledged for their potential to support learning while also raising concerns about plagiarism and diminished originality. While some studies proposed frameworks or policy recommendations to mitigate such risks, there was a general consensus on the need for clearer ethical guidelines, enhanced digital literacy, and interdisciplinary collaboration (Table 2).

Table 2: Distribution of benefits and challenges of AI in the personalization of learning in higher education, according to the results of the studies included in the study.

Author (year)	Benefits	Challenges	Outcomes	DOI
Aad and Hardey [28]	Time-saving, personalized student support, and faculty research enhancement	Ethical dilemmas, potential over-reliance, and academic dishonesty risks	Highlighting the dual nature of AI as a benefit and threat	10.1108/QAE-02-2024-0043
Abbasi et al. [29]	Personalized curricula, real-time feedback, and critical thinking development	Biases in AI, difficulty aligning with cultural contexts, and ethical concerns	Increased adoption of personalized AI curricula with mixed institutional readiness	10.1007/s10639-024-13113-z
Ajalo et al. [30]	Enhanced learning, personalized content, and accessible knowledge	Academic dishonesty and insufficient regulatory frameworks	High adoption of AI tools for assignments, research, and personal use	10.1371/journal.pone.0313776
Al-Ghonmein and Al-Moghrabi [31]	Critical thinking development, personalized learning, and resource optimization	Cybersecurity concerns, accuracy issues, and potential over-reliance	Potential improvement in learning outcomes through interactive AI use	10.11591/ijai.v13.i2.pp1206-1213
Almogren et al. [32]	Personalized skill development, real-time feedback, and adaptability	Scalability issues, limited faculty training, and infrastructure gaps	Improved engineering competency and enhanced adaptability	10.1109/ACCESS.2024.3498047
Al-Zahrani et al. [33]	Increased academic performance through personalized support	Over-reliance on digital tools and distraction risks	Increased knowledge retention and collaborative learning	10.1057/s41599-024-03432-4
Al-Zahrani and Alasmari [34]	Enhanced engagement, resource optimization, and personalized teaching	Privacy and security concerns, and algorithmic bias	Positive stakeholder perception and readiness for future AI	10.1007/s10639-024-13300-y
Azofeifa et al. [35]	Improved student support and engagement	Infrastructure limitations, digital divide, and unclear policies	Improved administrative planning and student outcomes	10.3389/feduc.2024.1412018

Valentyna et al. [36]	Personalized feedback and stakeholder engagement	Ethical concerns and content accuracy issues	Positive stakeholder acceptance with role-specific concerns	10.26803/ijlter.23.1.20
Chadha [37]	Enhanced learning experiences and improved academic outcomes	Ethical guidelines and over-reliance risks	Institutional policy recommendations for AI integration	10.32674/em2qsn46
Chan et al. [38]a	Immediate feedback and personalized writing assistance	Plagiarism and limited creativity development	Better writing efficiency and academic outcomes	10.1186/s41239-023-00411-8
Chan et al. [39]	Efficiency and personalized learning paths	Over-reliance and ethical considerations	Increased engagement but role-dependent concerns	10.1186/s40561-023-00269-3
Dahri et al. [40]	Improved cognitive skills and academic success	Misinformation and dependency on AI outputs	Enhanced academic success and career readiness	10.1007/s10639-024-13148-2
Demartini et al. [41]	Improved student support and adaptive curriculum	Data privacy concerns and integration complexity	Scalable adaptive learning model	10.3390/su16031347
Deri et al. [42]	Optimized learning pathways and reduced workload	Ethical concerns and accessibility challenges	Guidelines for effective AI integration	10.1344/REYD2024.30.45777
Elragal et al. [43]	Enhanced inclusion for students with disabilities	Infrastructure and training requirements	Customized learning support for diverse needs	10.1109/ACCESS.2024.3476953
Evangelista [44]	Higher-order learning and ethical AI usage	Plagiarism and over-reliance on AI	Framework for AI-driven assessments	10.30935/cedtech/15775
Firat [45]	Personalized learning and increased engagement	Bias, misinformation, and ethical issues	Policy recommendations and strategic guidelines	10.37074/jalt.2023.6.1.22
Francis et al. [46]	Enhanced student engagement and tailored learning experiences	Maintaining academic integrity and bias in AI-generated outputs	Recommendations for balancing innovation with academic rigor	10.3389/bjbs.2024.14048
Garay-Rondero et al. [47]	Improved problem-solving skills and personalized competency assessment	Evaluation complexity and over-reliance on AI tools	Improved engineering competency and problem-solving approaches	10.1080/2331186X.2024.2392424

(continued)

Table 2 (continued)

Author (year)	Benefits	Challenges	Outcomes	DOI
Gasaymeh et al. [48]	Enhanced creativity, personalized feedback, and academic efficiency	Plagiarism, misinformation, and technical access limitations	Moderate familiarity with AI and recognition of its benefits	10.3390/educsci14101062
Ghalia et al. [49]	Adaptive learning, increased accessibility, and engagement	Ethical concerns and lack of standardized evaluation	Diverse perceptions of AI's role in learning and teaching	10.62754/joe.v3i8.4807
Harb and Rowaished [50]	Personalized content delivery and enhanced skill development	Data privacy concerns, cost of AI tools, and skill gaps	Varied faculty engagement and effectiveness of AI in media education	10.24294/jipd7810
Hezam and Alkhateeb [51]	Enhanced motivation, engagement, and differentiated instruction	Teacher training gaps and quality of AI-generated content	Enhanced comprehension and motivation through AI integration	10.17507/tpls.1407.12
Huo and Siau [52]	Improved knowledge acquisition, co-ideation, and personalized learning	Trust issues, cognitive dependency, and policy gaps	Strategic recommendations for effective AI integration	10.4018/JGIM.364093
Ilieva et al. [53]	Real-time feedback, increased engagement, and improved learning outcomes	Dependence on chatbots and data privacy issues	Proposed framework for personalized chatbot-based education	10.3390/info14090492
Jin and Kamsin [54]	Improved engagement, personalized learning, and tailored content	Data privacy concerns, model overfitting, and dependence on data	Enhanced learning outcomes through tailored learning pathways	10.11591/ijict.v13i3.pp470-475
Kamalov et al. [55]	Enhanced collaborative learning, dynamic feedback, and time efficiency	AI bias, ethical concerns, and over-reliance on technology	Improved teaching efficiency and learner adaptability	10.3390/su151612451
Kayali et al. [56]	Enhanced access to information, time savings, and personalized learning	Accuracy of responses, privacy concerns, and technical issues	Positive perceptions of ChatGPT and increased learning efficiency	10.14742/ajet.8915

Reference	Benefits	Challenges/Risks	Outcomes/Recommendations	DOI
Kiryakova et al. [57]	Reduced teaching workload, personalized assessment, and student engagement	Risks of unethical use and reliance on AI-generated content	Time savings and positive adoption of ChatGPT in teaching	10.3390/educsci13101056
Li and Li [58]	Better language acquisition and adaptive content delivery	Complexity in system integration and maintaining data privacy	Increased student engagement and improved academic performance	10.4108/eetsis.5636
Nkedishu and Okonta [59]	Enhanced personalized learning and interdisciplinary collaboration	AI literacy gaps and limited interdisciplinary collaboration	Recommendations for AI centers and interdisciplinary workshops	10.47857/irjms.2024.05i04.01261
Noniashvili et al. [60]	Scalable and efficient learning continuity during disruptions	Lack of pre-established policies and rapid decision-making requirements	Sustainable and inclusive adaptive learning strategies	10.15549/jeecar.v11i4.1817
Núñez-Canal et al. [61]	Improved critical thinking, ethics training, and creative problem-solving	Resistance to AI integration, ethical concerns, and academic integrity	A framework for AI literacy and competency-based teaching models	10.31637/epsir-2024-1685
Opesemowo and Adekomaya [62]	Improved engagement, inclusivity, and academic performance	Resource constraints, infrastructural gaps, and AI skill shortages	Policy recommendations for national AI integration strategies	10.26803/ijlter.23.3.4
Orok et al. [63]	Enhanced academic performance and learning support	Distractions and academic dishonesty concerns	Recommendations for AI integration in curriculum development	10.1186/s12909-024-06255-8
Pavlovic et al. [64]	Increased motivation and support for individual study	Integrity of learning and over-reliance on AI	Frameworks for AI-supported personalized learning	10.22550/2174-0909.4160

(continued)

Table 2 (continued)

Author (year)	Benefits	Challenges	Outcomes	DOI
Pham et al. [65]	Enhanced engagement and adaptive support for practical tasks	Prompt dependency and variability of AI responses	Enhanced teaching strategies using ChatGPT in practical contexts	10.14742/ajet.8825
Pikhart et al. [66]	Improved vocabulary, listening, writing, and speaking skills	Connectivity issues and limited content diversity	Diverse applications of digital resources across contexts	10.3389/feduc.2024.1412377
Rapi et al. [67]	Culturally sensitive adaptive learning and immersive tools	Ethical concerns, access disparities, and balancing traditional teaching	Guidelines for culturally relevant AI integration	10.26803/ijlter.23.11.22
Rasul et al. [68]	Enhanced personalized learning, reduced workload, and creative assessments	Academic integrity concerns and dependency on AI-generated content	Recommendations for balanced AI use in assessment and learning	10.37074/jalt.2023.6.1.29
Rosales et al. [69]	Increased engagement, efficiency in learning, and improved academic support	Dependency risks, ethical concerns, and data privacy	Ethical guidelines for AI integration in academic contexts	NA
Saija et al. [70]	Improved engagement, flexible learning paths, and real-time assessment	Implementation hurdles, privacy issues, and adapting to diverse needs	Innovative frameworks for adaptive learning systems	10.3390/info15100596
Sanabria and Olivo [71]	Enhanced creativity, innovation, and problem-solving abilities	Ethical and privacy concerns and ensuring continuous feedback	Development of competency frameworks addressing 4IR megatrends	10.1108/ITSE-07-2023-0145

Study	Benefits	Challenges	Recommendations/Outcomes	DOI
Sarwanti et al. [72]	Enhanced productivity, writing support, and idea generation	Plagiarism risks and over-reliance on AI-generated content	Guidelines for responsible AI integration in academic settings	10.24059/olj.v28i3.4599
Shin et al. [73]	Improved personalized learning and efficient teaching methods	Ethical concerns and lack of standardization across universities	Proposals for AI-based teaching frameworks and policies	10.1109/ACCESS.2024.3447067
Tapalova and Zhiyenbayeva [74]	Tailored learning paths and real-time feedback	Data privacy concerns and uneven access to resources	Recommendations for scaling personalized learning systems	10.34190/ejel.20.5.2597
Ulla et al. [75]	Increased inclusivity and improved language engagement and skills	Academic integrity and dependency on AI-generated feedback	Policy recommendations for integrating GenAI in inclusive language education	10.1016/j.caeai.2024.100314
Wu [76]	Improved adaptability and reduced learning anxiety	Mental health issues and limited personalized feedback for all	Higher adaptability to AI-enabled learning environments	10.1007/s10803-023-06097-1
Xiao and Zhi [77]	Immediate feedback, enhanced writing skills, and idea generation	Dependence on AI-generated responses	Guidelines for ethical and effective AI usage	10.3390/languages8030212
Xu et al. [78]	Mitigation of digital overload and personalized feedback	System integration complexity and training needs	Policy recommendations for mitigating side effects in PLEs	10.1371/journal.pone.0295646
Xu [79]	Improved critical thinking and problem-solving	Dependence on technology and increased educational inequality	Improved test scores and cognitive performance	10.1007/s44217-024-00330-4
Yefymenko et al. [80]	Enhanced language proficiency and translation accuracy	Over-reliance on AI and lack of sufficient motivation	Increased success in language assessments and personalized learning outcomes	10.34069/AI/2024.74.02.25
Zhang et al. [81]	Enhanced knowledge creation, multilingual support, and reduced teaching workload	Ethical concerns, privacy risks, and potential over-reliance	Recommendations for digital literacy programs	10.1163/25902539-05030007

(continued)

Table 2 (continued)

Author (year)	Benefits	Challenges	Outcomes	DOI
Zhang et al. [82]	Improved language proficiency and cross-cultural communication skills	Dependence on AI-generated content and teacher adaptation	Guidelines for AI-enhanced curriculum design	10.1080/08839514.2023.2216051
Zhang [83]	Increased interactivity, personalized feedback, and autonomous learning	Resource access disparities and infrastructure limitations	Frameworks for optimizing blended learning through mobile interaction	10.3991/ijim.v18i23.52881
Zhou et al. [84]	Increased productivity, personalized instruction, and enhanced creativity	Over-reliance, lack of accuracy, and unclear usage guidelines	Policy suggestions for integrating AI in entrepreneurial education	10.53761/xzjprb23

3.7 SWOT Analysis of Generative AI in Personalized Learning

The strengths identified in the articles reflect the robust and diverse application of GenAI in higher education. A major strength is the broad and varied perspectives brought by faculty members, supported by longitudinal data that enables a deeper understanding of the evolution of AI in education. Wide faculty adoption, combined with structured feedback, demonstrates the system's effectiveness in adapting to different educational settings. Additionally, comprehensive analyses of AI's interdisciplinary impacts highlight its role in promoting digital literacy across disciplines. This strength is complemented by the diverse representation of both students and institutions, ensuring that AI solutions cater to various contexts. The presence of comprehensive models that address critical AI challenges further solidifies the system's potential in supporting scalable and targeted interventions in personalized learning.

Despite these strengths, the weaknesses highlight important limitations. Many studies lack input from students, which could limit the holistic assessment of AI's impact. Some institutions exhibit cautious adoption of AI due to concerns over ethical and practical risks, which hampers implementation speed. The generalizability of some findings is constrained, as studies focused on specific regions, such as Nigerian universities, may not translate globally. Moreover, some findings rely heavily on self-reported data from students, which could introduce biases or inconsistencies. A notable weakness is the limited empirical validation of several proposed models, leaving gaps in their practical applicability across different institutions.

The opportunities surrounding the integration of AI in personalized learning are vast and promising. Enhanced faculty training, coupled with AI-guided learning paths, provides a significant avenue for improving teaching efficacy and student outcomes. The development of comprehensive guidelines for AI integration ensures that institutions can implement AI-driven programs effectively and sustainably. The establishment of nationwide AI research centers could enhance collaboration and innovation within the educational ecosystem. Additionally, expanding digital literacy programs would equip students with critical skills for navigating AI-powered environments. Broader AI training across academic disciplines will further enable institutions to integrate technological advancements seamlessly.

On the other hand, several threats pose risks to the successful implementation of GenAI in personalized learning. Job insecurity among educators remains a significant concern, as some fear AI could replace or diminish their roles. Ethical issues, such as plagiarism and academic dishonesty, present challenges that require strict regulatory measures. Resistance to AI adoption, particularly in conservative academic environments, could delay progress and limit innovation. The rapid pace of technological change may outpace the development of policies needed to govern AI implementation effectively. Furthermore, an over-reliance on AI could undermine human-centric teaching approaches, potentially affecting critical thinking and creativity among students.

The strengths and opportunities demonstrate the immense potential for GenAI to transform personalized learning by creating adaptive and inclusive educational environments. The strengths show that institutions with diverse populations and faculties can benefit from AI-driven teaching and learning mechanisms, while opportunities highlight the pathway for large-scale implementation and continuous innovation.

However, the weaknesses and threats emphasize the importance of addressing underlying risks, such as ethical concerns, limited empirical validation, and resistance to change. Without addressing these aspects, the impact of AI on personalized learning may remain uneven and subject to significant disparities among institutions (Table 3).

Table 3: Summary of SWOT matrix analysis results extracted from research-included study results related to personalized learning through AI.

Strengths	Weaknesses	Opportunities	Threats
Broad perspective of the teaching staff and longitudinal data	Limited student perspectives in some studies	Advanced teacher training and AI-guided learning paths	Job insecurity for educators and ethical controversies
High adoption of AI among teachers, with structured feedback	Cautious adoption due to inherent risks	Development of comprehensive guidelines for AI implementation	Risks of plagiarism and academic dishonesty
Comprehensive analysis of interdisciplinary impacts on AI literacy	Lack of generalization in certain national studies	AI-powered research centers at the national level	Resistance to AI adoption in conservative academic environments
Diversity in the representation of students and institutions	Reliance on self-reported data by students	Expansion of digital literacy programs	Policy delays in the face of rapid technological change
Comprehensive models to address critical AI challenges	Scarce empirical validation of some proposed models	AI training for various academic disciplines	Over-reliance on AI that can affect human-centered education

As GenAI tools become more sophisticated, their role in redefining personalized learning will continue to expand. However, balancing technological innovation with ethical, pedagogical, and practical considerations remains paramount. Institutions should adopt a holistic approach, integrating AI systems as complements rather than replacements for traditional teaching methods. Additionally, regulatory frameworks and institutional policies need to evolve to ensure that AI technologies benefit diverse student populations equitably.

Moreover, the collaboration between AI developers, educators, and policymakers is critical to maximizing the benefits while mitigating the challenges. GenAI should not only be seen as a tool for efficiency but also as an opportunity to redefine pedagogical frameworks in a way that prioritizes critical thinking, creativity, and student agency.

4 Discussion

The systematic review conducted in this study revealed numerous benefits and challenges associated with the integration of GenAI into personalized learning environments in higher education. This discussion aims to contextualize these findings within current pedagogical frameworks, ethical considerations, and emerging trends in university teaching practices.

The analysis of the 57 studies included in this review illustrates a complex yet promising landscape regarding the deployment of GenAI in personalized learning settings. The literature consistently highlights a broad spectrum of advantages, with the most frequently reported being increased time efficiency, tailored content delivery, greater student engagement, and enhanced academic performance. These benefits are observed across a variety of disciplines, with particularly significant impact noted in technical fields such as engineering and language acquisition. Moreover, GenAI tools have demonstrated utility for academic staff by facilitating time management, enabling the design of adaptive curricula, and improving the efficiency of feedback processes.

Beyond individual-level benefits, the findings emphasize a growing trend of employing GenAI as a conduit between teaching and research activities, thereby contributing to the development of a more agile and interconnected academic ecosystem. In this regard, AI technologies are not only enhancing personalization but also enabling broader pedagogical and institutional transformation.

Nonetheless, the studies also underscore considerable challenges. A majority of the reviewed articles caution against ethical risks posed by GenAI, including the spread of misinformation, overdependence on automated tools, heightened risks of academic dishonesty, and concerns over academic integrity. Furthermore, structural issues are prevalent, such as inadequate digital infrastructure, limited faculty training, and the absence of comprehensive institutional policies to regulate AI use in educational contexts. The data also reveal pronounced regional disparities: while countries like China are leading in terms of scholarly output and technological implementation, other regions, including sub-Saharan Africa and parts of the MENA region, remain underrepresented pointing to systemic inequalities in access to and benefits from AI technologies.

The evidence suggests that while GenAI holds considerable transformative potential for personalized learning, it also introduces critical risks. If left unaddressed through robust policy frameworks, inclusive practices, and pedagogically sound integration strategies, these risks could worsen existing inequities and compromise educational quality.

This duality between promise and challenge serves as the foundation for the structured analysis that follows. In Section 4.1, we delve into the educational benefits of GenAI, exploring how it supports adaptive learning paths, fosters student autonomy, and enhances academic engagement. Section 4.2 then examines the technological advantages, with a focus on the system-level innovations that ease personalization on a scale. In contrast, Sections 4.3 and 4.4 address the limitations and challenges identified in the literature, distinguishing between ethical concerns such as data privacy, equity, integrity, and technical or pedagogical constraints like faculty readiness and infrastructural barriers. Finally, Section 4.5 presents a strategic synthesis of the findings through a SWOT analysis, offering actionable insights and recommendations for effective and equitable implementation. Together, these sections provide a comprehensive discussion of the opportunities and obstacles that define the evolving role of GenAI in higher education.

4.1 Convergence Between GenAI and Personalized Pedagogy

The documented benefits in this review support the notion that GenAI aligns with fundamental principles of personalized learning, such as adaptability, immediate feedback, and attention to individual differences among students. This finding is consistent with student-centered pedagogical models such as Vygotsky's [85] social constructivism, which emphasizes the role of social context and interaction in knowledge development.

Studies such as Almogren et al. [32] demonstrate that adaptive feedback generated by AI can enhance student motivation and engagement by providing more relevant and personalized learning paths. Similarly, Chan et al. [39] highlight how GenAI tools can assist in written production and language skill development, particularly among foreign language learners, fostering autonomous and constructive learning.

Furthermore, GenAI offers added value at the institutional level. Research by Aad and Hardey [28] underlines its usefulness for academic research and the reduction of administrative burdens, allowing educators to focus on more complex aspects of teaching such as the development of critical thinking and personalized interaction with students.

4.2 Ethical Tensions and Technological Dependence

However, alongside these educational promises, significant concerns emerge regarding the limitations and risks of these technologies. One of the most recurrent challenges in the reviewed literature is the ethical dimension, especially regarding data privacy, algorithmic bias, and academic plagiarism. Evangelista [44] warns that the indiscriminate use of tools such as ChatGPT may undermine academic integrity if clear regulatory frameworks and assessment strategies adapted to the new digital environment are not implemented.

Equally problematic is the phenomenon of "cognitive dependency," where students become overly reliant on AI-generated responses, diminishing their ability to think critically or generate original ideas [42, 65]. This not only affects learner autonomy but also the validity of learning assessment processes.

The review also revealed the lack of institutional policies and clear legislation regulating the use of these technologies, a shortfall that represents a direct threat to their ethical and sustainable implementation [34]. The lack of adequate faculty training and inequalities in technological infrastructure further exacerbate the situation, especially in low-resource settings [35, 50].

4.3 Reconfiguration of the Teaching Role and Pedagogical Relationships

One of the most sensitive debates around GenAI is its impact on the role of educators. While some studies point to improvements in teaching efficiency and the ability to personalize instruction [41, 55], others express concerns about the possible "dehumanization" of education. According to Bussey et al. [86], a sound ethical and pedagogical implementation of GenAI must treat it as a complementary tool rather than a substitute for the human role of the teacher.

The discussion also reflects that the development of digital and ethical competencies is crucial for both students and educators. Including AI training within university curricula could mitigate many of the current challenges, as proposed by Núñez-Canal et al. [87] in their integrative generative learning model.

4.4 Geographic Inequalities and Equity Challenges

A significant structural limitation noted in this review is the disparity in academic production and GenAI adoption across regions. For example, while China has shown remarkable research activity and technological adoption [82], regions such as sub-Saharan Africa or parts of the Arab world are underrepresented in the literature [33].

These disparities reflect not only asymmetries in technological infrastructure but also limited access to AI-enhanced educational resources. The lack of connectivity and public policies aimed at digital transformation in education may exacerbate the educational divide between the Global North and South, as suggested by Opesemowo and Adekomaya [62]. In this sense, GenAI risks reproducing or even amplifying existing educational inequalities unless it is integrated consciously and inclusively.

4.5 Practical Implications for Educational Policy

One of the main contributions of this systematic review lies in its implications for policymaking in higher education institutions. The evidence suggests that the effective adoption of GenAI requires clear regulatory frameworks that govern its ethical and pedagogical use, as well as governance structures that promote transparency and accountability [18, 21]. Without such frameworks, institutions risk falling into uncritical adoption of technologies that may violate student privacy or encourage unregulated automation of teaching.

Universities must also design institutional strategies to ensure that all educational actors, especially faculty and students, develop critical digital literacy. This includes not only the technical use of GenAI tools but also an understanding of their functioning, ethical implications, and inherent biases. Some studies recommend the incorporation of specific modules on artificial intelligence, ethics, and data analysis in all university disciplines, as part of a transversal approach to higher education [55, 59].

A significant investment in technological infrastructure and faculty training is also required. In contexts where these conditions are not assured, the adoption of GenAI could have a regressive effect, deepening pre-existing inequalities between institutions with high and low levels of digitization [35, 43].

4.6 Perspectives for Future Research

Despite the growing number of studies on GenAI and personalized learning, significant gaps remain in literature. First, many reviewed studies present methodological limitations related to the use of self-reported data, which can introduce social desirability bias or unreliable interpretations [40]. There is a need for greater incorporation of mixed methods, longitudinal studies, and experimental designs to assess the real impact of GenAI more rigorously on academic performance, motivation, and critical thinking skills.

Second, research must be expanded to underrepresented contexts, such as institutions in the Global South, rural universities, or those with nontraditional student populations. This will allow a more inclusive and contextualized understanding of the effects and challenges of these technologies [62].

In-depth investigation into the impact of GenAI on the pedagogical relationship is also recommended. How does technological mediation affect the construction of the student-teacher bond? What types of interactions are privileged or lost when AI is introduced into educational processes? These questions are key to ensuring that technological innovation does not displace the human and relational dimension of education.

4.7 Discussion of the SWOT Matrix: Institutional Impact and Strategic Lines of Action proposed

4.7.1 Strengths: Consolidating Institutional Capacities

The broad perspective of teaching staff and the availability of longitudinal data provide institutions with a solid foundation for designing evidence-informed personalized learning strategies. The high adoption of AI by educators, supported by structured feedback mechanisms, reflects an organizational culture open to innovation. Furthermore, the integration of comprehensive models addressing critical AI-related challenges, combined with the diversity in student and institutional representation, fosters a holistic understanding of the educational phenomenon.

Institutional impact: These strengths position institutions as adaptive environments capable of critically integrating emerging technologies, while promoting both quality and equity in education.

Strategic actions:

Consolidate inter-institutional networks to exchange best practices in AI-driven education.

Strengthen faculty monitoring systems to support reflective teaching and personalized learning.

Institutionalize longitudinal student data analysis systems to inform pedagogical decision-making.

4.7.2 Weaknesses: Internal Barriers to Transformation

The limited incorporation of student perspectives, the reliance on self-reported data, and the insufficient empirical validation of certain models reveal methodological and epistemological weaknesses. Additionally, the cautious adoption of AI due to perceived risks and the lack of generalizability in some national studies further constrain evidence-based implementation.

Institutional impact: These weaknesses may lead to biased or poorly informed decision-making, thereby undermining institutional credibility and hindering the scalability of technological innovations.

Strategic actions:

Design participatory research approaches that actively include student voices.

Integrate mixed-methods designs to enable data triangulation and enhance the robustness of evidence.

Promote pilot evaluations before large-scale deployment of AI-based models.

4.7.3 Opportunities: Catalysts for Strategic Transformation

The development of AI-guided learning pathways, the emergence of specialized research centers, and the expansion of digital literacy programs represent critical opportunities to accelerate educational innovation. In addition, the potential for implementing comprehensive AI teaching guidelines and integrating AI training across disciplines broadens the institutional scope of technological advancement.

Institutional impact: These opportunities can lead to sustainable competitive advantages, fostering a culture of innovation and positioning institutions as national or regional leaders in the ethical and pedagogical use of AI.

Strategic actions:

Create institutional programs for continuous AI training for both faculty and students.

Establish centers of excellence in AI for education with regional or national impact.

Develop internal normative frameworks to guide the ethical adoption of intelligent technologies.

4.7.4 Threats: Systemic Risks and Ethical-Political Challenges

The threats identified such as job insecurity among educators, academic dishonesty, institutional resistance, policy delays, and excessive reliance on AI highlight structural tensions that may destabilize teaching and learning processes and challenge the humanistic mission of higher education.

Institutional impact: These threats can generate internal conflicts, erode trust in AI, and undermine principles of human-centered education.

Strategic actions:

Implement policies that safeguard academic employment in the context of automation.

Develop integrity frameworks adapted to AI-mediated learning environments.

Promote inclusive deliberative processes to mitigate resistance and foster institutional consensus.

Design strategies that balance automation with the humanization of educational practices.

4.8 Evidence-Based Recommendations

Based on the findings of this review, the following recommendations are proposed for the effective, ethical, and equitable integration of GenAI in university settings:

Development of specific institutional policies on GenAI use, with emphasis on ethics, data privacy, and mechanisms for faculty supervision. These policies should be inclusive, participatory, and context-specific.

Ongoing faculty training that combines digital literacy with critical reflection on AI pedagogy and ethics. This training should be mandatory, discipline-contextualized, and supported by institutional resources.

Design of hybrid learning environments, where GenAI complements human pedagogical strategies rather than replacing them. This approach allows for leveraging technological advantages without losing the humanistic focus of learning.

Promotion of digital equity, ensuring that all students have access to devices, connectivity, and the competencies needed to engage meaningfully with AI tools.

Encouragement of applied and collaborative research, bringing together efforts from academic departments, innovation centers, and technology developers to create AI solutions tailored to the specific needs of educational communities.

5 Conclusion

This systematic review identified a wide range of benefits associated with the integration of GenAI into personalized learning environments in higher education. Chief among these were enhanced time efficiency, adaptive content delivery, increased student engagement, and improved academic performance. GenAI also proved valuable in supporting faculty with curriculum design, assessment, and research-related tasks. However, the review also highlighted significant challenges, including ethical concerns such as data privacy violations, algorithmic bias, academic dishonesty, and the risk of overreliance on AI-generated outputs. Infrastructural limitations, lack of faculty training, uneven policy development, and global disparities in AI adoption were identified as structural barriers to equitable implementation.

The implications for educational practice and institutional policy are substantial. For GenAI to be effectively and ethically integrated into university settings, institutions must adopt a holistic approach that prioritizes digital equity, inclusive policy frameworks, and the development of critical AI literacy for both students and faculty. AI should be seen not as a replacement for the educator but as a pedagogical tool that enhances the relational and humanistic dimensions of learning. Institutional strategies should include mandatory training programs, interdisciplinary AI ethics modules, and governance mechanisms that ensure transparency, accountability, and student protection.

In terms of future research, there is a pressing need for more methodologically diverse and context-sensitive studies. Longitudinal, experimental, and mixed-methods research designs are recommended to better assess the long-term effects of GenAI on learning outcomes, student autonomy, and critical thinking development. Furthermore, future investigations should target underrepresented regions and institutions, particularly those in the Global South, to develop a more globally inclusive understanding of AI's educational impact.

Finally, the review has certain methodological limitations that must be acknowledged. The exclusion of gray literature and the restriction to peer-reviewed journal articles may have omitted relevant insights from conference proceedings, institutional reports, or practitioner-based case studies. While this decision was made to ensure academic rigor, it may have limited the scope of perspectives captured. Additionally, most included studies relied on self-reported data, which could introduce bias or limit the objectivity of certain findings.

Despite these limitations, the review offers a robust and timely synthesis of the current state of research on GenAI in personalized learning. Its findings provide evidence-based guidance for educators, policymakers, and researchers seeking to harness the transformative potential of AI while safeguarding the pedagogical and ethical integrity of higher education.

References

[1]   Suyo-Vega J, Fernández-Bedoya V and Meneses-La-Riva M. Beyond traditional teaching: A systematic review of innovative pedagogical practices in higher education [version 1; peer review: 3 approved with reservations, 1 not approved]. *F1000Research*. 2024;13(22). https://doi.org/10.12688/f1000re search.143392.1.

[2]   Freeman S, Eddy SL, McDonough M, Smith MK, Okoroafor N, Jordt H, et al. Active learning increases student performance in science, engineering, and mathematics. *Proceedings of the National Academy of Sciences*. 2014;111(23):8410–8415. https://doi.org/10.1073/pnas.1319030111.

[3]   Venter A. Social media and social capital in online learning. *South African Journal of Higher Education*. 2019;33(3):241–257. https://doi.org/10.20853/33-3-3105.

[4]   Hannington M, Mbambo N, Lebelo M, Ramavhanda K, Johnson C, Lurani A, et al. Legitimacy, belonging and engagement: A qualitative exploration of safe learning spaces in a South African University. *The Clinical Teacher*. 2025;22(2):e70072. https://doi.org/10.1111/tct.70072.

[5]   Azram M, Hong M, Ahmad W and Sohail A. The impact of diversity experiences and innovative learning environments on the personal development of international students studying in China. *European Journal of Education*. 2024;59(3):e12655. https://doi.org/10.1111/ejed.12655.

[6]   Suazo Galdames I. From anatomy to algorithm: Scope of AI-assisted diagnostic competencies in health sciences education. *International Journal of Medical and Surgical Sciences*. 2024;11(3):1–24. https://doi.org/10.32457/ijmss.v11i3.2818.

[7]   Mabeza RM, Christophers B, Ederaine SA, Glenn EJ, Benton-Slocum ZP and Marcelin JR. Interventions associated with racial and ethnic University in US graduate medical education: A scoping review. *JAMA Network Open*. 2023;6(1):e2249335–e. https://doi.org/10.1001/jamanetworkopen.2022.49335.

[8]   Khan FA, Graf S, Weippl ER and Tjoa AM. Implementation of affective states and learning styles tactics in web-based learning management systems. *2010 10th IEEE International Conference on Advanced Learning Technologies*. 2010: 734–735.

[9]   Sfenrianto S and Hasibuan ZA. Step-function approach for E-learning personalization. *TELKOMNIKA (Telecommunication Computing Electronics and Control)*. 2017;15(3):1362–1367. http://doi.org/10.12928/telkomnika.v15i3.4457.

[10] Wood D. Problematizing the inclusion agenda in higher education: Towards a more inclusive technology enhanced learning model. *First Monday*. 2015;20(9). https://doi.org/10.5210/fm. v20i9.6168.

[11] Pasupuleti MK. AI-powered learning: Transformative innovations in smart education technology. Advanced AI-driven EdTech: Transformative innovations in smart learning. In: *International Journal of Academic and Industrial Research Innovations (IJAIRI):* National Education Services; 2024. pp. 292–310.

[12] Imran M, Almusharraf N, Abdellatif MS and Abbasova MY. Artificial intelligence in higher education: Enhancing learning systems and transforming educational paradigms. *International Journal of Interactive Mobile Technologies*. 2024;18(18):34–48. https://doi.org/10.3991/ijim.v18i18.49143.

[13] Valentini A. Educación superior, inteligencia artificial y transformación digital en América Latina y el Caribe. *SciComm Report*. 2025;5(1):1–13. https://doi.org/10.32457/scr.v5i1.2830.

[14] Vashishth TK, Sharma V, Kk S, Kumar B, Panwar R and Chaudhary S. AI-driven learning analytics for personalized feedback and assessment in higher education. In: Nguyen TVT and Vo NTM, editors. *Using Traditional Design Methods to Enhance AI-Driven Decision Making*. Hershey, PA, USA: IGI Global; 2024. pp. 206–230.

[15] Denga EM and Denga SW. Revolutionizing education: The power of technology. Revolutionizing curricula through computational thinking, logic, and problem solving. *IGI Global*. 2024: 167–188.

[16] Busch F, Hoffmann L, Truhn D, Palaian S, Alomar M, Shpati K, et al. International pharmacy students' perceptions towards artificial intelligence in medicine – A multinational, multicentre cross-sectional study. *British Journal of Clinical Pharmacology*. 2024;90(3):649–661. https://doi.org/10.1111/bcp.15911.

[17] Abdekhoda M and Dehnad A. Adopting artificial intelligence driven technology in medical education. *Interactive Technology and Smart Education*. 2024;21(4):535–545. https://doi.org/10.1108/ITSE-12-2023-0240.

[18] Singh P. Artificial intelligence and student engagement: Drivers and consequences. In: Gierhart AR editor. *Cases on Enhancing P-16 Student Engagement with Digital Technologies*. Hershey, PA, USA: IGI Global; 2025. pp. 201–232.

[19] Zheng S and Han M. The impact of AI enablement on students' personalized learning and countermeasures – A dialectical approach to thinking. *Journal of Infrastructure, Policy and Development, [Sl]*. 2024. https://doi.org/10.24294/jipd10274.

[20] Suazo Galdames I. Inteligencia artificial en investigación científica. *SciComm Report*. 2023;3(1):1–3. https://doi.org/10.32457/scr.v3i1.2149.

[21] Chaudhry M and Goswami N. Shaping tomorrow's minds: The future of education in an AI (artificial intelligence)-augmented world. In: Doshi R, Dadhich M, Poddar S and Hiran KK editors. *Integrating Generative AI in Education to Achieve Sustainable Development Goals*. Hershey, PA, USA: IGI Global; 2024. pp. 386–407.

[22] Issa WB, Shorbagi A, Al-Sharman A, Rababa M, Al-Majeed K, Radwan H, et al. Shaping the future: Perspectives on the Integration of Artificial Intelligence in health profession education: A multi-country survey. *BMC Medical Education*. 2024;24(1):1166. https://doi.org/10.21203/rs.3.rs-4396289/v1.

[23] Hong QN, Gonzalez-Reyes A and Pluye P. Improving the usefulness of a tool for appraising the quality of qualitative, quantitative and mixed methods studies, the Mixed Methods Appraisal Tool (MMAT). *Journal of Evaluation in Clinical Practice*. 2018;24(3):459–467. https://doi.org/10.1111/jep.12884.

[24] Al-Abdullatif FA, Alnasib BN, Alruwaili HAS, Al Katam MH, Al Hasan SA, Alsager JS, et al. The use and challenges of artificial intelligence among university students: The case of Saudi Arabia. *Eurasian Journal of Educational Research*. 2024;2024(112):219–238. https://doi.org/10.14689/ejer.2024.112.022.

[25] Ray P and PP R. Large language models in laparoscopic surgery: A transformative opportunity. *Laparoscopic Endoscopic and Robotic Surgery*. 2024;7(4):174–180. https://doi.org/10.1016/j.lers.2024. 07.002.

[26] Oluwad K, Adeoti A, Agodirin S, Nottidge T, Usman M, Gali M, et al. Exploring artificial intelligence in the Nigerian medical educational space: An online cross-sectional study of perceptions, risks and benefits among students and lecturers from ten universities. *The Nigerian Postgraduate Medical Journal*. 2023;30(4):285–292. https://doi.org/10.4103/npmj.npmj_186_23.

[27] Wei C. A study of Piano Timbre teaching in the context of artificial intelligence interaction. *Computational Intelligence and Neuroscience*. 2021;2021:4920250. https://doi.org/10.1155/2021/4920250.

[28] Aad S and Hardey M. Generative AI: Hopes, controversies and the future of faculty roles in education. *Quality Assurance in Education*. 2024. https://doi.org/10.1108/QAE-02-2024-0043,

[29] Abbasi BN, Wu Y and Luo Z. Exploring the impact of artificial intelligence on curriculum development in global higher education institutions. *Education and Information Technologies*. 2024. https://doi.org/10.1007/s10639-024-13113-z.

[30] Ajalo E, Mukunya D, Nantale R, Kayemba F, Pangholi K, Babuya J, et al. Widespread use of ChatGPT and other Artificial Intelligence tools among medical students in Uganda: A cross-sectional study. *PloS One*. 2025;20(1):e0313776. https://doi.org/10.1371/journal.pone.0313776.

[31] Al-Ghonmein AM and Al-Moghrabi KG. The potential of ChatGPT technology in education: Advantages, obstacles and future growth. *IAES International Journal of Artificial Intelligence*. 2024;13 (2):1206–1213. https://doi.org/10.11591/ijai.v13.i2.pp1206-1213.

[32] Almogren A, Al-Rahmi W, Dahri N, Almogren AS, Al-Rahmi WM and Dahri NA. Integrated technological approaches to academic success: Mobile learning, social media, and AI in visual art education. *IEEE Access*. 2024;12:175391–175413. https://doi.org/10.1109/ACCESS.2024.3498047.

[33] Al-Zahrani A, Alasmari T, Al-Zahrani AM and Alasmari TM. Exploring the impact of artificial intelligence on higher education: The dynamics of ethical, social, and educational implications. *Humanities & Social Sciences Communications*. 2024;11(1). https://doi.org/10.1057/s41599-024-03432-4.

[34] Al-Zahrani AM and Alasmari TM. A comprehensive analysis of AI adoption, implementation strategies, and challenges in higher education across the Middle East and North Africa (MENA) region. *Education and Information Technologies*. 2025. https://doi.org/10.1007/s10639-024-13300-y.

[35] Azofeifa J, Rueda-Castro V, Camacho-Zuñiga C, Chans G, Membrillo-Hernández J, Caratozzolo P, et al. Future skills for Industry 4.0 integration and innovative learning for continuing engineering education. *Frontiers in Education*. 2024;9. https://doi.org/10.3389/feduc.2024.1412018.

[36] Bobrytska VI, Krasylnykova HV, Beseda NA, Krasylnykov SR and Skyrda TS. Artificial intelligence (AI) in Ukrainian higher education: A comprehensive study of stakeholder attitudes, expectations and concerns. *International Journal of Learning, Teaching and Educational Research*. 2024;23(1):400–426. https://doi.org/10.26803/ijlter.23.1.20.

[37] Chadha A. Transforming higher education for the digital age: Examining emerging technologies and pedagogical innovations. *Journal of Interdisciplinary Studies in Education*. 2024;13:53–70. https://doi.org/10.32674/em2qsn46.

[38] Chan C, Hu W, Chan CKY and Hu W. Students' voices on generative AI: Perceptions, benefits, and challenges in higher education. *International Journal of Educational Technology in Higher Education*. 2023;20(1). https://doi.org/10.1186/s41239-023-00411-8.

[39] Chan C, Lee K, Chan CKY and Lee KKW. The AI generation gap: Are Gen Z students more interested in adopting generative AI such as ChatGPT in teaching and learning than their Gen X and millennial generation teachers? *Smart Learning Environments*. 2023;10(1). https://doi.org/10.1186/s40561-023-00269-3.

[40] Dahri NA, Yahaya N and Al-Rahmi WM. Exploring the influence of ChatGPT on student academic success and career readiness. *Education and Information Technologies*. 2024. https://doi.org/10.1007/s10639-024-13148-2.

[41] Demartini C, Sciascia L, Bosso A, Manuri F, Demartini CG, Sciascia L, et al. Artificial intelligence bringing improvements to adaptive learning in education: A case study. *Sustainability*. 2024;16(3). https://doi.org/10.3390/su16031347.

[42] Deri M, Singh A, Zaazie P, Anandene D, Deri MN, Singh A, et al. Leveraging artificial intelligence in higher educational institutions: A comprehensive overview. *Revista de Educacion y Derecho-educational and Law Review*. 2024;(30). https://doi.org/10.1344/REYD2024.30.45777.

[43] Elragal A, Awad AI, Andersson I and Nilsson J. A conversational AI Bot for efficient learning: A prototypical design. *IEEE Access*. 2024;12:154877–154887. https://doi.org/10.1109/ACCESS.2024.3476953.

[44] Evangelista EDL. Ensuring academic integrity in the age of ChatGPT: Rethinking exam design, assessment strategies, and ethical AI policies in higher education. *Contemporary Educational Technology*. 2025;17(1). https://doi.org/10.30935/cedtech/15775.

[45] Firat M. What ChatGPT means for universities: Perceptions of scholars and students. *Journal of Applied Learning and Teaching*. 2023;6(1):57–63. https://doi.org/10.37074/jalt.2023.6.1.22.

[46] Francis NJ, Jones S and Smith DP. Generative AI in higher education: Balancing innovation and integrity. *British Journal of Biomedical Science*. 2024;81:14048. https://doi.org/10.3389/bjbs.2024.14048.

[47] Garay-Rondero C, Castillo-Paz A, Gijón-Rivera C, Domínguez-Ramírez G, Rosales-Torres C, Oliart-Ros A, et al. Competency-based assessment tools for engineering higher education: A case study on complex problem-solving. *Cogent Education*. 2024;11(1). https://doi.org/10.1080/2331186X.2024.2392424.

[48] Gasaymeh A, Beirat M, Abu QA, Gasaymeh A-M-M, Beirat MA and Abu Qbeita A. University students' insights of generative artificial intelligence (AI) writing tools. *Education Sciences*. 2024;14(10). https://doi.org/10.3390/educsci14101062.

[49] Ghalia N, Isami E, Bsoul M, Aabed S, Haeb RS and Mustafa M. Impact of artificial intelligence in education: Insights from students and faculty members at Yarmouk University. *Journal of Ecohumanism*. 2024;3(8):1278–1289. https://doi.org/10.62754/joe.v3i8.4807.

[50] Harb WA and Rowaished S. Employing artificial intelligence technology in developing practical content for media specialization – A case study of Palestine Technical University, Kadoorie. *Journal of Infrastructure, Policy and Development*. 2024;8(13). https://doi.org/10.24294/jipd7810.

[51] Hezam AMM and Alkhateeb A. Short stories and AI tools: An exploratory study. *Theory and Practice in Language Studies*. 2024;14(7):2053–2062. https://doi.org/10.17507/tpls.1407.12.

[52] Huo X and Siau KL. Generative artificial intelligence in business higher education: A focus group study. *Journal of Global Information Management*. 2024;32(1). https://doi.org/10.4018/JGIM.364093.

[53] Ilieva G, Yankova T, Klisarova-Belcheva S, Dimitrov A, Bratkov M, Angelov D, et al. Effects of generative chatbots in higher education. *Information*. 2023;14(9). https://doi.org/10.3390/info14090492.

[54] Jin Z and Kamsin A. Personalized learning model based on machine learning algorithms. *International Journal of Informatics and Communication Technology*. 2024;13(3):470–475. https://doi.org/10.11591/ijict.v13i3.pp470-475.

[55] Kamalov F, Santandreu Calonge D and Gurrib I. New era of artificial intelligence in education: Towards a sustainable multifaceted revolution. *Sustainability (Switzerland)*. 2023;15(16). https://doi.org/10.3390/su151612451.

[56] Kayalı B, Yavuz M, Balat Ş and Çalışan M. Investigation of student experiences with ChatGPT-supported online learning applications in higher education. *Australasian Journal of Educational Technology*. 2023;39(5):20–39. https://doi.org/10.14742/ajet.8915.

[57] Kiryakova G, Angelova N, Kiryakova G and Angelova N. ChatGPT-A challenging tool for the university professors in their teaching practice. *Education Sciences*. 2023;13(10). https://doi.org/10.3390/educsci13101056.

[58] Li Y and Li Y. The digital transformation of college english classroom: Application of artificial intelligence and data science. *EAI Endorsed Transactions on Scalable Information Systems*. 2024;11 (5):12. https://doi.org/10.4108/eetsis.5636 EA APR 2024.

[59] Nkedishu VC and Okonta V. Unpacking optimism versus concern: Tertiary students' multidimensional views on the rise of artificial intelligence (AI). *International Research Journal of Multidisciplinary Scope*. 2024;5(4):362–377. https://doi.org/10.47857/irjms.2024.05i04.01261.

[60] Noniashvili M, Matchavariani L, Noniashvili M and Matchavariani L. Opportunities and challenges for using artificial intelligence in academic continuity: Case of Georgia. *Journal of Eastern European and Central Asian Research*. 2024;11(4):813–827. https://doi.org/10.15549/jeecar.v11i4.1817.

[61] Núñez-Canal M, Fernández-Ardavín A, Díaz Marcos L and Aguado Tevar O. Integrative generative learning: A higher educational model for artificial intelligence challenges. *European Public and Social Innovation Review*. 2024;9. https://doi.org/10.31637/epsir-2024-1685.

[62] Opesemowo OAG and Adekomaya V. Harnessing artificial intelligence for advancing sustainable development goals in South Africa's higher education system: A qualitative study. *International Journal of Learning, Teaching and Educational Research*. 2024;23(3):67–86. https://doi.org/10.26803/ijl ter.23.3.4.

[63] Orok E, Okaramee C, Egboro B, Egbochukwu E, Bello K, Etukudo S, et al. Pharmacy students' perception and knowledge of chat-based artificial intelligence tools at a Nigerian University. *BMC Medical Education*. 2024;24(1):1237. https://doi.org/10.1186/s12909-024-06255-8.

[64] Pavlovic D, Soler-Adillon J and Stanisavljevic-Petrovic Z. A full-time professor dedicated to you: ChatGPT from university students' perspective. *Revista Espanola de Pedagogia*. 2024;82(289):563–584. https://doi.org/10.22550/2174-0909.4160.

[65] Pham T, Nguyen B, Ha S, Ngoc T, Pham T, Nguyen B, et al. Digital transformation in engineering education: Exploring the potential of AI-assisted learning. *Australasian Journal of Educational Technology*. 2023;39(5):1–19. https://doi.org/10.14742/ajet.8825.

[66] Pikhart M, Klimova B and Al-Obaydi LH. Exploring university students' preferences and satisfaction in utilizing digital tools for foreign language learning. *Frontiers in Education*. 2024;9. https://doi.org/10.3389/feduc.2024.1412377.

[67] Rapi M, Rusdi M and Idris R. Challenges and opportunities of artificial intelligence adoption in Islamic education in Indonesian higher education institutions. *International Journal of Learning, Teaching and Educational Research*. 2024;23(11):423–443. https://doi.org/10.26803/ijlter.23.11.22.

[68] Rasul T, Nair S, Kalendra D, Robin M, Santini FO, Ladeira WJ, et al. The role of ChatGPT in higher education: Benefits, challenges, and future research directions. *Journal of Applied Learning and Teaching*. 2023;6(1):41–56. https://doi.org/10.37074/jalt.2023.6.1.29.

[69] Rosales V, Estupiñán S, Mora D, Quinde Rosales VX, Garcia Estupinan SB and Tenelanda Mora DB. Artificial intelligence and its usefulness in the academic field: An analysis from the perspective of university students. *Revista Conrado*. 2024;20(99):187–193.

[70] Sajja R, Sermet Y, Cikmaz M, Cwiertny D, Demir I, Sajja R, et al. Artificial intelligence-enabled intelligent assistant for personalized and adaptive learning in higher education. *Information*. 2024;15 (10). https://doi.org/10.3390/info15100596.

[71] Sanabria-Z J and Olivo PG. AI platform model on 4IR megatrend challenges: Complex thinking by active and transformational learning. *Interactive Technology and Smart Education*. 2024;21(4):571–587. https://doi.org/10.1108/ITSE-07-2023-0145.

[72] Sarwanti S, Sariasih Y, Rahmatika L, Islam M, Riantina E, Sarwanti S, et al. Are they literate on ChatGPT? University language students' perceptions, benefits and challenges in higher education learning. *Online Learning*. 2024;28(3):105–130. https://doi.org/10.24059/olj.v28i3.4599.

[73] Shin C, Seo D, Jin S, Lee S, Park H, Shin C, et al. Educational technology in the university: A comprehensive look at the role of a professor and artificial intelligence. *IEEE Access*. 2024;12:116727–116739. https://doi.org/10.1109/ACCESS.2024.3447067.

[74]   Tapalova O and Zhiyenbayeva N. Artificial intelligence in education: AIEd for personalised learning pathways. *Electronic Journal of e-Learning*. 2022;20(5):639–653. https://doi.org/10.34190/ejel.20. 5.2597.

[75]   Ulla MB, Advincula MJC, Mombay CDS, Mercullo HMA, Nacionales JP and Entino-Señorita AD. How can GenAI foster an inclusive language classroom? A critical language pedagogy perspective from Philippine university teachers. *Computers and Education: Artificial Intelligence*. 2024;7. https://doi.org/ 10.1016/j.caeai.2024.100314.

[76]   Wu L. Students' foreign language learning adaptability and mental health supported by artificial intelligence. *Journal of Autism and Developmental Disorders*. 2024;54(10):3921–3932. https://doi.org/ 10.1007/s10803-023-06097-1.

[77]   Xiao Y and Zhi Y. An exploratory study of EFL learners' use of ChatGPT for language learning tasks: Experience and perceptions. *Languages*. 2023;8(3). https://doi.org/10.3390/languages8030212.

[78]   Xu X, Wang X, Zhang Y, Zheng R, Xu X, Wang X, et al. Applying ChatGPT to tackle the side effects of personal learning environments from learner and learning perspective: An interview of experts in higher education. *Plos One*. 2024;19(1). https://doi.org/10.1371/journal.pone.0295646.

[79]   Xu Q. Action research plan: A methodology to examine the impact of artificial intelligence (AI) on the cognitive abilities of university students. *Discover Education*. 2024;3(1). https://doi.org/10.1007/ s44217-024-00330-4.

[80]   Yefymenko T, Bilous T, Zhukovska A, Sieriakova I, Moyseyenko I, Yefymenko T, et al. Technologies for using interactive artificial intelligence tools in the teaching of foreign languages and translation. *Amazonia Investiga*. 2024;13(74):299–307. https://doi.org/10.34069/AI/2024.74.02.25.

[81]   Zhang X, Li D, Wang C, Jiang Z, Ngao AI, Liu D, et al. From ChatGPT to China' Sci-Tech: Implications for Chinese higher education. *Beijing International Review of Education*. 2023;5(3):296–314. https://doi.org/10.1163/25902539-05030007.

[82]   Zhang X, Sun J and Deng Y. Design and application of intelligent classroom for english language and literature based on artificial intelligence technology. *Applied Artificial Intelligence*. 2023;37(1). https://doi.org/10.1080/08839514.2023.2216051.

[83]   Zhang Y. Exploring blended learning models enhanced by mobile interactive technology in higher education. *International Journal of Interactive Mobile Technologies*. 2024;18(23):15–29. https://doi.org/ 10.3991/ijim.v18i23.52881.

[84]   Zhou X, Zhang J, Chan C, Zhou X, Zhang J and Chan C. Unveiling students' experiences and perceptions of Artificial Intelligence usage in higher education. *Journal of University Teaching and Learning Practice*. 2024;21(6):1–20. https://doi.org/10.53761/xzjprb23.

[85]   Vygotsky LS. *Mind in Society: The Development of Higher Mental Processes (E. Rice, Ed. & Trans.).* Cambridge, MA: Harvard University Press; (Original work published 1930, 1933 ... ; 1978).

[86]   Bussey MP, Inayatullah S and Milojevic I, editors. *Alternative Educational Futures: Pedagogies for an Emergent World*. Netherlands: Sense Publishers; 2008.

[87]   Nuñez-Canal M, Fernandez Ardavin A, Díaz-Marcos L and Aguado Tevar O. Aprendizaje Generativo integral: Un modelo para la educación superior ante los desafíos de la inteligencia artificial. *European Public and Social Innovation Review*. 2024;9:1–21. https://doi.org/10.31637/epsir-2024-1685.

Talip Gönülal

AI in Higher Education: When Do University Students Consider It Ethical or Unethical?

Abstract: This chapter investigated the ethical dimensions of artificial intelligence (AI) use in higher education through the perspective of university students. Using a qualitative research method, this study analyzed responses elicited from a group of English Language Teaching (ELT) students to explore how they conceptualized ethical boundaries when using AI in academic tasks. The findings revealed that students developed a foundational ethical framework distinguishing between assistive AI use (e.g., brainstorming and proofreading) and substitutive ones (e.g., completing assignments or homework). Most ELT students identified a threshold of 25–30% as an acceptable level of AI assistance in academic tasks. Their ethical judgments were affected by such factors as the amount of personal contribution, the nature of tasks, the intent of using AI, and transparency considerations. Students perceived ethical AI use as something that supported rather than replaced their intellectual engagement, while they defined unethical AI use as direct copying and complete reliance on AI tools. This chapter, therefore, contributes to our understanding of the recent ethical discussion of AI use in higher education and offers targeted recommendations for students, academics, and institutions.

Keywords: AI ethics, ethical AI use, unethical AI use, higher education, academic integrity, student perceptions

1 Introduction

The emergence of artificial intelligence (AI) platforms such as ChatGPT in late 2022 marked a significant turning point for many people. As these technologies become increasingly sophisticated and accessible, they have quickly influenced higher education by changing how teachers teach and how students learn [1–5]. AI tools offer students several affordances, from language learning and research to writing and editing [6–9]. However, as these technologies can now produce human-like text that is increasingly difficult to distinguish from content created by humans themselves [10], their use in academic tasks raises ethical concerns regarding academic honesty and integrity, pushing these issues to the forefront of educational discussions [11–15]. While some academics and students perceive AI as valuable academic support that

Talip Gönülal, Department of Foreign Languages Education, Erzincan Binali Yıldırım University, Turkey, e-mail: talip.gonulal@erzincan.edu.tr

https://doi.org/10.1515/9783112206393-004

improves their learning and productivity, others view overreliance on AI-generated outputs as a form of academic misconduct that harms the fundamental purposes of education, leading to a devaluation of academic degrees [12, 16].

These different views highlight the need to explore how students themselves conceptualize and deal with the ethical boundaries of AI usage in academic contexts. Although AI use in higher education is growing rapidly, institutions and universities have been slower to develop clear policies and guidelines. As a result, students, often the earliest adopters, are left to make ethical decisions on their own, without much direction from their institutions. Therefore, understanding how students distinguish between ethical and unethical AI use is essential not only for revealing the current state of AI use in higher education but also for shaping effective policies and encouraging responsible AI use moving forward. Although AI ethics has been widely studied in broader societal contexts [17–21], research specifically addressing students' ethical reasoning in academic AI use remains limited.

This chapter aims to address this research gap by examining Turkish university students' ethical orientations toward AI use in academic tasks, focusing on their perceptions, concerns, and decision-making processes. The Turkish higher education context provides a valuable perspective, as the country has rapidly embraced AI in higher education and established nationwide ethical guidelines for its use [22]. This study explores how Turkish university students define what constitutes ethical and unethical AI use within academic contexts and the various factors that influence their ethical judgments when utilizing AI tools for academic tasks. These insights can inform institutional discussions on AI policies and contribute to the development of clearer, more effective guidelines for responsible AI integration in higher education.

1.1 Growing Use of AI Tools in Educational Contexts

In recent years, the integration of AI in higher education has expanded drastically, with a wide range of tools becoming increasingly accessible to students and academics. These technologies range from simple grammar checkers and plagiarism detection software to sophisticated AI platforms capable of producing complex written content, solving problems, and engaging in human-like conversations [1, 6, 7, 23].

Among the most widely used AI tools by students are large language models such as ChatGPT, which has seen rapid and widespread adoption since its launch [24, 25]. In addition to ChatGPT, students also commonly used writing assistants like Grammarly, QuillBot, Wordtune, and Jenni [15, 26]. Research shows that these tools are primarily used for specific academic purposes. For instance, Gruenhagen et al. [13] found that Australian university students commonly used AI to search for information on a topic (60%), to gain a clearer understanding of subjects (48%), and to analyze texts or questions (34%). Furthermore, 37% of students reported using AI to support them in completing their assessments.

Besides general usage trends, several studies have explored the impact of AI tools on academic writing. Allen and Mizumoto [27], for example, examined how Japanese students used ChatGPT for editing and proofreading compared to traditional writing groups. Their study revealed a general preference for AI tools in revising academic work. Similarly, Wang et al. [28] studied Chinese medical students and found that ChatGPT not only reduced writing time but also improved the quality of scientific writing and increased students' satisfaction with the writing process. In the Turkish context, Yılmaz-Virlan and Tomak [29] investigated how Turkish English as foreign language (EFL) learners perceived AI tools in writing. Their findings showed that students primarily valued AI for generating ideas and enhancing language skills. In the same vein, Malik et al. [26] conducted a large-scale study with Indonesian undergraduate students, showing that AI-powered tools had a positive impact on academic writing by enhancing grammar accuracy, plagiarism detection, language translation, and essay organization. The study found that these tools improved students' overall writing abilities, self-efficacy, and understanding of academic integrity. However, students also expressed concerns about potential negative effects on creativity, critical thinking, and ethical writing. Importantly, Malik et al. [26] emphasized the need for a balanced integration of AI and human authorship, highlighting that AI should not replace human originality and critical analysis in academic essay writing.

Additionally, research across various countries [3–5] has also identified key factors influencing students' intentions to use AI tools. These studies have consistently found that performance expectancy and habit are strong predictors of AI adoption in academic contexts. While some regional differences exist – such as the role of hedonic motivation in Nordic countries and social influence in the USA – the anticipated academic benefits of AI remain a common driver. These findings highlight the need to consider students' motivations and usage patterns when evaluating the growing role of AI in higher education.

1.2 Conceptualizing AI Ethics in Higher Education

In tandem with the increased use of AI tools in higher education, scholarly work on AI ethics has received significant attention. Recent research has emphasized several key ethical principles guiding responsible AI use, including transparency, fairness, accountability, non-maleficence, privacy, accessibility, and bias [3, 17–20]. These principles can provide a framework for evaluating and addressing the ethical implications of AI technologies in educational contexts. For example, transparency in AI ethics refers to the disclosure of AI use in academic work, explicitly stating how AI tools are used, and their roles throughout the academic work [21]. In line with this, fairness requires that AI tool use does not disadvantage specific student populations [31]. In educational contexts, accountability requires determining who should be responsible for AI-generated content or decisions – whether it be developers, institutions, teachers, or students [20].

The ethical dimensions of AI use in higher education have given rise to new forms of potential misconduct. One emerging concept is "AI-giarism," a blend of AI and plagiarism [32]. While traditional academic ethics have focused on plagiarism, commonly defined as using someone else's work without giving them credit [33], plagiarism itself involves complex judgments influenced by factors such as the quantity of copied content, citation practices, writer intent, academic level, and language context [34–37].

The increased use of AI in education further complicates plagiarism discussions. Chan [11] described AI-giarism as "the unethical practice of using AI technologies, particularly generative language models, to generate content that is plagiarized either from original human-authored work or directly from AI-generated content, without appropriate acknowledgment of the original sources or AI's contribution" (p. 3). This definition points out two key ethical concerns: the misuse of human-authored content through AI tools and the misrepresentation of AI-generated text as original student work. Hirmiz [38] added a valuable distinction between assistive AI, which supports users without replacing them, and substitutive AI, which performs tasks in place of the user. This distinction offers a valuable lens for examining the ethical implications of various AI applications in academic settings.

Research on student perceptions of AI use in academic settings reveals complex and sometimes contradictory attitudes toward ethical boundaries. Several studies indicate that students generally recognize AI's potential benefits for improving learning while also expressing concerns about potential misuse [11, 13, 23, 24, 39]. For instance, Bayraktar Balkır and Zehir Topkaya's [39] investigation on how Turkish EFL students understood and responded to the ethical challenges of using AI in academic contexts showed that students were moving beyond seeing AI use as simply right or wrong. Instead, they were developing more thoughtful and context-based views, often stressing the importance of using AI in moderation, giving proper credit, and maintaining personal effort. Many students perceived AI as a useful learning tool when it helped rather than replaced their own work. At the same time, they worried about the risks of plagiarism and relying too much on AI, which they thought could harm their learning progress. On the other hand, a study by Gruenhagen et al. [13] reported that more than one-third of Australian university students had used chatbots for assessment assistance without perceiving this use as a breach of academic integrity. This suggests that there might be a gap between institutional policies and what students think is acceptable AI use. Similarly, Chan [11] reported that students had complex and varied views on the ethics of using AI. Although they clearly disapproved of using AI to directly create work and present it as their own, they were less sure about the ethical problems of using AI in less direct ways. Specifically, students found it hard to decide what was acceptable when using AI for things like generating ideas, making outlines, or improving language. This mix of attitudes showed that there is a strong need for further research in this area and clearer guidelines regarding how AI should be used properly in higher education.

However, current institutional policies on AI in higher education often remain vague and overly broad, probably due to the limited evidence on actual AI usage patterns [40]. In response, several scholars [14, 41, 42] suggested that AI-related policies should name specific tools and clearly define what uses are acceptable and unacceptable. They recommended that any restrictions or requirements for using AI tools should be clear and easy to understand, with real examples to help both students and academics understand how to use them responsibly. Furthermore, they suggested that users should be provided with good AI literacy training to help them use these tools responsibly and ethically.

Notwithstanding the aforementioned research on AI technologies in higher education and the ongoing discussions surrounding ethical AI use, a significant gap exists in our understanding of how students themselves interpret and apply these principles in real academic settings. While existing studies have examined general ethical principles related to AI use and suggested AI policies, less attention has been paid to students' ethical decision-making processes when engaging with AI tools for academic purposes. Specifically, there is a lack of understanding of how students deal with the ethical challenges of AI use, including their definitions of ethical versus unethical AI use and the factors that influence their ethical judgments in different academic contexts. This study, therefore, aimed to investigate how university students defined ethical versus unethical uses of AI in their academic work. The goal was to offer information that can help create better university policies and teaching practices.

2 Methodology

Considering the purposes of this study, a qualitative research approach was employed because it allowed for an in-depth exploration of participants' perspectives, experiences, and reasoning processes [43]. The study utilized thematic analysis as the primary analytical framework, following Braun and Clarke's [44] guidelines.

Participants in this study were chosen through purposive sampling, as they possessed specific characteristics relevant to the research focus. The sample consisted of 31 fourth-year students (65% of whom were female) from an English Language Teaching (ELT) program at a state university in Turkey. These students had substantial experience with using AI tools in various academic contexts. Importantly, at the time of the study, 71% of them were enrolled in a specialized course on AI-enhanced ELT, which provided them with structured opportunities to engage with different AI technologies in diverse academic tasks. Their frequent and varied use of AI tools meant that they often encountered situations requiring ethical consideration, whether consciously or unconsciously. This made them particularly valuable participants for this study to examine how ethical issues surrounding AI use in academic tasks were perceived and discussed by university students.

Data were collected through an 18-item questionnaire containing open-ended questions to elicit detailed responses about participants' experiences with and perceptions of AI use in academic tasks. The questionnaire was developed based on a review of relevant literature and refined through consultation with two experts in educational technology and research methodology. The questionnaire included questions addressing students' background information, AI usage patterns, ethical perceptions, specific AI use cases, and ethical reasoning. The questionnaire was administered face-to-face, and participants were free to respond to the questions either in English or Turkish. This helped them share their thoughts and experiences in a more detailed and meaningful way. Additionally, the open-ended nature of the questions allowed participants to detail their perspectives, providing rich qualitative data for analysis.

To increase methodological rigor, several strategies were implemented during data collection and analysis. Thematic analysis followed Braun and Clarke's [44] six-step model, beginning with familiarization and followed by systematic coding, theme generation, review, definition, and reporting. To improve coding reliability, the dataset was initially coded by the researcher and then reviewed by a qualitative research expert. Acknowledging the potential influence of the researcher's dual role as instructor and investigator, participant anonymity was maintained, and responses were coded inductively to capture emergent themes rather than to impose preexisting assumptions.

Ethical approval for the study was obtained from the Ethics Committee of Erzincan Binali Yıldırım University. All participants were informed about the purpose of the study and signed consent forms prior to participation. Participation was voluntary, and students were assured of confidentiality and anonymity in data reporting.

3 Results

3.1 AI Usage Patterns

Participants in this study showed a moderate to high level of AI familiarity, with an average familiarity level of 3.48. The frequency of AI use among participants closely mirrored their familiarity levels, with an average usage frequency level of 3.42 (Table 1). Additionally, participants generally perceived AI tools as providing an academic advantage in their courses, with an average score of 3.65. Additionally, participants believed, on average, that approximately 28% of AI use was acceptable in academic tasks. However, the relatively high standard deviation suggests that students had varying perceptions of what constituted an acceptable threshold, indicating a lack of consensus on the ethical boundaries of AI use in academic settings.

Table 1: AI usage patterns.

	Min.	Max.	Median	*M*	SD
AI familiarity	2	5	3	3.48	.72
AI use frequency	2	5	3	3.42	.72
Gaining academic advantage with AI	2	5	4	3.65	.75
Acceptable AI use percentage	5	55	30	28.39	11.79

In looking at the specific AI tools used by participants (Table 2), out of 17 different AI tools mentioned, ChatGPT emerged as the most commonly used AI tool, used by 87% of the students. Other notable tools included Grammarly (26%) and Gemini (23%).

When asked about their primary motivations for using AI tools, participants reported being motivated primarily by the desire to generate ideas and enhance creativity (73%), to save time and be more efficient (57%), and to get help with language, especially to improve their written English (47%). Less frequently mentioned motivations included easier access to desired information (17%).

Table 2: AI usage behaviors.

		N	%
AI tools used			
	ChatGPT	27	87.1
	Grammarly	8	25.8
	Gemini	7	22.6
	QuillBot	3	9.7
	Other tools	7	22.6
Reasons for using AI			
	Creativity/idea generation	22	73.3
	Time-saving/efficiency	17	56.7
	Language support	14	46.7
	Practicality/access to information	5	16.7
Academic stages AI used			
	Brainstorming	17	56.7
	Research	14	46.7
	Proofreading	11	36.7
	Editing and revising	8	26.7
	Drafting and writing	4	13.3
	Getting feedback	2	6.7

Participants reported using AI tools for a variety of academic tasks, with brainstorming being the most common purpose (57%), followed by research (47%), proofreading

(37%), and editing and revising their work (27%). Interestingly, drafting and writing content were mentioned by only 13% of participants. This shows that students mostly use AI to prepare for their work or to improve it at the end, not to create the main content from scratch.

3.2 Student Definitions of Ethical and Unethical AI Use

When participants were asked to define ethical and unethical AI use in academic tasks, they mentioned several key dimensions of AI use (Table 3). First, participants characterized ethical AI use by significant human involvement, with AI serving as a supplementary tool that enhances rather than replaces their effort. That is, participants emphasized maintaining human agency and responsibility in the learning process, with AI playing a supportive role. In contrast, according to the participants, unethical AI use included minimal human involvement, with complete delegation of academic responsibilities to AI tools. The ethical-unethical boundary appears to be determined largely by the balance between human and AI contributions to the final product. For example, participants reported that up to almost 30% of AI use is acceptable in academic tasks.

Table 3: Student definitions of ethical and unethical AI use.

Dimensions	Student definitions
Degree of human involvement	Ethical: – Using AI for brainstorming or drafting where the AI contribution remains minimal. – Minimal AI use in academic tasks is ethical. Unethical: – If AI completes the entire work, it is unethical. – Having AI complete all the work is entirely unethical. We cannot transfer the burden and responsibility to AI.
Purpose and context of use	Ethical: – Using AI for brainstorming before starting an assignment, for feedback after completing the assignment, and for proofreading to identify typos and errors is ethical. Unethical: – Using AI during examinations or having AI complete the whole assignments is not ethical.

Table 3 (continued)

Dimensions	Student definitions
Transparency and attribution	Ethical: – Verifying sources, confirming their reliability, and properly crediting AI assistance is completely ethical. Unethical: – Presenting AI-generated work as your own without proper attribution.
Engagement with AI outputs	Ethical: – Not solely relying on AI and critically evaluating AI-generated output. – Cross-checking would be an excellent example. In other words, using AI as an assistant rather than the primary source. Unethical: – Copying and pasting AI-generated information without verifying its accuracy. – Using AI-generated output without any modification or evaluation.

Participants also distinguished between ethical and unethical AI use based on purpose and context. Using AI for idea generation, proofreading, or as a reference tool was generally considered ethical, whereas using AI to complete entire assignments or exams was considered unethical. In addition, participants emphasized the importance of transparency about AI assistance and appropriate attribution when AI-generated content is used.

Also, participants stressed the importance of critically engaging with AI outputs rather than accepting them blindly. Participants' ethical AI use involved verification, cross-checking, and critical evaluation of AI-generated content, whereas unethical use involved direct copying of AI output without any critical engagement.

3.3 Factors Influencing Ethical Judgments

Apart from asking participants to provide their definitions of ethical and unethical AI use, they were also asked to describe cases where they used AI tools and they believed that their AI use was completely ethical, completely unethical but they used it anyways, or they were not sure whether it was ethical or unethical.

As presented in Table 4, participants' ethical judgments were influenced by multiple factors. When explaining why they considered certain AI uses ethical, their reasoning often centered on maintaining significant personal contribution, using AI for enhancement rather than replacement of their own thinking, and viewing it as a

Table 4: Sample cases where AI use was considered ethical, unethical, or uncertain.

Level of Ethicality	Cases	Ethical Reasoning
Ethical AI use	– Once I used AI to find some creative ways to organize my presentation for a class.	Because it gives me inspiration and does not replace my own work.
	– I usually get immediate assistance on any problems during preparing my homework. I think this ethical.	Because I am getting only advice and feedback on my problems.
	– I wrote my assignment myself but used AI to check my grammar and word choices.	Because I use it as a good grammar checker. It's not changing the content or ideas that were entirely mine.
	– After finishing my essay, I asked AI to review it and give me suggestions for improvement.	Because it is very similar to get feedback from a more knowledgeable peer.
Unethical AI use	– I was once tasked with reading a paragraph and writing what I have learned about it. I used AI to do it, effectively cheating my way out of a homework.	Because I learned nothing which was the whole point of that task. Yet, the assignment was still regarded as passable.
	– I used ChatGPT to write a significant portion of my final paper	Because I submitted AI-generated content as my own work.
	– I used it during an online exam when we were not supposed to.	Because my use of AI violated the rules of the exam.
Ethically uncertain AI use	– I had to paraphrase some texts, so I had ChatGPT do it for me. I did not know it was ethical or not though. In order to paraphrase I had to do a full reading but I didn't.	Because I was uncertain about the ethical aspects of using AI to paraphrase texts without doing the full reading while it saved time.
	– In a part-time job application, I used AI to create a quick CV for me. I inserted a few details about me and AI drafted a really good CV for me that I could never create such a CV myself	I was able to get the job because they were impressed by my achievements in the CV. I was happy but at the same time I was not sure if I could get the same job if I had prepared the cv myself.

form of secondary assistance. For instance, one participant described his use of AI to better organize his presentation for a class and justified it as ethical by stating "Because it gives me inspiration and does not replace my own work." Many participants emphasized that using AI for brainstorming, feedback, or language correction was ethical as long as they did most of the work and used AI minimally or treated its suggestions merely as suggestions.

Factors contributing to ethical uncertainty included the potential for unfair advantage, unclear boundaries, and concerns about plagiarism. For example, one participant expressed uncertainty about using AI tools to paraphrase where the personal contribution was limited or minimal, even if their intention was not to cheat. Other participants worried about the unfair advantage they gained compared to those who did not use AI tools. Some expressed a fear of unintentionally cheating, which highlights the anxiety surrounding the risk of inadvertently crossing ethical lines.

When explaining why they considered certain AI uses unethical, participants mentioned factors such as the perceived lack of learning, minimal personal effort, dishonesty, and violation of academic regulations. One participant stated "Because I learned nothing, which was the whole point of that task . . ." showing that undermining the educational purpose of assignments was a key determinant in their ethical judgments. Another participant simply stated "My use of AI violated the rules of the exam" emphasizing academic dishonesty and rule-breaking as clear indicators of unethical use.

4 Discussion and Conclusion

This study shows that university students have developed some basic ethical frameworks for evaluating AI use in academic contexts. Rather than viewing AI use as inherently ethical or unethical, participants demonstrated a contextual understanding that distinguishes between different applications, purposes, and degrees of personal contribution. This aligns with recent research suggesting that ethical judgments about AI use are highly context-dependent and multifaceted [11, 26, 30, 38, 41].

Students generally consider AI use most ethical when it serves supportive functions (e.g., brainstorming, proofreading, and feedback) rather than substitutive ones (e.g., writing complete assignments). This distinction closely echoes Hirmiz's [38] conceptualization of assistive versus substitutive AI tools, suggesting that students intuitively recognize the ethical implications of different types of AI assistance. The emphasis on using AI for early-stage work (e.g., brainstorming and doing initial research) and refinement (e.g., proofreading and editing) rather than content production indicates a preference for maintaining intellectual ownership while using AI to increase productivity and creativity.

Although there is no clear consensus among the students in this study, they generally consider 25–30% AI assistance as the threshold beyond which its use in academic work becomes ethically questionable. This perceived threshold reflects their intuitive sense of proportionality and provides a quantitative measure for checking acceptable levels of AI integration from the student perspective. To the best of my knowledge, the existing literature, while discussing the issue of overreliance on AI [12, 16], does not offer comparable thresholds. However, this finding aligns with Chan's [11] discussion of AI-giarism, which suggests that ethical concerns arise when AI-generated content begins to overshadow human originality. Similarly, Malik et al. [26] emphasize that academic work should remain primarily the product of human intellectual effort.

However, it is important to note here that the proportionality of AI use, while significant, is not the only factor that influences students' ethical judgments. In fact, they consider multiple points when evaluating AI use, including the degree of intellectual development, task specificity, fairness, time efficiency, and transparency. This multifaceted approach to ethical reasoning suggests that students are balancing competing values and considerations rather than applying simplistic rules or guidelines. When participants described instances of AI use they considered unethical, they frequently mentioned that the use of AI undermined the learning purpose of the assignment (e.g., "I learned nothing, which was the whole point of that task"). This suggests that students' ethical frameworks are anchored in an understanding of education as a process of intellectual development rather than merely the production of academic outputs. To some extent, this aligns with the concerns raised by several scholars [13, 26] that AI, if used improperly, may hinder critical thinking and learning.

The findings of this study have significant implications for academic integrity policies in higher education. First, they suggest that simplistic prohibitions or permissions regarding AI use may be insufficient to address the complex ethical judgments that students are already making. Instead, policies might be more effective if they provide clear guidelines for appropriate AI use in different academic contexts, recognizing that some uses are more ethically acceptable than others. The emphasis participants placed on personal contribution suggests that policies should focus on maintaining meaningful intellectual engagement rather than simply detecting or preventing AI use. This might involve redesigning assessments to emphasize process over product, requiring students to document their thinking processes, or incorporating more personal reflections that cannot be easily generated by AI. As Chan [11] noted, assessment design that emphasizes critical thinking, application, and reflection may be more effective than technological detection in promoting academic integrity in the age of AI. The acceptable thresholds identified by participants could inform specific guidelines regarding appropriate levels of AI assistance. While exact percentages may be difficult to enforce, policies could establish general expectations that academic work should remain primarily the product of human intellectual effort, with AI serving a supportive rather than substitutive role [38].

The importance of transparency and acknowledgment in participants' ethical frameworks suggests that policies should emphasize proper citation and disclosure of AI assistance [17, 20, 21]. This aligns with emerging practices in academic publishing, where journals increasingly require authors to disclose AI use in manuscript preparation. Educational institutions could adopt similar requirements for student work, promoting a culture of honesty rather than secrecy around AI use. Finally, the findings highlight the importance of educating students about the ethical dimensions of AI use rather than simply imposing rules. The fact that participants in this study showed basic ethical reasoning that balanced multiple considerations shows that AI training that helps them develop critical AI literacy may be more effective in the long term [41].

However, several limitations of this study should be acknowledged. First, the small sample size limits the generalizability of the findings to other student populations. Future research should explore diverse perspectives across different disciplines, educational levels, and cultural contexts to develop a more comprehensive understanding of student ethical frameworks. Second, the study captured a snapshot of student perceptions at a specific moment in a rapidly evolving technological environment. Longitudinal studies investigating how ethical perceptions change as AI technologies develop and become more integrated into academic practices would provide valuable insights into the evolution of ethical standards. Third, the study focused exclusively on student perspectives, without direct comparison to institutional policies or academic views. Future research should examine alignment between student ethical frameworks and institutional guidelines to identify potential gaps or inconsistencies.

Based on the findings of this study, several recommendations can be offered for educational practice in the age of AI. These recommendations aim to support ethical, transparent, and educationally meaningful AI use by addressing the responsibilities of students, academics, and educational institutions.

Recommendations for students:

1. *Practice transparency:* Clearly disclose the use of AI tools in academic work. Use available institutional templates or acknowledgment statements to specify when and how AI contributed to an assignment.

2. *Maintain intellectual ownership:* Use AI tools as assistive aids (e.g., for brainstorming ideas, doing research, or proofreading) rather than as substitutes (e.g., generating entire assignments). Ensure that your academic work reflects your original thinking and efforts.

3. *Verify AI-generated content*: Critically evaluate information produced by AI tools. Cross-check information with reliable sources and remain alert to potential AI hallucinations or biases.

4. *Develop AI literacy*: Actively engage in AI literacy training opportunities provided by your institution. Understand the affordances and risks of generative AI tools, especially in relation to academic honesty, bias, and data privacy.

Recommendations for academics:
1. *Redesign assessments:* In your assignments and assessments, shift from product-only evaluations to process-oriented assessments (e.g., drafts, revision logs, and reflective statements). Require students to document their thought process and any AI involvement.
2. *Clarify expectations:* Provide clear, assignment-specific guidelines on acceptable and unacceptable uses of AI. Co-create AI use guidelines with students to build shared responsibility and mutual understanding around ethical academic practices.
3. *Integrate AI literacy into instruction:* Include structured activities that teach ethical AI use, such as critical analysis of AI output, citation exercises, and classroom debates on emerging dilemmas.
4. *Model ethical AI use:* Use AI tools responsibly in your own teaching and research. Disclose your use of AI in syllabi, presentations, and feedback to normalize ethical practices.

Recommendations for institutions:
1. *Promote AI literacy:* Offer workshops, courses, or online training that can educate students, academics, and staff about the ethical, practical, and disciplinary implications of AI use in higher education.
2. *Develop clear AI policies*: Rather than straightforward prohibitions or permissions, create policies that acknowledge the diverse range of AI applications, from ethically acceptable assistive uses (e.g., brainstorming and proofreading) to more questionable substitutive uses (e.g., full assignment generation). Involve academics, students, and ethics committees in developing these policies to ensure they reflect a broad, inclusive understanding of academic integrity in the age of AI.
3. *Design practical AI-ethics evaluation tools:* Create rubrics, checklists, or matrices to help instructors and administrators assess whether student AI use is ethical. These tools could consider dimensions such as AI-human contribution ratio, intent, attribution, and task relevance.
4. *Encourage transparency and disclosure practices:* Create institutional templates for disclosing AI use in assignments, theses, or research papers, fostering a culture of academic honesty and reflection.
5. *Differentiate between AI tools and use cases:* Recognize that not all AI tools serve the same purpose. For example, Grammarly may support surface-level corrections, while ChatGPT can generate full-text responses. Policies should reflect these distinctions in permitted and prohibited uses.

6. *Establish AI ethics committees or advisory boards:* Form special committees or boards to oversee AI-related policy development, review emerging AI use cases, and provide ongoing guidance as AI tools and their uses continue to evolve.

In summary, this study reveals that Turkish university students engage in thoughtful and context-sensitive ethical reasoning, differentiating between assistive uses of AI (e.g., brainstorming and proofreading) and substitutive uses (e.g., full assignment generation). These findings not only contribute to the discourse on academic integrity but also offer practical insights for shaping institutional AI policies and teaching practices. In line with the book's theme of AI in higher education, this chapter emphasizes the importance of aligning AI use with pedagogical goals, student agency, and transparency. Future research could adopt longitudinal designs to track how students' ethical perceptions evolve as AI policies mature and as generative AI tools become more integrated into academic workflows. Additionally, cross-cultural comparative studies would further enhance our understanding of both shared patterns and culturally specific differences in students' approaches to ethical AI use.

Disclosure of AI use: GPT-4.1 mini and Gemini 2.5 flash were used for grammar and style edits of this chapter. Manus AI was also used to edit the reference list in the Vancouver referencing style. All content and intellectual contributions remain the sole responsibility of the author.

References

[1] Ahmad N, Murugesan S and Kshetri N. Generative artificial intelligence and the education sector. *Computer*. 2023;56(6):72–76. https://doi.org/10.1109/MC.2023.3263576.

[2] Zou B, Reinders H, Thomas M and Barr D. Using artificial intelligence technology for language learning. *Frontiers in Psychology*. 2023;14:1287667. https://doi.org/10.3389/fpsyg.2023.1287667.

[3] Faraon M, Rönkkö K, Milrad M and Tsui E. International perspectives on artificial intelligence in higher education: An explorative study of students' intention to use ChatGPT across the Nordic countries and the USA. *Education and Information Technologies*. 2025;1–46. https://doi.org/10.1007/s10639-025-13492-x.

[4] Lavidas K, Voulgari I, Papadakis S, Athanassopoulos S, Anastasiou A, Filippidi A, Komis V and Karacapilidis N. Determinants of humanities and social sciences students' intentions to use artificial intelligence applications for academic purposes. *Information*. 2024;15(6):314. https://doi.org/10.3390/info15060314.

[5] Ishmuradova II, Chistyakov AA, Brodskaya TA, Kosarenko NN, Savchenko NV and Shindryaeva NN. Latent profiles of AI learning conditions among university students: Implications for educational intentions. *Contemporary Educational Technology*. 2025;17(2):ep565. https://doi.org/10.30935/cedtech/15907.

[6] Crompton H and Burke D. Artificial intelligence in higher education: The state of the field. *International Journal of Educational Technology in Higher Education*. 2023;20:22. https://doi.org/10.1186/s41239-023-00392-8.

[7]   Gayed JM, Carlon MKJ, Oriola AM and Cross JS. Exploring an AI-based writing assistant's impact on English language learners. *Computers and Education: Artificial Intelligence*. 2022;3:100055. https://doi.org/10.1016/j.caeai.2022.100055.

[8]   Sari N. The role of artificial intelligence (AI) in developing English language learners' communication skills. *Journal on Education*. 2023;6(1):750–757. https://doi.org/10.31004/joe.v6i1.2990.

[9]   Song C and Song Y. Enhancing academic writing skills and motivation: Assessing the efficacy of ChatGPT in AI-assisted language learning for EFL students. *Frontiers in Psychology*. 2023;14:1260843. https://doi.org/10.3389/fpsyg.2023.1260843.

[10] Casal JE and Kessler M. Can linguists distinguish between ChatGPT/AI and human writing? A study of research ethics and academic publishing. *Research Methods in Applied Linguistics*. 2023;2 (3):100068. https://doi.org/10.1016/j.rmal.2023.100068.

[11] Chan CKY. Students' perceptions of 'AI-giarism': Investigating changes in understandings of academic misconduct. *Education and Information Technologies*. 2024. https://doi.org/10.1007/s10639-024-13151-7.

[12] Eke DO. ChatGPT and the rise of generative AI: Threat to academic integrity?. *Journal of Responsible Technology*. 2023;13:100060. https://doi.org/10.1016/j.jrt.2023.100060.

[13] Gruenhagen JH, Sinclair PM, Carroll JA, Baker PR, Wilson A and Demant D. The rapid rise of generative AI and its implications for academic integrity: Students' perceptions and use of chatbots for assistance with assessments. *Computers and Education: Artificial Intelligence*. 2024;7:100273. https://doi.org/10.1016/j.caeai.2024.100273.

[14] Roe J, Renandya WA and Jacobs GM. A review of AI-powered writing tools and their implications for academic integrity in the language classroom. *Journal of English and Applied Linguistics*. 2023;2(1):3. https://doi.org/10.59588/2961-3094.1035.

[15] Yeo MA. Academic integrity in the age of artificial intelligence (AI) authoring apps. *TESOL Journal*. 2023;14(3):e716. https://doi.org/10.1002/tesj.716.

[16] DR C, PA C and Shipway JR. Chatting and cheating: Ensuring academic integrity in the era of ChatGPT. *Innovations in Education and Teaching International*. 2024;61(2):228–239. https://doi.org/10.1080/14703297.2023.2190148.

[17] Floridi L, Cowls J, Beltrametti M, Chatila R, Chazerand P, Dignum V, et al. AI4People – An ethical framework for a good AI society: Opportunities, risks, principles, and recommendations. *Minds and Machines*. 2018;28:689–707. https://doi.org/10.1007/s11023-018-9482-5.

[18] Jobin A, Ienca M and Vayena E. The global landscape of AI ethics guidelines. *Nature Machine Intelligence*. 2019;1(9):389–399. https://doi.org/10.1038/s42256-019-0088-2.

[19] Holmes W, Porayska-Pomsta K, Holstein K, Sutherland E, Baker T, Shum SB, et al. Ethics of AI in education: Towards a community-wide framework. *International Journal of Artificial Intelligence in Education*. 2022. https://doi.org/10.1007/s40593-021-00239-1.

[20] Huang C, Zhang Z, Mao B and Yao X. An overview of artificial intelligence ethics. *IEEE Transactions on Artificial Intelligence*. 2022;4(4):799–819. https://doi.org/10.1109/TAI.2022.3194503.

[21] Weaver K. The artificial intelligence disclosure (AID) framework: An introduction. *College and Research Libraries News*. 2023;84(9):392–395. https://doi.org/10.5860/crln.85.10.407.

[22] YÖK. Yükseköğretim kurumları bilimsel araştırma ve yayın faaliyetlerinde üretken yapay zekâ kullanımına dair etik rehber [Internet]. Ankara: Yükseköğretim Kurulu; 2024 [cited 2025 Apr 26]. Available from: https://www.yok.gov.tr/Documents/2024/yapay-zeka-kullanimina-dair-etik-rehber.pdf

[23] Farhi F, Jeljeli R, Aburezeq I, Dweikat FF, Al-shami SA and Slamene R. Analyzing the students' views, concerns, and perceived ethics about ChatGPT usage. *Computers and Education: Artificial Intelligence*. 2023;5:100180. https://doi.org/10.1016/j.caeai.2023.100180.

[24] Ngo TTA. The perception by university students of the use of ChatGPT in education. *International Journal of Emerging Technologies in Learning*. 2023;18(17):4. https://doi.org/10.3991/ijet.v18i17.39019.

[25]   Zhai X. ChatGPT user experience: Implications for education. *SSRN*. 2022. https://doi.org/10.2139/ssrn.4312418.

[26]   Malik AR, Pratiwi Y, Andajani K, Numertayasa IW, Suharti S and Darwis A. Exploring artificial intelligence in academic essay: Higher education students' perspective. *International Journal of Educational Research Open*. 2023;5:100296. https://doi.org/10.1016/j.ijedro.2023.100296.

[27]   Allen TJ and Mizumoto A. ChatGPT over my friends: Japanese English-as-a-Foreign-Language learners' preferences for editing and proofreading strategies. *RELC Journal*. 2024. https://doi.org/10.1177/00336882241262533.

[28]   Wang J, et al. The impact of using ChatGPT on academic writing among medical undergraduates. *Annals of Medicine*. 2024. https://doi.org/10.1080/07853890.2024.2426760.

[29]   Yılmaz Virlan A and Tomak B. A Q method study on Turkish EFL learners' perspectives on the use of AI tools for writing: Benefits, concerns, and ethics. *Language Teaching Research*. 2024. https://doi.org/10.1177/13621688241308836.

[30]   Ryan M and Stahl BC. Artificial intelligence ethics guidelines for developers and users: Clarifying their content and normative implications. *Journal of Information Communications and Ethics in Society*. 2020;19(1):61–86. https://doi.org/10.1108/JICES-12-2019-0138.

[31]   Baker RS and Hawn A. Algorithmic bias in education. *International Journal of Artificial Intelligence Education*. 2022. https://doi.org/10.1007/s40593-021-00285-9.

[32]   Khalaf MA. Does attitude towards plagiarism predict aIgiarism using ChatGPT?. *AI and Ethics*. 2025;5(1):677–688. https://doi.org/10.1007/s43681-024-00426-5.

[33]   Park C. In other (people's) words: Plagiarism by university students–literature and lessons. *Assessment and Evaluations in Higher Education*. 2003;28(5):471–488. https://doi.org/10.1080/02602930301677.

[34]   Armstrong JDII. Plagiarism: What is it, whom does it offend, and how does one deal with it?. *American Journal of Roentgenology*. 1993;161(3):479–484. https://doi.org/10.2214/ajr.161.3.8352091.

[35]   Childers D and Bruton S. Should it be considered plagiarism? Student perceptions of complex citation issues. *Journal of Academic Ethics*. 2016;14:1–17. https://doi.org/10.1007/s10805-015-9250-6.

[36]   Sowden C. Plagiarism and the culture of multilingual students in higher education abroad. *ELT Journal*. 2005;59(3):226–233. https://doi.org/10.1093/elt/cci042.

[37]   Wager E. Defining and responding to plagiarism. *Learned Publishing*. 2014;27:33–42. https://doi.org/10.1087/20140105.

[38]   Hirmiz R. *The Ethical Implications of AI in Healthcare [Dissertation]*. Toronto: York University; 2024.

[39]   Bayraktar Balkir N and Zehir Topkaya E. Evolving perceptions of AI use and academic integrity: Insights from EFL learners in Turkish higher education. *Journal of Academic Ethics*. 2025. https://doi.org/10.1007/s10805-025-09632-0.

[40]   Chan CKY. A comprehensive AI policy education framework for university teaching and learning. *International Journal of Educational Technology in Higher Education*. 2023;20:38. https://doi.org/10.1186/s41239-023-00408-3.

[41]   Perkins M. Academic integrity considerations of AI large language models in the post-pandemic era: ChatGPT and beyond. *Journal of University Teaching and Learning Practice*. 2023;20(2):1–24. https://doi.org/10.53761/1.20.02.07.

[42]   Ugwu NF, Igbinlade AS, Ochiaka RE, Ezeani UD, Okorie NC, Opele JK, et al. Clarifying ethical dilemmas in using artificial intelligence in research writing: A rapid review. *Higher Learning Research Communications*. 2024;14(2):29–47. doi: 10.18870/hlrc.v14i1.1549.

[43]   Creswell JW and Poth CN. *Qualitative Inquiry and Research Design: Choosing among Five Approaches*. 4th ed. Thousand Oaks (CA): Sage Publications; 2016.

[44]   Braun V and Clarke V. Using thematic analysis in psychology. *Qualitative Research in Psychology*. 2006;3(2):77–101. https://doi.org/10.1191/1478088706qp063oa.

Rama Yusvana*, Azli Nawawi

How Gamified Pedagogies and AI Tools Influence Student Engagement and Retention in Higher Education Learning Environments

Abstract: Higher education institutions face ongoing challenges in sustaining student engagement and retention within digital and hybrid learning environments, worsened by the COVID-19 pandemic and unequal resource access. Traditional pedagogical approaches often fail to address these complexities, requiring the integration of emerging technologies with inclusive strategies. This chapter synthesizes the transformative potential of gamified pedagogies and artificial intelligence (AI)-driven tools, such as generative AI, virtual reality (VR), and adaptive learning platforms, in improving engagement and retention. Based on self-determination theory and behavioral psychology, the analysis bridges theoretical frameworks with empirical case studies across disciplines, including AI-mediated collaborative environments, VR-based experiential learning, and hybrid massive open online courses. Key findings reveal that gamification elements (e.g., badges and leaderboards) and AI personalization significantly enhance intrinsic motivation, peer collaboration, and self-directed learning. However, ethical dilemmas emerge, including risks to academic integrity, content homogenization, and over-reliance on automated systems, alongside infrastructural barriers such as digital divides and inadequate faculty training. The chapter advocates for scalable implementation strategies emphasizing faculty professional development, equitable infrastructure investment, and policy frameworks that balance innovation with accountability. By proposing interdisciplinary approaches to mitigate ethical risks and systemic inequities, this work contributes a roadmap for institutions to harness AI and gamification effectively while prioritizing pedagogical rigor and inclusivity. Implications extend to policymakers, educators, and EdTech developers, underscoring the need for collaborative efforts to align technological advancements with equitable, student-centered learning ecosystems.

Keywords: Gamified pedagogies, AI-driven tools, student engagement, higher education, generative AI

*Corresponding author: Rama Yusvana, Faculty of Engineering Technology, Tun Hussein Onn University of Malaysia (UTHM), Pagoh Higher Education Hub, Panchor 84600, Johor, Malaysia, e-mail: rama@uthm.edu.my

Azli Nawawi, Faculty of Engineering Technology, Tun Hussein Onn University of Malaysia (UTHM), Pagoh Higher Education Hub, Panchor 84600, Johor, Malaysia, e-mail: azle@uthm.edu.my

https://doi.org/10.1515/9783112206393-005

1 Introduction

Higher education institutions struggle with ongoing challenges in maintaining student engagement and retention within digital or hybrid learning environments. Traditional pedagogical methods often ineffectively address the integration of virtual and in-person learners, resulting in participation inequalities, as academic staff reveal limited proficiency in student-centered strategies [1] and digital competencies [1, 2]. The COVID-19 pandemic strengthened these challenges, revealing cognitive overload, technical drawbacks, and socioeconomic inequities, particularly for students requiring essential resources such as stable internet connectivity or acceptable study environments [3, 4]. Additionally, Prabakaran et al. [5] testified that hybrid classrooms face miscommunication and coordination issues during collaborative activities, further delaying justifiable learning outcomes [5]. These issues highlight the urgency for institutions to prioritize digital transformation while tackling systemic barriers to access and inclusiveness.

In response, emerging studies emphasize the transformative possibility of gamified pedagogies and artificial intelligence (AI)-driven tools in stimulating engagement. According to Naseer et al. [6], adaptive learning platforms incorporating game mechanics, such as badges and leveling systems, combined with AI-driven personalization, have exhibited significant improvements in academic performance and motivation through personalized feedback [6]. Similarly, immersive technologies such as augmented reality (AR) and metaverse-based environments encourage interactivity and accommodate diverse learning styles, boosting emotional regulation and self-efficacy [5, 7]. Nevertheless, effective implementation requires institutional investments in technological infrastructure, continuous professional development for educators, and strategies to alleviate digital divides, as uneven access and technical barriers continue [8, 9]. Scalable solutions such as chatbots and learning analytics systems further simplify administrative tasks and individualize support, but their success hinges on addressing inclusive gaps and familiarizing them with resource-constrained contexts [9, 10]. Collectively, these innovations emphasize the dual imperative for higher education: leveraging gamification and AI to enhance engagement while confirming equitable access to reduce systemic inequalities in hybrid learning ecosystems.

This chapter aims to integrate the impact of gamified pedagogies and AI-driven tools in concentrating on the engagement and retention challenges within higher education. By bridging theoretical frameworks, such as self-determination theory (SDT) and behavioral psychology, with practical applications across various case studies, such as virtual reality (VR) environments, AI tutors, and hybrid massive open online courses (MOOCs), it seeks to evaluate their efficacy in fostering motivation, collaboration, and personalized learning. Additionally, the chapter analytically investigates ethical dilemmas, such as academic integrity risks and content homogenization, alongside infrastructural and equity barriers hindering equitable access. The objectives are therefore threefold: (1) to evaluate the theoretical foundations and empirical impacts of gamification and AI on student outcomes; (2) to investigate ethical, pedagogical, and institutional challenges in execution; and (3) to recommend strategies for scalable adoption, highlighting faculty development, policy frameworks, and inclusive infrastructure. Through this integrated approach, the chapter highlights the need to balance innovation with responsibility to ensure these technologies improve educational equity and rigor in the evolving learning environments.

2 Gamified Pedagogies: Theory, Case Studies, and Impact

2.1 Theoretical Foundations of Gamification

The theoretical foundations of gamification in education are deeply ingrained in motivation theory and behavioral psychology, specifically through the lens of SDT and reward systems. Fundamental to SDT are the psychological needs of autonomy, competence, and relatedness, which enhance motivation and engagement. For instance, Mohanty and Christopher [11] revealed that gamification elements such as experience points and progress bars directly stimulate intrinsic motivation, improving outcomes, whereas extrinsic motivation showed adverse effects. Similarly, Xie et al. [12] discovered that users' intentions to adopt new technologies were significantly influenced by perceived competence, autonomy, and relatedness, supporting SDT principles. These findings highlight the importance of integrating autonomy-supportive gamified structures, as demonstrated by Dantzer and Perry [13], who applied relationship motivation theory in cross-age mentoring, highlighting language and practices that cultivate mentees' psychological needs.

Behavioral psychology further reveals how reward systems within gamification catalyze engagement and retention. Zeng et al. [14] discovered that online class-related enjoyment in some way enhanced academic achievement through the facilitating roles of school motivation and learning engagement, emphasizing the relationship between affective states and behavioral outcomes. Similarly, Gao et al. [15] observed that psychological needs fulfillment influenced excessive WeChat use via affective states and usage concentration, indicating that well-designed reward mechanisms must equalize the need fulfillment to counteract counterproductive behaviors. The role of accountability, as noted by Yang et al. [16], extends this discussion, demonstrating that organizational users' intensified responsiveness to SDT-embedded appeals stems from diverse responsibilities, a principle relevant to educational contexts where coordinated rewards and progress tracking strengthen commitment.

Collectively, these studies highlight the need to embed SDT-aligned gamification strategies (Figure 1) into pedagogical frameworks. Schürmann et al. [17] further supported specific training in motivation theory for educators, as preservice teachers with seminar exposure displayed higher perceived knowledge and relevance of these concepts. This synthesis points out that gamified pedagogies, when supported with SDT and behavioral reward systems, can effectively tie together the intrinsic motivation, tone down disengagement, and cultivate sustained academic participation.

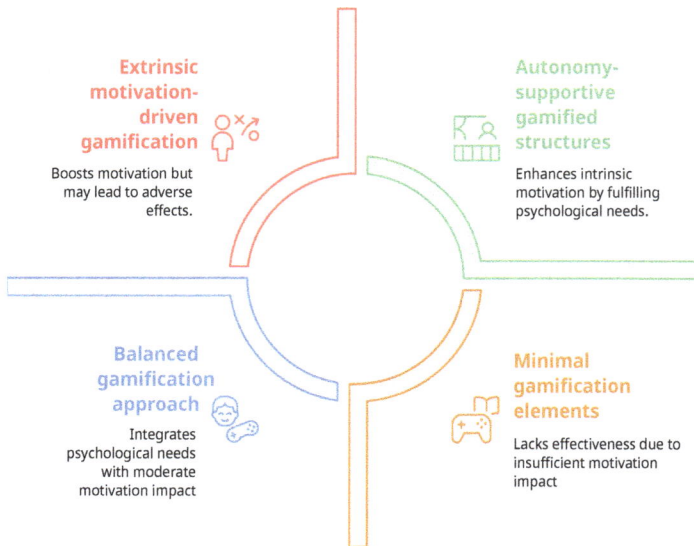

Figure 1: Gamification strategies in education. (Graphics by Napkin AI)

2.2 Case Studies of Successful Integration

The integration of gamified pedagogies and AI tools in higher education has been demonstrated through various case studies, exhibiting their efficacy in nurturing engagement and retention. The following cases, drawn exclusively from published literature [18–23], demonstrate successful integrations of gamified pedagogies and AI tools. Aslan et al. [18] developed Kid Space, a multimodal AI system at Purdue University, which merges conversational agents and virtual environments to improve collaborative problem-solving (CPS) among learners. Quantitative results showed major correlations between CPS behaviors, joint engagement, and pedagogical interventions, indicating that AI-mediated environments can support interactive learning when designed to align with developmental needs. Similarly, Richter and Langesee [19] executed AI tools such as ChatGPT-4 in a multinational program at TU Dresden, concentrating on virtual collaborative learning. The study draws attention to AI's role in overcoming logistical barriers, formulating tasks, and enabling cross-border cooperation, particularly in a smart city project, thereby improving critical thinking and collaborative skills.

Immersive learning platforms, though not exactly like Roblox, prove analogous principles. Subramanian et al. [20] make use of VR at a high school in Manila (Philippines) to reimagine mythological education through nonlinear storytelling and AI-driven interactions, encouraging cultural understanding and critical thinking. In parallel, Gao et al. [21] at the University of Michigan incorporated generative AI (GenAI) avatars into photogrammetry-based VR environments for sustainability education, imitating real-world settings like Naxos, Greece. This approach facilitated natural language interactions with AI tutors, replicating study-abroad experiences while lowering the costs. These cases highlight the potential of immersive technologies to construct experiential learning spaces, akin to role-playing environments, which intensify engagement through interactivity.

AI-assisted hybrid models further exhibit adaptableness across disciplines. Iacono et al. [22] developed an extended reality-based MOOC at the Università Degli Studi di Genova, encompassing gamification and ChatGPT for Italian language instruction. The study highlighted GenAI's capacity to reorganize content creation and enable individualized, interactive lessons, bridging skill gaps among educators. Similarly, Kok et al. [23] evaluated AI-driven gamification frameworks, documenting their capability to personalize content and deliver real-time feedback, as seen in engineering education case studies. These implementations emphasize how AI tutors and gamified modules can transform traditional curricula into dynamic, adaptive systems, concentrating on ethical and accessibility challenges while focusing on learner-centric design. Together, the studies acknowledge that strategic integration of AI and gamification promotes immersive, collaborative, and personalized educational experiences, essential for upholding student motivation and academic success (Figure 2).

These cases were identified through a literature search (2019–2025) across Scopus and Web of Science, using keywords aligned with the theoretical framework (e.g., "gamified pedagogies," "AI-driven tools," "student engagement," and "higher education"). Inclusion criteria prioritized includes empirical studies demonstrating measurable outcomes (e.g., engagement, retention, and collaboration metrics), disciplinary and geographical diversity (e.g., engineering, humanities; Global North/South contexts), and innovative use of AI/gamification in hybrid/digital environments. Synthesis employed thematic coding grounded in SDT (autonomy, competence, and relatedness) and behavioral psychology (reward systems), with cases categorized by technological modality (e.g., VR, chatbots and adaptive platforms) to map efficacy patterns.

2.3 Impact on Engagement and Retention

Gamified pedagogies have demonstrated significant potential in enhancing student engagement and retention within higher education. For instance, the implementation

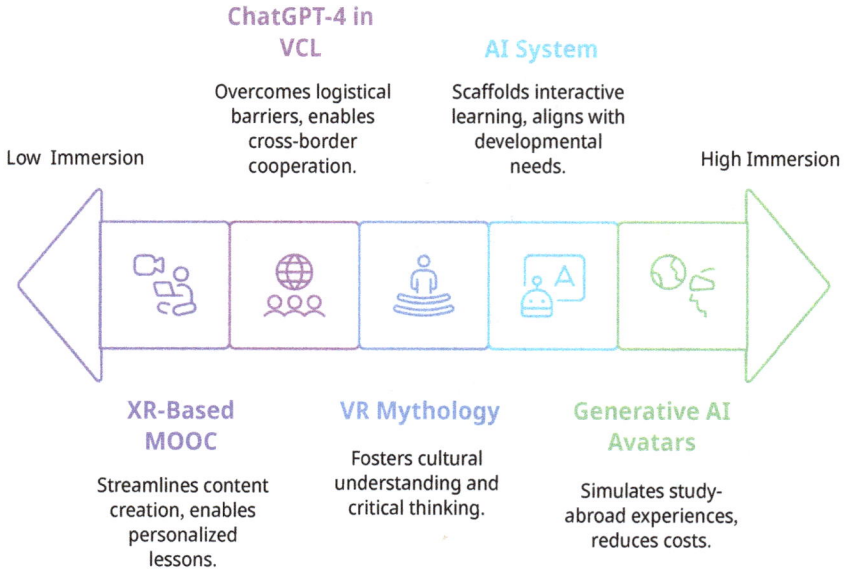

Figure 2: Integration of AI and gamification shows a spectrum of immersion. (Graphics by Napkin AI)

of the participation competition, a semester-long group-based activity, was found to elevate academic learning, cognitive engagement, and peer interaction in marketing classrooms, as evidenced by higher scores on the Higher Education Student Engagement Scale (HESES) compared to control groups [24]. Similarly, student-designed games in physical education fostered autonomy and perceived empowerment, with quantitative results revealing statistically significant improvements in these areas [25]. The integration of online student response systems, particularly game-based variants, further amplified participation and engagement by leveraging instant feedback and interactive elements [26]. These findings collectively underscore the role of gamification in reducing dropout rates by sustaining student interest and motivation.

In addition, gamified frameworks have proven effective in fostering community through collaborative mechanisms. Interactive didactic games, when structured with clear guidelines and age-appropriate narratives, not only enhanced cognitive development but also encouraged sustained attention and problem-solving incentives, thereby promoting peer cooperation [27]. Similarly, team challenges within student-designed games were noted to increase peer engagement, though limitations such as occasional lack of flow were acknowledged [25]. Social value derived from leaderboards and freemium models further reinforced community bonds, as perceived social benefits positively correlated with sustained platform use and collaborative participation [28]. These strategies highlight the dual capacity of gamification to cultivate academic resilience and social cohesion, thereby addressing engagement and retention challenges in higher education environments (Figure 3).

Figure 3: Gamification's impact on higher education. (Graphics by Napkin AI)

2.4 Gamified Pedagogies and Student Outcomes in Western Higher Education

Recent empirical studies from Western higher education contexts provide robust evidence for the efficacy of gamified pedagogies in enhancing student engagement and retention [29]. A longitudinal study involving over 1,000 higher education students in Greece demonstrated that gamified learning significantly improved success rates (39% over online, 13% over traditional), excellence rates (130% over online, 23% over traditional), and average grades (24% over online, 11% over traditional) in a demanding computer science course. Furthermore, gamified learning boosted student retention by 42% compared to online and 36% compared to traditional methods. Student perceptions underscored these quantitative gains, with a vast majority believing gamification increased educational effectiveness (80.54%), improved learning productivity (81.62%), and offered more enjoyable experiences (65.41%), while promoting motivation (83.78%) and active participation (81.08%).

Beyond general gamification, adaptive gamification has shown particular promise in Western educational settings, notably in addressing gender-based learning disparities [30]. A study conducted in Greece with 9-year-old science students revealed that adaptive gamification led to significantly higher post-test scores compared to traditional inquiry-based learning. Crucially, while female students initially scored lower than males, the adaptive gamification intervention resulted in no significant post-test score difference between genders, indicating equal benefit. In contrast, traditional inquiry-based learning in the control group only showed significant improvement for male students, with female students not demonstrating statistically significant progress. This suggests that adaptive gamification can actively contribute to closing initial gender-based achievement gaps, fostering greater parity in science and STEM (science, technology, engineering, and mathematics) fields.

3 AI-Driven Tools: Personalization and Ethical Dilemmas

3.1 Generative AI in Education

GenAI tools have emerged as pivotal resources for adaptive content generation, enabling personalized learning experiences through automated quizzes, simulations, and tailored educational materials. For instance, educators in Bhutan leveraged ChatGPT for lesson planning, content creation, and generating assessment questions, highlighting its role in reducing workload and enhancing teaching efficiency [31]. Correspondingly, students at Sultan Qaboos University used GenAI tools for academic support, involving brainstorming and personalized learning, while academics utilized them to create customized learning content [32]. Preservice teachers also incorporated ChatGPT-4 into curriculum development, streamlining theoretical concepts and envisioning pedagogical strategies through iterative prompt improvement [33]. These applications exhibit GenAI's capacity to simplify content creation and familiarize instructional methods to various learner needs.

However, the integration of GenAI tools presents risks related to standardization and content reliability. Studies uncover that while GenAI boosts individual creativity, it exposes standardizing outputs, as surveyed in short story writing where AI-generated ideas grow in similarity across works, possibly dampening collective novelty [34]. Concerns about over-reliance on AI-generated content were also recorded, with academics voicing apprehensions regarding weakened critical thinking and academic integrity due to errors and biases in outputs [35, 36]. In music education, AI tools democratize creativity but elevate ethical dilemmas around ingenuity and cultural bias, requiring frameworks to balance innovation with accountability [37]. These findings highlight the dual nature of GenAI: while it improves efficiency, its unrestricted use risks grinding down the essential academic skills and diversity of thought.

Tackling these challenges involves structured interventions. For example, cultivating AI literacy and developing assessment frameworks that acknowledge collaborative human-AI contributions can lessen the risks [32, 37]. Moreover, guaranteeing rigorous validation of AI-generated content and encouraging ethical guidelines are essential to sustaining academic rigor and creativity in AI-augmented educational environments [31, 36]. By offsetting adaptability with accountability, institutions can utilize GenAI's potential while preserving educational integrity (Figure 4).

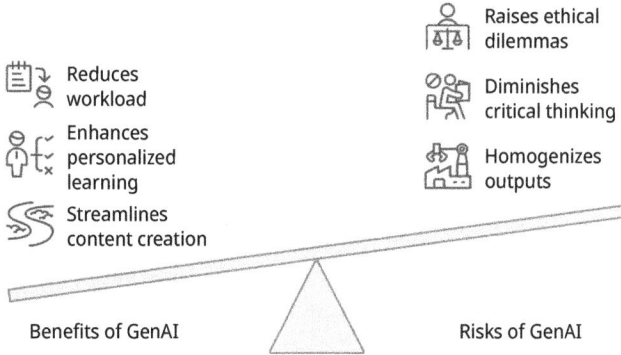

Figure 4: Balancing GenAI benefits and risks in education. (Graphics by Napkin AI)

3.2 Agent-Based AI and Self-Directed Learning

Agent-based AI tools have demonstrated potential in fostering student autonomy by enabling self-directed learning through adaptive and interactive systems. For instance, GenAI chatbots like ERNIE Bot in second language education were found to enhance emotional engagement by fulfilling students' psychological needs for autonomy and competence, with teacher support mediating behavioral and cognitive engagement [38]. Similarly, large language models acting as testing assistants provided developers with conversational frameworks to autonomously generate test cases, leveraging their capacity for creative "hallucinations" to identify software vulnerabilities [39]. In team-based learning environments, adaptive AI agents dynamically adjusted their autonomy levels according to task phases, enhancing human-AI collaboration while maintaining team cohesion [40]. These examples illustrate how AI tools empower learners to navigate complex tasks independently, aligning with self-determination principles.

However, the autonomy granted by such tools necessitates careful oversight to mitigate risks of over-reliance. Studies highlight that while AI systems like automated grading reduce educator workload, their reliability and ethical implications demand rigorous validation frameworks to ensure assessment accuracy and fairness [41]. The distinction between autonomy-as-agency and autonomy-as-authenticity further highlights the need for balanced integration, as over-dependence on AI-driven decisions risks corroding critical thinking and authenticity in learning outcomes [42]. Trust dynamics in AI systems also reveal that user-based steerability diminishes reliance on creator-aligned outputs, suggesting the importance of designing tools that prioritize user goals to prevent inaccurate acceptance [43]. Consequently, educators must adopt hybrid approaches that combine AI-driven autonomy with structured guidance, ensuring tools complement rather than replace human oversight. By integrating transparency in AI operations and fostering interdisciplinary collaboration, institutions can cultivate environments where agent-based AI enhances self-directed learning while safeguarding academic rigor and ethical standards.

3.3 Ethical and Pedagogical Concerns

The integration of GenAI tools in higher education has raised concerns regarding academic integrity, particularly with the increase of AI-generated essays and assessments. For instance, studies reveal that AI-generated outputs often contain inaccuracies and risks of plagiarism, undermining the authenticity of academic work [44]. Similarly, educators report that over-reliance on AI tools like ChatGPT can weaken critical thinking and lead to academic dishonesty, as students may prioritize practicality over intellectual engagement [45, 46]. These concerns are further intensified by findings that advanced GenAI models, such as ChatGPT-4, can perform competently in diverse assessment formats, heightening risks of undetected cheating [47]. Such challenges necessitate robust policies, including revised assessment designs and enhanced plagiarism detection mechanisms, to uphold academic standards.

Another critical issue involves the potential of prioritizing AI-driven entertainment features over substantive learning outcomes. While gamified AI tools enhance engagement, their emphasis on user-friendly interfaces and instant feedback may unconsciously shift focus from deep learning to superficial interaction [48]. For example, educators highlight that students' preference for AI-generated convenience can reduce their motivation to engage in critical analysis, fostering dependency on automated solutions [49]. Additionally, ethical frameworks proposed by scholars stress the need to balance technological allure with pedagogical rigor, ensuring that AI tools complement rather than surpass core educational objectives [50]. Addressing these dual challenges requires a strategic approach that harmonizes innovation with academic integrity, ensuring AI serves as a supportive tool rather than a disruptive force in higher education.

In addition, AI-generated outputs may contain inaccuracies and pose substantial risks of plagiarism, thereby undermining the authenticity of academic work [29]. Advanced GenAI models, such as ChatGPT-4, can perform competently across diverse assessment formats, making it increasingly difficult for educators to discern human-authored from AI-generated content and heightening risks of undetected cheating. Furthermore, an over-reliance on AI tools can weaken critical thinking skills, as students may prioritize convenience over intellectual engagement, potentially leading to a decrease in originality and diversity of thought by reducing novelty.

Beyond academic integrity, ethical AI integration in higher education must address critical equity gaps and the risks of algorithmic bias. Unequal access to AI-driven tools and immersive platforms, particularly in under-resourced regions, exacerbates existing educational inequalities and limits the democratization of these technologies. The inherent risks of biased AI models, if not rigorously audited and transparently deployed, can perpetuate or even amplify existing societal biases, leading to discriminatory educational outcomes for marginalized groups. Therefore, ethical deployment necessitates transparent policies to govern tool deployment, strategic budgeting for robust digital infrastructure, and fostering partnerships to expand connectivity and enhance tool accessibility, ensuring that AI serves all learners fairly and inclusively.

All case studies cited herein were sourced from peer-reviewed literature, ethical compliance – including participant consent, data anonymization, and institutional review – aligns with the protocols reported in the original works.

4 Implementation Strategies and Institutional Challenges

4.1 Overcoming Faculty Resistance

Professional development programs are important in addressing faculty resistance to AI adoption, as they enhance educators' technical competencies and ethical understanding. Kumar and Sharma [51] stressed the necessity of comprehensive training to equip teachers with skills for integrating GenAI, while Ng et al. [52] demonstrated that workshops foster positive attitudes toward AI tools among educators. Similarly, Al-Abdullatif [53] identified AI literacy as a critical factor influencing adoption, suggesting that hands-on training bridges gaps in perceived ease of use. For instance, Lu et al. [54] revealed that preservice teachers engaged in AI-assisted training exhibited improved self-efficacy, highlighting the role of experiential learning in reducing skepticism. Nevertheless, challenges such as ethical concerns persist, requiring continuous professional development to align technological integration with pedagogical goals.

Demonstrating efficacy through pilot studies further mitigates resistance by providing empirical evidence of AI's benefits. Lu et al. [54] utilized a quasi-experimental design to show significant improvements in preservice teachers' higher-order thinking and self-efficacy through AI-assisted training. Similarly, Nadim and Fuccio [55] advocated for balanced AI integration to preserve critical thinking, highlighting the importance of evidence-based approaches. Kaplan-Rakowski et al. [56] reinforced this by linking frequent AI use to positive faculty perspectives, suggesting that pilot initiatives can shift attitudes by showcasing practical applications. These findings collectively emphasize the value of localized, data-driven experiments in validating AI's educational utility.

Institutional support and policy frameworks are equally critical in fostering faculty acceptance. Hazzan-Bishara et al. [57] expanded the technology acceptance model to highlight how institutional backing and credible information directly influence

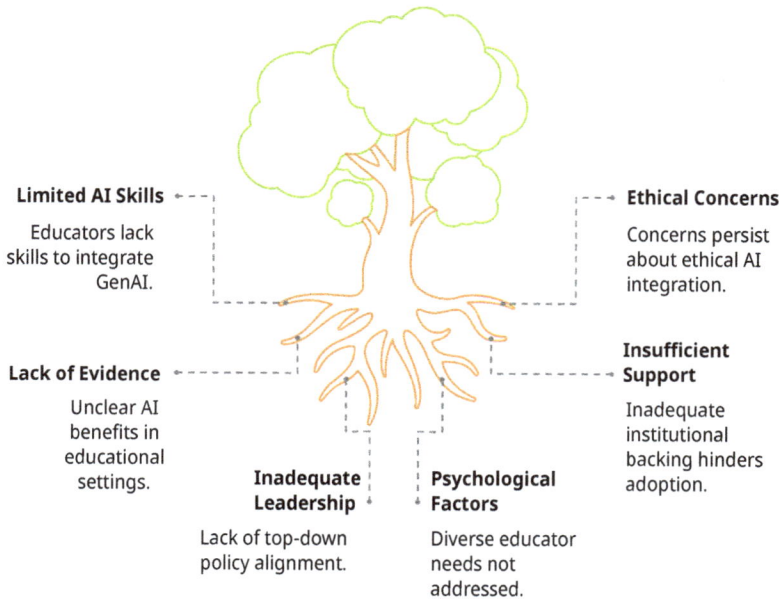

Limited AI Skills

Educators lack skills to integrate GenAI.

Ethical Concerns

Concerns persist about ethical AI integration.

Lack of Evidence

Unclear AI benefits in educational settings.

Insufficient Support

Inadequate institutional backing hinders adoption.

Inadequate Leadership

Lack of top-down policy alignment.

Psychological Factors

Diverse educator needs not addressed.

Figure 5: Resistance of AI adoption in education by faculty. (Graphics by Napkin AI)

adoption intentions. Jogezai et al. [58] noted that organizational guidelines, though indirectly impactful, promote pedagogically relevant AI use when paired with literacy initiatives. Conversely, Cheah and Kim [59] identified resistance among STEM teachers due to inadequate leadership awareness, stressing the need for top-down policy alignment. These insights underscore the importance of clear ethical guidelines, technical support, and leadership involvement in sustaining faculty engagement.

Finally, tailoring strategies to psychological/contextual factors for diverse educator needs. Chen et al. [60] found that secondary teachers exhibited greater reluctance toward AI tools compared to elementary educators, suggesting grade-specific training programs may be needed. Gao et al. [61] further linked preservice teachers' personality traits and prior experience to AI acceptance, indicating the need for personalized professional development. Kaplan-Rakowski et al. [56] corroborated this by demonstrating that hands-on AI use fosters optimism, indicating that experiential learning aligns with individual psychological profiles. Such complex approaches, combining training, evidence, and contextual adaptation, are essential for overcoming resistance and fostering sustainable AI integration in higher education (Figure 5).

4.2 Infrastructural and Equity Considerations

Equity gaps in access to AI-driven tools and immersive platforms remain a critical barrier, particularly in underdeveloped regions. Oyelere and Aruleba [62] highlighted the prominent digital divide in sub-Saharan Africa, where unequal access to resources limits the democratization of AI tools like ChatGPT for programming education. Similarly, Singh et al. [63] identified unequal technology access as a key challenge in adopting AI-enabled cloud computing, necessitating scalable solutions such as device loan programs to ensure universal access. In Peru, Dávila Cisneros et al. [64] found that students faced disparities in AI resource availability, intensifying inequalities in higher education. To address these gaps, strategic budgeting for infrastructure and partnerships with EdTech firms are essential. For instance, El Koshiry and Tony [65] suggested for public-private collaborations to expand connectivity and resource access, particularly in regions with aging infrastructure.

Partnerships with EdTech developers could alleviate cost difficulties and enhance tool accessibility. Weligodapola and Kumarapperuma [66] emphasized the role of institutional-policymaker collaboration in overcoming infrastructural limitations in developing countries, while Oyelere and Aruleba [62] proposed tailored AI-driven strategies for Kenya, Nigeria, and South Africa to align tools with local needs. Nevertheless, infrastructural challenges persist, as noted by Mbangeleli and Funda [67], whose study on South African universities revealed that high costs and data insecurity hinder AI integration despite its potential to enhance academic support. Dritsas and Trigka [68] further stressed the need for investments in immersive technologies like AR/VR, which require robust digital infrastructure to ensure equitable implementation.

Ethical and logistical concerns further complicate the digital divide issue. For example, Ilayaraja et al. [69] highlighted the risks of biased AI models and data privacy issues in low-income countries, urging transparent policies to govern tool deployment. Similarly, Chanpradit [70] cautioned against over-reliance on GenAI in academic writing without addressing cultural aspects and access disparities. These challenges necessitate holistic strategies, combining infrastructure development, ethical guidelines, and stakeholder collaboration to foster inclusive, equitable AI adoption in higher education (Figure 6).

4.3 Policy Frameworks for Ethical Deployment

The development of comprehensive guidelines for AI transparency, data privacy, and bias mitigation is important to ensure ethical GenAI deployment in higher education. Jin et al. [71] highlighted that universities are globally prioritizing academic integrity and equity by establishing ethical guidelines, such as requiring instructor consent and proper AI-generated content attribution. Similarly, Ullah et al. [72] analyzed insti-

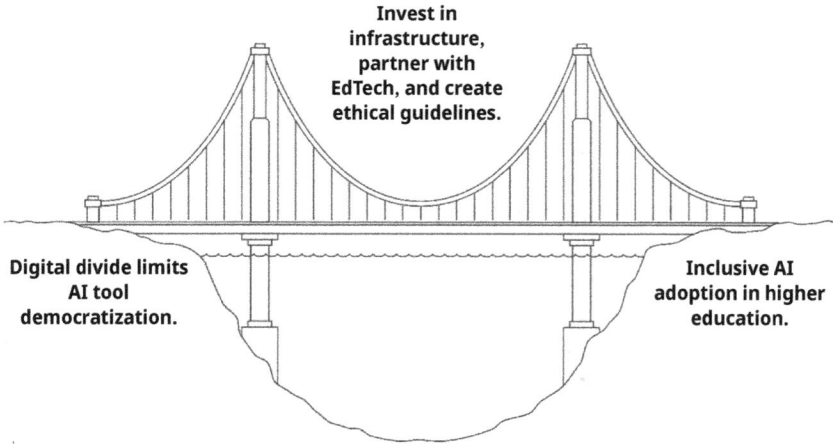

Invest in infrastructure, partner with EdTech, and create ethical guidelines.

Digital divide limits AI tool democratization.

Inclusive AI adoption in higher education.

Figure 6: Bridging the AI equity gap in education. (Graphics by Napkin AI)

tutional policies and found that while most universities permit GenAI use under specific conditions, gaps persist in explaining AI algorithms and documenting prompts, underscoring the need for clearer transparency protocols. Wang et al. [73] further emphasized ethical concerns like data privacy and accuracy, noting that US institutions often provide syllabus templates and workshops to guide responsible AI integration [74]. However, Ghimire and Edwards [75] revealed that many educational administrators lack specialized policies addressing algorithmic transparency, leaving student privacy vulnerabilities unaddressed. These findings collectively stress the urgency of iterative, institution-specific guidelines to mitigate biases and safeguard sensitive information.

Institutional audits are equally important to align AI tools with pedagogical objectives and ensure compliance with ethical standards. For example, Jin et al. [71] proposed ongoing policy evaluations and stakeholder collaboration to adapt GenAI frameworks to evolving educational needs. Wang et al. [73] supported this by encouraging discipline-specific audits, as generic guidelines may fail to address unique disciplinary challenges. Conversely, Ullah et al. [72] identified limited mechanisms for reporting misconduct in existing policies, suggesting that audits should incorporate accountability measures. To bridge these gaps, universities must adopt complex evaluation strategies, such as integrating AI literacy training into curricula and revising assessment designs to deter misuse. Such audits not only reinforce ethical practices but also enhance the alignment of GenAI tools with institutional learning outcomes.

4.4 Practicalities of EdTech Integration and Policy Needs

The practical integration of adaptive gamification in Western educational contexts, as observed through teacher perspectives in Greece, highlights both significant benefits and notable challenges [29]. Teachers consistently reported improved student understanding and motivation, positive learning attitudes, and enhanced student cooperation, alongside high perceived usability of the adaptive gamification environment. However, educators faced difficulties in accurately distinguishing embedded learning strategies and assessing student progress, as well as challenges in initial instruction. Concerns also arose regarding the cognitive levels and age appropriateness of content for younger students, coupled with behavioral issues like impatience during experiments. Frequent technical problems, often linked to inadequate school infrastructure, further compounded these implementation challenges.

Effective policy alignment and robust professional development are critical for successful EdTech adoption in higher education, as evidenced by teacher feedback on training programs in Greece [76]. While professional development based on the Technological Pedagogical and Science Knowledge (TPASK) model was generally well-received for familiarizing teachers with adaptive gamification, specific areas for improvement were identified. Teachers requested more time for adaptation processes, additional hands-on practice, and improved interaction among trainees and trainers. Crucially, there was a consistent demand for increased focus on troubleshooting technical problems and a deeper understanding of the instructional strategies embedded within the applications. These insights underscore that successful integration requires not only pedagogical training but also robust IT infrastructure and continuous, context-specific support to bridge the gap between theoretical potential and practical classroom realities.

5 Strategic Roadmap for Ethical AI and Gamification Implementation

Table 1 provides an actionable roadmap (summary or checklist) for institutions to guide their implementation efforts as discussed in the previous sections.

5.1 Faculty Professional Development

As mentioned previously, overcoming faculty resistance and limited proficiency require comprehensive, ongoing professional development programs. These must move beyond basic technical training to equip educators with AI literacy, foster positive attitudes, and build self-efficacy through hands-on, experiential learning. Crucially, pro-

Table 1: Actionable roadmap/checklist for ethical AI and gamification deployment in higher education.

Thematic area	Specific action/ recommendation	Rationale/benefit	Key considerations
Faculty professional development	1. Implement comprehensive AI and gamification training programs	Equips educators with necessary skills and fosters positive attitudes toward new tools, bridging perceived ease-of-use gaps.	Programs should be continuous, experiential, and tailored (grade-specific and personalized) to address diverse faculty needs and ethical concerns.
	2. Integrate AI literacy into curricula	Develops educators' and students' critical understanding of AI's capabilities, limitations, biases, and ethical implications.	Should go beyond technical skills to encompass critical thinking, responsible use, and ethical considerations for AI outputs.
Equitable infrastructure investment	3. Prioritize strategic budgeting for digital infrastructure	Addresses equity gaps in access to AI tools and immersive platforms, preventing the amplification of existing digital divides.	Includes ongoing maintenance, upgrades, and planning for robust infrastructure required by immersive technologies (AR/VR).
	4. Foster public-private partnerships and device loan programs	Expands connectivity and resource access, alleviates cost difficulties, and ensures universal access to essential technology.	Focus on regions with aging infrastructure and implement scalable solutions to bridge unequal technology access.
Policy frameworks for ethical deployment	5. Develop comprehensive and iterative ethical guidelines	Ensures AI transparency, data privacy, bias mitigation, and upholds academic integrity and equity.	Guidelines must be institution-specific, iterative, and include clear protocols for AI algorithms, prompt documentation, and attribution.
	6. Conduct regular institutional and discipline-specific audits	Aligns AI tools with pedagogical objectives, ensures compliance with ethical standards, and incorporates accountability measures.	Audits should be ongoing, involve stakeholder collaboration, and be tailored to unique disciplinary challenges.
	7. Revise assessment designs and implement accountability measures	Deters misuse of AI tools, maintains academic rigor, and ensures fair assessment practices.	Focus on designing assessments that require critical thinking, human oversight, and transparent policies for reporting misconduct.

Table 1 (continued)

Thematic area	Specific action/ recommendation	Rationale/benefit	Key considerations
Interdisciplinary approaches and stakeholder collaboration	8. Promote interdisciplinary research and collaboration	Addresses cultural, technical, and ethical complexities of emerging technologies and fosters adaptive, student-centered ecosystems.	Involves collaboration among educators, policymakers, EdTech developers, and researchers to align advancements with equitable learning.
	9. Balance innovation with pedagogical rigor and accountability	Ensures AI tools complement rather than surpass core educational objectives and values.	Requires continuous dialogue and strategic decision-making to harness AI's potential while prioritizing educational equity and rigor.

fessional development should empower educators to critically engage with, adapt, and innovate using AI within their pedagogical contexts, shifting institutional culture to value such innovation. Given varying levels of reluctance, prior experience, and differing needs across teaching levels and disciplines, a uniform approach is ineffective. Instead, professional development must be differentiated, personalized, and potentially grade-specific, utilizing mentorship, communities of practice, and tailored resources. Continuous, iterative professional development is essential to align AI integration with evolving pedagogical goals, address ethical concerns, and support experimentation and reflection. Integrating AI literacy directly into university curricula is also recommended to prepare future educators.

5.2 Equitable Infrastructure Investment

The persistent digital divide, particularly in underdeveloped regions, critically hinders equitable AI adoption in education. Strategic, sustained investment in robust digital infrastructure – including initial deployment, ongoing maintenance, upgrades for demanding technologies like AR/VR, reliable internet, technical support, and software – is paramount. Addressing fundamental equity gaps also necessitates fostering EdTech partnerships and public-private collaborations to alleviate costs and expand connectivity, alongside scalable solutions like device loan programs to ensure universal student access. Crucially, equitable infrastructure is an ethical imperative; without foundational access, AI ethics discussions are irrelevant for excluded populations. Policymakers and institutions must therefore view this investment not as optional but as core to educational equity and ethical AI strategy, requiring long-term funding models and global collaboration to address disparities. Sustainable progress demands contin uous budgeting for the evolving digital ecosystem, moving beyond one-time hardware procurement.

5.3 Policy Frameworks for Ethical Deployment

Universities require dynamic, iterative, and institution-specific policy frameworks to ensure the ethical deployment of GenAI, prioritizing transparency (explaining algorithms and documenting prompts), data privacy, bias mitigation, academic integrity, and equity. These frameworks must move beyond static rules to proactive governance, incorporating continuous audits – especially discipline-specific ones – to align tools with pedagogical goals and ethical standards, and including robust accountability measures. Establishing dedicated ethics committees with diverse stakeholders is crucial for ongoing policy evaluation and updates, fostering ethical foresight. Furthermore, effective frameworks necessitate a tiered approach: setting high-level institutional principles for consistency and compliance, while allowing decentralized development of field-specific guidelines to accommodate diverse academic practices and applications. Transparent policies are particularly urgent for deployment in low-income countries.

5.4 Interdisciplinary Approaches and Stakeholder Collaboration

The complex integration of AI and gamification demands collaborative, interdisciplinary approaches across organizational silos. Sustainable and ethical implementation, specially for emerging technologies like AR/VR and metaverse platforms, requires 'istic research addressing cultural, technical, and ethical dimensions. Success 'es on active stakeholder collaboration – among educators, policymakers, and Ed-developers – to co-create adaptive, equitable, student-centered ecosystems. This tates moving beyond top-down mandates to foster cross-departmental initia-'mal feedback channels, and viewing integration as an ongoing co-creation 'rucially, a dynamic, co-evolutionary relationship is needed: educators and 'rs must proactively engage with developers to shape tools, embedding edu-'lues and ethics from the conceptual stage, ensuring technology consistently 'gical goals rather than merely reacting to its progress.

n

; the innovative potential of gamified pedagogies and AI-driven 'lenges of student engagement and retention within higher education. By synthesizing theoretical frameworks such as SDT with empirical case studies, it reveals how game mechanics (e.g., badges and leaderboards) and AI personalization foster intrinsic motivation, collaborative learning, and self-directed autonomy. However, these innovations are challenged by ethical dilemmas – including risks to academic integrity, content homogenization, and infrastructural inequities – that demand careful oversight.

The interplay between technological advancement and pedagogical responsibility emerges as a central theme. While gamification and AI catalyze personalized, immersive learning experiences, their success hinges on balancing innovation with ethical imperatives: equitable access, educator agency, and safeguarding academic rigor. This synthesis resolves the tension between technological promise and systemic barriers, advocating for scalable strategies that prioritize inclusivity and faculty empowerment.

The chapter's insights hold critical implications for institutional policies, urging investments in equitable infrastructure, professional development, and transparent ethical frameworks to align AI integration with pedagogical goals. Looking ahead, emerging technologies such as AR/VR and metaverse-based platforms present fertile ground for deepening experiential learning, but their adoption necessitates interdisciplinary research to address cultural, technical, and ethical complexities. Stakeholder collaboration among educators, policymakers, and EdTech developers will be pivotal in designing adaptive, student-centered ecosystems.

Limitations include reliance on secondary case studies; therefore, future work should directly investigate longitudinal ethical implications. However, this work positions gamification and AI not merely as tools but as transformative forces in higher education, capable of reshaping learning landscapes when embedded. The path forward demands a commitment to harmonizing technological progress with the enduring values of education.

References

[1]   Fleur JL and Dlamini R. Towards learner-centric pedagogies: Technology-enhanced teaching and learning in the 21st century classroom. *Journal of Education*. 2022 Oct;88:1–17. doi: 10.17159/2520-9868/i88a01.

[2]   Van Wyk H and Rosa C. An exploratory study to evaluate the teaching strategies in the hybrid higher education classroom. *South African Journal of Higher Education*. 2024;38(3):211–229. doi: 10.20853/38-3-6371.

[3]   Bashir A, Bashir S, Rana K, Lambert P and Vernallis A. Post-COVID-19 adaptations; the shifts towards online learning, hybrid course delivery and the implications for biosciences courses in the higher education setting. *Frontiers in Education*. 2021;6. doi: 10.3389/feduc.2021.711619.

[4]   Lotfi FZ, Suwartono T, Maziane B, Nurhayati S, Laajan Y and Nachit B. Collaborative concept maps in higher education: Pedagogical contributions, cognitive challenges, and optimization strategies for

interactive isual learning. *Educational Process: International Journal*. 2025;14. doi: 10.22521/edupij.2025.14.85.

[5] Prabakaran N, Patrick HA and Kareem J. Enhancing English language proficiency and digital literacy through metaverse-based learning: A mixed-methods study in higher education. *Journal of Information Technology Education: Research*. 2025;24. doi: 10.28945/5484.

[6] Naseer F, Khan MN, Addas A, Awais Q and Ayub N. Game mechanics and artificial intelligence personalisation: A framework for adaptive learning systems. *Education Sciences*. 2025 Feb;15(3):301. doi: 10.3390/educsci15030301.

[7] Kaur DP, Kumar A, Dutta R and Malhotra S. The role of interactive and immersive technologies in higher education: A survey. *Journal of Engineering Education Transformations*. 2022;36(2):79–86. doi: 10.16920/jeet/2022/v36i2/22156.

[8] Lu B and Hanim RN. Enhancing learning experiences through interactive visual communication design in online education. *Eurasian Journal of Educational Research*. 2024;2024(109):134–157. doi: 10.14689/ejer.2024.109.009.

[9] Popescu R-I, Sabie OM and Trușcă MI. The contribution of artificial intelligence to stimulating the innovation of educational services and university programs in public administration. *Transylvanian Review of Administrative Sciences*. 2023;2023(70):85–108. doi: 10.24193/tras.70E.5.

[10] Celik I, Gedrimiene E, Silvola A and Muukkonen H. Response of learning analytics to the online education challenges during pandemic: Opportunities and key examples in higher education. *Policy Futures in Education*. 2023;21(4):387–404. doi: 10.1177/14782103221078401.

[11] Mohanty S and Christopher BP. A study on role of gamification elements in training outcomes: Comparing the mediating effect of intrinsic and extrinsic motivation. *Learning Organization*. 2023;30(4):480–500. doi: 10.1108/TLO-08-2022-0098.

[12] Xie Y, Zhou R, Chan AHS, Jin M and Qu M. Motivation to interaction media: The impact of automation trust and self-determination theory on intention to use the new interaction technology in autonomous vehicles. *Frontiers in Psychology*. 2023;14. doi: 10.3389/fpsyg.2023.1078438.

[13] Dantzer B and Perry N. Co-constructing knowledge with youth: What high-school aged mentors say and do to support their mentees' autonomy, belonging, and competence. *Educational Action Research*. 2023;31(2):195–212. doi: 10.1080/09650792.2021.1968457.

[14] Zeng Y, Zhang W, Wei J and Zhang W. The association between online class-related enjoyment and academic achievement of college students: A multi-chain mediating model. *BMC Psychology*. 2023;11(1). doi: 10.1186/s40359-023-01390-1.

[15] Gao Q, et al. What links to psychological needs satisfaction and excessive WeChat use? The mediating role of anxiety, depression and WeChat use intensity. *BMC Psychology*. 2021;9(1). doi: 10.1186/s40359-021-00604-8.

[16] Yang N, Singh T and Johnston AC. A replication study of user motivation in protecting information security using protection motivation theory and self-determination theory. *AIS Transactions on Replication Research*. 2021;7. doi: 10.17705/1atrr.00053.

[17] Schürmann L, Gaschler R and Quaiser-Pohl C. Motivation theory in the school context: Differences in preservice and practicing teachers' experience, opinion, and knowledge. *European Journal of Psychology of Education*. 2021;36(3):739–757. doi: 10.1007/s10212-020-00496-z.

[18] Aslan S, et al. An early investigation of collaborative problem solving in conversational AI-mediated learning environments. *Computers and Education: Artificial Intelligence*. 2025;8. doi: 10.1016/j.caeai.2025.100393.

[19] Richter TT and Langesee L-M. Supporting virtual learning and international academic visits with AI. *IDIMT 2024: Changes to ICT, Management, and Business Processes through AI – 32nd Interdisciplinary Information Management Talks*. 2024;141–146. doi: 10.35011/IDIMT-2024-141.

[20] Subramanian M, Gupta RK, Karthic RV, Jadhav KP, Tabuena AC and Tabuena YMH. Reimagining mythology with virtual reality: Postmodernist techniques for immersive retellings in educational

platforms. *1st International Conference on Advances in Computer Science, Electrical, Electronics, and Communication Technologies, CE2CT 2025*. 2025;858–861. doi: 10.1109/CE2CT64011.2025.10939597.

[21] Gao H, Lindquist M and Vergel RS. AI-driven avatars in immersive 3D environments for education workflow and case study of the Temple of Demeter, Greece. *Journal of Digital Landscape Architecture*. 2024;2024(9):640–651. doi: 10.14627/537752059.

[22] Iacono S, Zolezzi D and Vercelli G. Metaverse, A year after: Evolution of XR tools and generative-AI. *2023 IEEE International Conference on Metrology for eXtended Reality, Artificial Intelligence and Neural Engineering, MetroXRAINE 2023 – Proceedings*. 2023;34–39. doi: 10.1109/MetroXRAINE58569.2023.10405700.

[23] Kok CL, Koh YY, Ho CK, Teo TH and Lee C. Enhancing learning: Gamification and immersive experiences with AI. *IEEE Region 10 Annual International Conference, Proceedings/TENCON*. 2024;1853–1856. doi: 10.1109/TENCON61640.2024.10902870.

[24] Peña P, Riley J and Davis N. Increasing student engagement & contributions: Introducing the semester-long participation competition. *Marketing Education Review*. 2024;34(2):99–106. doi: 10.1080/10528008.2024.2337345.

[25] Fernandez-Rio J and Morales-Sallés P. Student-designed games in secondary education. Effects and perspectives from students and teachers. *Journal of Educational Research*. 2020;113(3):204–212. doi: 10.1080/00220671.2020.1778614.

[26] Wang W, Ran S, Huang L and Swigart V. Student perceptions of classic and game-based online student response systems. *Nurse Educator*. 2019;44(4):E6–E9. doi: 10.1097/NNE.0000000000000591.

[27] Baikulova A, Akimbekova S, Kerimbayeva R, Arzymbetova S and Moldagali B. Leveraging digital interactive didactic games to enhance cognitive development in preschool education. *E-Learning and Digital Media*. 2024. doi: 10.1177/20427530241261294.

[28] Hamari J, Hanner N and Koivisto J. 'Why pay premium in freemium services?' A study on perceived value, continued use and purchase intentions in free-to-play games. *International Journal of Information Management*. 2020;51. doi: 10.1016/j.ijinfomgt.2019.102040.

[29] Lampropoulos G and Sidiropoulos A. Impact of gamification on students' learning outcomes and academic performance: A longitudinal study comparing online, traditional, and gamified learning. *Education Sciences*. 2024 Apr;14(4):367. doi: 10.3390/educsci14040367.

[30] Zourmpakis A-I, Kalogiannakis M and Papadakis S. The effects of adaptive gamification in science learning: A comparison between traditional inquiry-based Learning and gender differences. *Computers*. 2024 Dec;13(12):324. doi: 10.3390/computers13120324.

[31] Wangdi T and Rigdel KS. ChatGPT and education: Bhutanese teachers' knowledge, perceptions, and practices. *Pertanika Journal of Social Sciences and Humanities*. 2025;33(2):779–801. doi: 10.47836/pjssh.33.2.13.

[32] Alshamy A, Al-Harthi ASA and Abdullah S. Perceptions of generative AI tools in higher education: Insights from students and academics at Sultan Qaboos University. *Education Sciences*. 2025;15(4). doi: 10.3390/educsci15040501.

[33] Biberman-Shalev L. Prompting theory into practice: Utilizing ChatGPT-4 in a curriculum planning course. *Education Sciences*. 2025;15(2). doi: 10.3390/educsci15020196.

[34] Doshi AR and Hauser OP. Generative AI enhances individual creativity but reduces the collective diversity of novel content. *Science Advances*. 2024;10(28). doi: 10.1126/sciadv.adn5290.

[35] Bozkurt A, *et al*. The manifesto for teaching and learning in a time of generative AI: A critical collective stance to better navigate the future. *Open Praxis*. 2024;16(4):487–513. doi: 10.55982/openpraxis.16.4.777.

[36] Hughes L, Malik T, Dettmer S, Al-Busaidi AS and Dwivedi YK. Reimagining higher education: Navigating the challenges of generative AI adoption. *Information Systems Frontiers*. 2025. doi: 10.1007/s10796-025-10582-6.

[37] Cheng L. The impact of generative AI on school music education: Challenges and recommendations. *Arts Education Policy Review*. 2025. doi: 10.1080/10632913.2025.2451373.

[38] Li Y and Chiu TKF. The mediating effects of needs satisfaction on the relationship between teacher support and student engagement with generative artificial intelligence (GenAI) chatbots from a self-determination theory (SDT) perspective. *Education and Information Technologies*. 2025. doi: 10.1007/s10639-025-13574-w.

[39] Feldt R, Kang S, Yoon J and Yoo S. Towards autonomous testing agents via conversational large language models. *Proceedings – 2023 38th IEEE/ACM International Conference on Automated Software Engineering, ASE 2023*. 2023;1688–1693. doi: 10.1109/ASE56229.2023.00148.

[40] Hauptman AI, Schelble BG, McNeese NJ and Madathil KC. Adapt and overcome: Perceptions of adaptive autonomous agents for human-AI teaming. *Computers in Human Behavior*. 2023;138. doi: 10.1016/j.chb.2022.107451.

[41] Huang X, Chang L-H, Veermans K and Ginter F. Breakpoints in iterative development and interdisciplinary collaboration of AI-driven automated assessment. *2024 21st International Conference on Information Technology Based Higher Education and Training, ITHET 2024*. 2024. doi: 10.1109/ITHET61869.2024.10837673.

[42] Prunkl C. Human autonomy at risk? An analysis of the challenges from AI. *Minds and Machines*. 2024;34(3). doi: 10.1007/s11023-024-09665-1.

[43] Saffarizadeh K, Keil M and Maruping L. Relationship between trust in the artificial intelligence creator and trust in artificial intelligence systems: The crucial role of artificial intelligence alignment and steerability. *Journal of Management Information Systems*. 2024;41(3):645–681. doi: 10.1080/07421222.2024.2376382.

[44] Guleria A, Krishan K, Sharma V and Kanchan T. ChatGPT: Ethical concerns and challenges in academics and research. *Journal of Infection in Developing Countries*. 2023;17(9):1292–1299. doi: 10.3855/jidc.18738.

[45] Chavez JV, *et al*. Discourse analysis on the ethical dilemmas on the use of AI in academic settings from ICT, science, and language instructors. *Forum for Linguistic Studies*. 2024;6(5):349–363. doi: 10.30564/fls.v6i5.6765.

[46] Williams RT. The ethical implications of using generative chatbots in higher education. *Frontiers in Education*. 2023;8. doi: 10.3389/feduc.2023.1331607.

[47] Nikolic S, *et al*. ChatGPT, copilot, gemini, SciSpace and Wolfram versus higher education assessments: An updated multi-institutional study of the academic integrity impacts of generative artificial intelligence (GenAI) on assessment, teaching and learning in engineering. *Australasian Journal of Engineering Education*. 2024;29(2):126–153. doi: 10.1080/22054952.2024.2372154.

[48] Quince Z and Nikolic S. Student identification of the social, economic and environmental implications of using generative artificial intelligence (GenAI): Identifying student ethical awareness of ChatGPT from a scaffolded multi-stage assessment. *European Journal of Engineering Education*. 2025 Mar;1–20. doi: 10.1080/03043797.2025.2482830.

[49] Kamali J, Alpat MF and Bozkurt A. AI ethics as a complex and multifaceted challenge: Decoding educators' AI ethics alignment through the lens of activity theory. *International Journal of Educational Technology in Higher Education*. 2024;21(1). doi: 10.1186/s41239-024-00496-9.

[50] Stahl BC and Eke D. The ethics of ChatGPT – Exploring the ethical issues of an emerging technology. *International Journal of Information Management*. 2024;74. doi: 10.1016/j.ijinfomgt.2023.102700.

[51] Kumar R and Sharma S. Secondary school teachers' perspectives on GenAI proliferation: Generating advanced insights. *International Journal for Educational Integrity*. 2025;21(1). doi: 10.1007/s40979-025-00180-z.

[52] Ng DTK, Chan EKC and Lo CK. Opportunities, challenges and school strategies for integrating generative AI in education. *Computers and Education: Artificial Intelligence*. 2025;8. doi: 10.1016/j.caeai.2025.100373.

[53] Al-Abdullatif AM. Modeling teachers' acceptance of generative artificial intelligence use in higher education: The role of AI literacy, intelligent TPACK, and perceived trust. *Education Sciences*. 2024;14 (11). doi: 10.3390/educsci14111209.

[54] Lu J, Zheng R, Gong Z and Xu H. Supporting teachers' professional development with generative AI: The effects on higher order thinking and self-efficacy. *IEEE Transactions on Learning Technologies*. 2024;17:1279–1289. doi: 10.1109/TLT.2024.3369690.

[55] Nadim MA and Fuccio RD. Unveiling the potential: Artificial intelligence's negative impact on teaching and research considering ethics in higher education. *European Journal of Education*. 2025;60 (1). doi: 10.1111/ejed.12929.

[56] Kaplan-Rakowski R, Grotewold K, Hartwick P and Papin K. Generative AI and teachers' perspectives on its implementation in education. *Journal of Interactive Learning Research*. 2023;34(2):313–338.

[57] Hazzan-Bishara A, Kol O and Levy S. The factors affecting teachers' adoption of AI technologies: A unified model of external and internal determinants. *Education and Information Technologies*. 2025. doi: 10.1007/s10639-025-13393-z.

[58] Jogezai N, Baloch FA, Jaffar M and Khilji G. From technology to pedagogy: Determinants of university faculty's pedagogically relevant use of generative AI. *Quality Assurance in Education*. 2025. doi: 10.1108/QAE-12-2024-0255.

[59] Cheah YH and Kim J. STEM teachers' perceptions, familiarity, and support needs for integrating generative artificial intelligence in K-12 education. *School Science and Mathematics*. 2025. doi: 10.1111/ssm.18334.

[60] Chen R, Lee VR and Lee MG. A cross-sectional look at teacher reactions, worries, and professional development needs related to generative AI in an urban school district. *Education and Information Technologies*. 2025. doi: 10.1007/s10639-025-13350-w.

[61] Gao M, Zhang H, Dong Y and Li J. Embracing generative AI in education: An experiential study on preservice teachers' acceptance and attitudes. *Educational Studies*. 2025. doi: 10.1080/03055698.2025.2483831.

[62] Oyelere SS and Aruleba K. A comparative study of student perceptions on generative AI in programming education across Sub-Saharan Africa. *Computers and Education Open*. 2025;8. doi: 10.1016/j.caeo.2025.100245.

[63] Singh M, Rana S, Upadhyay S and Jha AK. AI-enabled cloud computing application in the education sector. *Establishing AI-Specific Cloud Computing Infrastructure*. 2025;71–102. doi: 10.4018/979-8-3693-9694-0.ch003.

[64] Cisneros JDD, et al. Adjustment of Peruvian University students to artificial intelligence. *Artseduca*. 2023;36:237–248. doi: 10.6035/artseduca.3615.

[65] Koshiry AE and Tony MAA. Enabling smart education: An overview of innovations and challenges in modern learning. *International Journal of Innovative Research and Scientific Studies*. 2025;8(2):3184–3200. doi: 10.53894/ijirss.v8i2.5962.

[66] Weligodapola M and Kumarapperuma CU. Transformative impact of predictive and generative AI on education workforce in developing countries. *2025 5th International Conference on Advanced Research in Computing: Converging Horizons: Uniting Disciplines in Computing Research through AI Innovation, ICARC 2025 – Proceedings*. 2025. doi: 10.1109/ICARC64760.2025.10963062.

[67] Mbangeleli NBA and Funda V. Mapping the evidence around the use of AI-powered tools in South African Universities: A systematic review. *Proceedings of the International Conference on Education Research, ICER 2024*. 2024;159–167 [Online]. Available: https://www.scopus.com/record/display.uri?eid=2-s2.0-85216104609&origin=resultslist&sort=plf-f&src=s&sot=b&sdt=b&s=TITLE%28%22Mapping+the+Evidence+Around+the+Use+of+AI-Powered+Tools+in+South+African+Universities%3A+A+Systematic+Review%22%29.

[68] Dritsas E and Trigka M. Methodological and technological advancements in E-learning. *Information (Switzerland)*. 2025;16(1). doi: 10.3390/info16010056.

[69] Ilayaraja L, Yukthika SS, Prabakaran S and Pavithra S. Integrating artificial intelligence and educational technology for inclusive and ethical education practices. *3rd International Conference on Intelligent Data Communication Technologies and Internet of Things, IDCIoT 2025*. 2025;1152–1157. doi: 10.1109/IDCIOT64235.2025.10914832.

[70] Chanpradit T. Generative artificial intelligence in academic writing in higher education: A systematic review. *Edelweiss Applied Science and Technology*. 2025;9(4):889–906. doi: 10.55214/25768484. v9i4.6128.

[71] Jin Y, Yan L, Echeverria V, Gašević D and Martinez-Maldonado R. Generative AI in higher education: A global perspective of institutional adoption policies and guidelines. *Computers and Education: Artificial Intelligence*. 2025;8. doi: 10.1016/j.caeai.2024.100348.

[72] Ullah M, Naeem SB and Boulos MNK. Assessing the guidelines on the use of generative artificial intelligence tools in universities: A survey of the world's top 50 universities. *Big Data and Cognitive Computing*. 2024;8(12). doi: 10.3390/bdcc8120194.

[73] Wang H, Dang A, Wu Z and Mac S. Generative AI in higher education: Seeing ChatGPT through universities' policies, resources, and guidelines. *Computers and Education: Artificial Intelligence*. 2024;7. doi: 10.1016/j.caeai.2024.100326.

[74] Zhu W, *et al*. Could AI ethical anxiety, perceived ethical risks and ethical awareness about AI influence university students' use of generative AI products? An ethical perspective. *International Journal of Human-Computer Interaction*. 2025;41(1):742–764. doi: 10.1080/10447318.2024.2323277.

[75] Ghimire A and Edwards J. *From Guidelines to Governance: A Study of AI Policies in Education*. 2024. pp. 299–307. doi: 10.1007/978-3-031-64312-5_36.

[76] Sofia K and Kalogiannakis M. Teachers' perspectives on integrating adaptive gamification applications into science teaching. *Journal of Electrical Systems*. 2024;20(11s):2593–2600.

Ivy Shen

Intelligent Learning: Generative AI in Higher Education and Authentic Learning

Abstract: This chapter intends to demonstrate how artificial intelligence (AI) can play a vital role in enhancing authentic learning in higher education. It provides a summary of the history of AI, its current status, and its applications in higher education. Then, the chapter proceeds to explore the potential of the technology in assisting authentic learning. Specifically, the chapter proposes that AI can be integrated into college classrooms to assist project-based learning, interdisciplinary learning, personalized learning, and authentic assessment.

Keywords: Artificial intelligence, higher education, machine learning, natural language processing, authentic learning, authentic assessment

1 Introduction

Imagine this: your alarm goes off at 7 am. Your smart home system adjusts the room temperature and slowly opens the curtain to let in sunlight, according to the setting you made the night before. You ask Alexa for the day's weather forecast, and it tells you to expect rain, so it suggests bringing an umbrella. It also reminds you of your meetings and syncs your calendar with real-time traffic updates to inform you when you should leave for work. Feeling confident, you walk into the kitchen and command the Wi-Fi-enabled voice-controlled coffee maker to brew a cup of morning joe. As you enjoy the refreshing aroma, your smart refrigerator alerts you that you are low on milk, suggesting an online grocery order. You confirm the purchase with a voice command, and the order will be delivered to your door later in the evening. Feeling hungry, you use face ID to unlock your phone and upload a picture of the contents in your pantry to an app, and it generates a recipe using the ingredients you have. As you lay out these ingredients on the counter, your phone reads the cooking instructions to you step by step. On your commute to work, your GPS analyzes real-time traffic conditions, offering the quickest route to avoid congestion. You see a self-driving car as you wait for the traffic light. At work, you receive a long email from your boss that you do not have time to read as you are swamped with multiple tasks. An artificial intelligence (AI) program summarizes the long email for you and highlights the key points from the message, saving much of your precious time. During lunch break, your fitness app notifies you that you need to go for a short walk and stay hydrated

Ivy Shen, Southeast Missouri State University, e-mail: yshen@semo.edu

https://doi.org/10.1515/9783112206393-006

based on your health goals. Later, when you have a video conference, the AI in your webcam automatically adjusts the lighting and background for a professional look. As you are getting ready to leave work, you have a few minutes to spare to check in with your financial app, which tracks your spending and suggests investment opportunities. After a long day at work, you return home to relax. Your streaming service recommends a new show based on your viewing history while your smart home system adjusts the lighting and temperature to your preferred evening setting. Exhausted from work, you ask Alexa to provide a list of available orders from Grubhub as you sit on your comfortable sofa. The day ends as you ask your virtual assistant to set the alarm for the next morning.

This is not science fiction but a regular day of a typical individual living in the modern world. Technology is deeply woven into nearly every aspect of modern life, making it more prevalent than ever before. Most people carry powerful computers in their pockets, rely on GPS to navigate, and use digital platforms for shopping, banking, and social interaction. In schools and workplaces, digital tools are essential for productivity and collaboration. Higher education, like many other sectors, has shown a great deal of interest in AI, as the technology becomes more and more sophisticated. The use of AI in higher education has risen rapidly in the past few years, with a concomitant proliferation of new AI tools available [1]. Universities employ AI to plan resource allocation and automate routine tasks. Students and faculty members use AI for academic and educational purposes. AI will likely become more prevalent in higher education in the years to come.

For college students, authentic learning has always been a hot pursuit because they desire to be professionally prepared and gainfully employed after their academic journey. A recent survey of 4,880 college students reveals that 61% of respondents are interested in skill-based learning opportunities to advance their careers [2]. Authentic learning is an approach that focuses on practical skills. In authentic learning, students tackle tasks that are reflective of the challenges they will face outside the classroom [3]. Rather than focusing solely on theoretical knowledge, authentic learning encourages students to apply what they have learned to solve practical problems, collaborate with others, and explore topics that have real-world applications. This chapter aims to demonstrate how AI can play a vital role in enhancing authentic learning. It provides a summary of the history of AI, its current status, and its applications in higher education. Then the chapter proceeds to explore the potential of the technology in assisting authentic learning. Specifically, the author proposes that AI can be integrated into college classrooms to assist project-based learning (PBL), interdisciplinary learning, personalized learning, and authentic assessment. The author begins this chapter with an overview of the technology and an explanation of how it works. Specifically, the author focuses on machine learning (ML), natural language processing (NLP), and smart automation, which are among the most widely used fields of AI in higher education. Then the author proceeds to argue why and how AI can be integrated to improve authentic learning in higher education.

2 What Is Artificial Intelligence?

Technological advancements vary in scale and impact. Some are small like a software update. Others are groundbreaking innovations that significantly reshape how we live and work. Over the past few decades, advancements in fields such as computing, AI, and telecommunications have transformed the norm of everyday life. These innovations not only redefine daily life but also create new opportunities. Technology today has revolutionized the way we live. What was once considered science fiction is now a reality. Many of the smart devices and productivity programs are powered by AI, making them intelligent and responsive. AI is defined as a machine agent that is smart enough to think and act [4]. AI technology allows the agent to perform tasks that typically require human intelligence. This integration has revolutionized everything from how we manage our homes to how we work and communicate. The global AI market is expected to grow from USD 638.23 billion in 2024 to USD 3,680.47 billion by 2034 [5]. AI has become increasingly prevalent in everyday life. From voice assistants like Siri and Alexa to personalized recommendations on streaming platforms, AI is integrated into daily routines, often without users even realizing it. Whether it is navigating traffic with GPS, using facial recognition to unlock phones, or interacting with chatbots for customer service, AI-powered systems are simplifying tasks. In addition to personal devices, AI plays a significant role in broader sectors such as healthcare, finance, and retail. In healthcare, AI algorithms are helpful in many ways, such as diagnosing diseases, analyzing medical images, and even predicting patient outcomes [6]. In finance, AI is used for fraud detection, investment analysis, and automated customer support [7]. Retailers employ AI to analyze consumer behavior, optimize inventory, and offer personalized shopping experiences [8].

There are two types of AI: narrow AI and general AI [9]. Artificial narrow intelligence (ANI) carries out "intelligent" behaviors in specific contexts. If the context of behavior specification is changed in a narrow AI system, some level of human reprogramming is necessary to enable the system to retain its level of intelligence [10]. Virtual assistants like Siri or Alexa use ANI to understand the user's voice commands and provide answers or perform actions like playing music or setting reminders. ANI can displace humans in low-level cognitive demand tasks that are repetitive, but it cannot perform high-level cognitive demand tasks [11]. ANI can often perform these kinds of tasks more efficiently or accurately than humans, but its abilities are limited to predefined problems. Artificial general intelligence (AGI) can understand, learn, and carry out a wide range of tasks, similar to human intelligence [12]. AGI is largely theoretical and does not currently exist [13]. The AI currently being implemented is mostly ANI.

2.1 A Brief History of Artificial Intelligence

AI research began soon after World War II. During the 1956 Dartmouth Summer Research Conference, American computer scientist John McCarthy first introduced the term "artificial intelligence" [14]. AI is a branch of computer science focused on creating systems capable of performing tasks that typically require human intelligence. Equipped with AI technology, chatbots are computer programs that can simulate human conversations. Joseph Weizenbaum, a computer scientist at the Massachusetts Institute of Technology (MIT), created the first chatbot, ELIZA, in 1966. ELIZA was designed to simulate conversation using pattern-matching techniques to respond to user input [15]. ELIZA looks for the keyword in the user input and generates the most relevant response using predefined scripts. Interactions with ELIZA are likely clunky and even nonsensical, given the limited database. ELIZA is one of the first NLP computer programs. NLP is a field of AI that focuses on enabling computers to understand, interpret, and generate human language. NLP combines computational linguistics, ML, and deep learning techniques to process and analyze large amounts of natural language data, such as text or speech [16]. In 1972, American psychiatrist Kenneth Colby created PARRY, a modification of ELIZA that simulates the thought processes of a person with paranoid schizophrenia, enabling it to "converse" with others [17]. Like ELIZA, PARRY used NLP to interact with users. It analyzed the user input, identified key phrases, and generated responses that aligned with its paranoid personality. Jabberwacky, a chatbot created by British programmer Rollo Carpenter in 1988, was the first AI program capable of simulating natural human conversation. By collecting and storing inputted phrases in its database, Jabberwacky made interactions feel more natural compared to its predecessors. With its ability to tell jokes and use slang, Jabberwacky provided users with engaging and entertaining conversations [18].

In the twenty-first century, chatbots began to feature ML and advanced algorithms. ML, a branch of AI, utilizes data and algorithms to automate analytical models by processing input and output data, uncovering patterns within the data, and mimicking human learning processes [19]. This approach enhances the accuracy of responses. SmarterChild, one of the first chatbots designed to interact with users in a conversational manner, appeared on online instant messenger platforms such as AOL Instant Messenger (AIM), MSN Messenger, and Yahoo Messenger. SmarterChild can engage in basic conversations with users, responding to questions and comments on news, sports, and weather. *NPR News* [20] exclaimed that the chatbot "has to have a soul" since it responds like a human. Soon SmarterChild was outsmarted by IBM Watson, a question-answering computer system. Watson was developed to run on a cluster of computers and contains information equivalent to approximately 1 million books. Watson gained international fame when it competed on the TV quiz show *Jeopardy!*, where it faced former champion players and won the top prize of $1 million [21].

Chatbots continued to advance in the 2010s, when virtual assistants, or intelligent assistants, began to perform tasks that previously required human touch, such as sending text messages, placing phone calls, setting timers, and looking up the internet. Virtual assistants are AI-driven software programs designed to perform tasks, provide information, or assist with various functions through natural language interactions. They use a combination of NLP, ML, and sometimes voice recognition to understand and respond to user requests. Soon, these sophisticated chatbots became popular across various industries due to their ability to simulate human conversations. To cut costs and minimize labor, many companies have deployed chatbots on customer service and messaging platforms. Additionally, some conversational chatbots are available in multiple languages, facilitating seamless communication with international users. Industries where consumers expect 24/7 service are especially focused on developing virtual assistants to deliver continuous customer care. For example, Progressive Insurance [22] utilizes the chatbot Flo to enhance the efficiency of its customer service operations. Flo can manage numerous inquiries at once and deliver relevant responses in real time. Engaging with Flo online feels similar to a human conversation, as Flo can chat about a wide range of topics with customers, from estimated quotes to bill payments to insurance bundling. For another example, Bank of America's virtual financial assistant Erica has had 32 million clients and over 1 billion client interactions since its launch in 2018. The chatbot impressively handles over a whopping 1.5 million interactions every day [23].

2.2 How Artificial Intelligence Works

AI is defined by a set of capabilities, and there are various approaches to develop these capabilities. The two major approaches to develop AI systems are handcrafted knowledge and ML [24]. Handcrafted knowledge AI refers to AI systems that rely on human-created rules or knowledge structures rather than data-driven learning approaches like ML. In handcrafted knowledge systems, experts manually design and encode knowledge into the AI, enabling it to perform specific tasks or reason within a certain domain [24, 25]. The aforementioned chatbots, ELIZA and PARRY, use a rule-based system to simulate a "conversation" without a true understanding of the process. The intelligence from such systems comes from the data input. Tax preparation software programs such as TurboTax and FreeTaxUSA are examples of handcrafted knowledge systems. Users provide tax information according to prespecified data formats. The software then processes that data using the formally programmed rules of the tax code. The output can be good enough to pass an IRS audit [24]. While handcrafted knowledge AI played a critical role in early AI development, it has largely been supplanted by ML. Unlike handcrafted knowledge AI, which requires human rules, ML systems can generate their own rules. It can be far easier to train a system by showing it examples of desired input-output behavior than to program it manually by anticipating the desired response for all possible inputs [26]. ML systems learn from training data rather than being programmed explicitly. Training data can either be structured (e.g., databases) or unstructured (e.g., images and audio). The following section provides a more detailed explanation of ML systems.

3 Machine Learning

3.1 What Is Machine Learning?

ML is a branch of AI where computers learn from data to make predictions or decisions without being explicitly programmed to perform specific tasks [27]. The program can recognize patterns, make sense of new information, or even predict future events based on past data with minimal or no human instructions. ML involves feeding data to a computer. The driving force of ML is data, which can be words, numbers, and images. The computer then uses this data to find patterns. The more diverse and accurate the training data, the better the machine can identify patterns, leading to more accurate results. Thus, the core principle of ML is that machines learn by receiving a large sum of accurate data. For example, the user can train the computer to recognize pictures of cats by feeding it thousands of images labeled as "cat" or "not cat." Over time, the computer learns common features of a typical cat, like fur (sorry, hairless cats), whiskers, a tail, and pointy ears (sorry, Scottish folds), enabling it to recognize a cat on its own. ML methods process data points generated within a specific application domain. Some examples of domains are healthcare, education, energy, entertainment, and finance. Each data point is defined by various characteristics, which can be categorized into two groups: features and labels [28]. Features are properties that can be easily measured or computed automatically. Labels, on the other hand, are properties that are not easily measurable and often represent a higher-level fact or quantity of interest, the identification of which typically requires human expertise. Features are used to train the computer and predict the label. In a dataset predicting a student's learning performance, features might include the student's attendance rate, assignment completion and quality, quiz scores, and participation frequency. Labels would be the final course grade, pass/fail outcome, and dropout likelihood. ML learns to predict the label of a data point based on the features of this data point. The main task is to enable the program to predict the label of an object given by the set of features [28]. Once trained, the user can give the computer new data, and it will predict the label. Using the examples mentioned above, after the training phase, when the computer receives thousands of images of cats and not-cats, it will be able to recognize the new cats in the new images. The computer will also be able to predict whether a student is struggling academically or at risk of dropping out of school.

There are three types of ML algorithms: supervised, unsupervised, and reinforcement [29]. In supervised learning, the computer is given input-output pairs. The supervision refers to the feeding of the information to the algorithm to help it learn. For example, if a university wants to identify the characteristics of an at-risk student, it can provide the machine with profiles of at-risk and not-at-risk students. The program can generate common characteristics such as class attendance rate, homework completion, and participation frequency. The input "supervises" the program to figure out the information users are seeking. In unsupervised learning, the computer is fed data without any explicit labels. The goal is to identify patterns within the data on its own. Unlike supervised learning, where the algorithm learns from labeled data, unsupervised learning involves the model discovering hidden patterns or groupings in the data, such as clustering similar items together. Online retailers regularly utilize this feature to make product recommendations based on purchase history. Streaming services make movie recommendations based on viewing history. Higher education institutions can leverage pattern recognition to predict the future performance of current students. The third type, reinforcement learning, enables the algorithm to learn to make decisions by interacting with an environment. It involves getting a positive reward or a negative reward. The program learns through trial-and-error interactions with a dynamic environment [30]. Over time, the program begins taking the optimal actions to earn positive rewards. If the algorithm classifies a student as at-risk judging by the profile and the student indeed drops out, the algorithm gets a positive reward. If the student remains enrolled, the algorithm gets a negative reward. The machine can learn in both cases and better understand the problem and the environment.

3.2 Application in Higher Education

Many universities have begun to use ML to identify and support at-risk students before academic issues escalate. By analyzing data such as attendance, grades, engagement in online platforms, and patterns of communication, ML models can flag students who may be struggling. Once identified, institutions can proactively reach out with targeted interventions, such as academic advising, tutoring, or mental health support. This approach allows universities to move from reactive to preventive strategies, helping students stay on track and improving overall retention and graduation rates. For example, in Indiana, Ivy Tech Community College launched a pilot study using AI to analyze data from 10,000 course sections. Within the first 2 weeks of the semester, the system identified 16,000 students at risk of failing. Outreach staff were then assigned to personally call each student and offer support. By the end of the term, 3,000 of those students avoided failure, with 98% of contacted students earning a grade of C or higher. Through its ongoing initiative, Project Student Success, the college has now supported a total of 34,712 students [31]. For another example, Western Governors University in Utah used ML predictive modeling to improve student retention. The university employed the program to identify at-risk students, and then developed early-intervention programs. With the help of AI, the university raised the 4-year graduation rate by 5 percentage points between 2018 and 2020 [32].

In addition to being used to retain students, ML models predict future student enrollment trends, helping universities manage resources and staffing more effectively. As discussed, ML models look for patterns in input data. Many universities employ ML models to analyze historical and real-time data and forecast student enrollment trends. This information allows the institution to predict classroom and facility usage and adjust staffing and scheduling accordingly. For example, ML can help determine peak hours for library usage, dining halls, or campus Wi-Fi, enabling universities to allocate resources like staff, utilities, and space more effectively. It can also optimize energy consumption by learning patterns in building usage and adjusting heating, cooling, and lighting systems to reduce costs. Scholars [33, 34] also propose to optimize cloud computing resources according to actual demand and to reduce the cost of cloud services.

Some universities are now using ML to determine the optimal use of physical and virtual infrastructures. For example, the School of the Art Institute of Chicago (SAIC) deploys ML to gauge which students are most likely to accept its offers. Applicant data are entered into the ML system to process. The ML system can parse more than 100 factors, including the number of SAIC events the applicant attended, the types of programs they are interested in, and where they went to high school. It then reports the likelihood that a student would accept the admissions offer [35]. Oftentimes, institutions see "summer melt" from students who accept an enrollment offer but do not end up attending. This technology offers a fairly accurate prediction of the incoming student population, allowing the institution to think ahead about allocating resources. ML has also come in handy to assist admissions. By analyzing historical application data such as test scores, academic records, and extracurricular activities, ML algorithms can help identify strong candidates more quickly and fairly. This analysis enables universities to predict applicant success based on patterns from previous cohorts.

4 Natural Language Processing

4.1 What Is Natural Language Processing?

NLP is another branch of AI. It is a collection of computational techniques for automatic analysis and representation of human languages [36]. Research on computer language processing dates back to the late 1940s. Machine translation (MT) was the first computer-based application related to natural language. Early MT systems search the dictionary for appropriate words for translation and reorder the words to make sense of the translated text using word-order rules of the target language [37]. The approach fails to take into account the lexical ambiguity inherent in natural language. A large percentage of English words have multiple meanings. The word "bank" can mean a financial institution, where money is kept and managed or the side of a river. The same sentence may have different meanings in different contexts. The sentence "she is really something" can be construed as a compliment to the person who just gave an outstanding speech that deserves admiration or a negative comment on the person who behaved inappropriately in a social setting. Not able to detect these linguistic features, the 1940s MTs were frequently inaccurate. In 1952, two English biophysicists, Alan Hodgkin and Andrew Huxley, developed the Hodgkin-Huxley model that shows how the brain uses neurons to form an electrical network, consequently inspiring the idea of NLP [38]. The Hodgkin-Huxley model describes how electrical impulses (action potentials) travel along neurons by modeling the flow of ions across the neuron's membrane, forming neural networks [39]. Various families of neural networks have been used in the study and development of the field of AI for their computing properties [40].

In 1957, American linguist Noam Chomsky published a short monograph titled *Syntactic Structures*, in which he treated language as a formal system, similar to mathematics, where sentences are generated according to a set of rules [41]. This view is central to the idea of generative grammar, which aims to provide a set of formal rules that can predict all the grammatical sentences in a language. Chomsky further introduced the phrase structure grammar, which is a system of rules that break down sentences into their "constituent parts" (like noun phrases and verb phrases) [41, p. 26]. Chomsky's theory of generative grammar introduced formal systems for understanding the syntax of natural languages [41]. In NLP, one of the fundamental tasks is syntactic parsing, where machines analyze sentences to understand their grammatical structure. NLP systems use grammar rules (similar to those in Chomsky's syntactic structures) to break down sentences into their components, or constituent parts, in a way similar to Chomsky's phrase structure grammar. Through the combination of linguistic, statistical, and AI methods, NLP can be used either to determine the meaning of a text or to produce a human-like response [42]. The goal of NLP is to allow computers to process and analyze large amounts of natural language data (spoken or written language) and to perform tasks like translation, sentiment analysis, and question answering. Humans use language to communicate ideas, thoughts, and emotions. NLP allows computers to process this language, which means they can better understand what human users want.

4.2 Application in Higher Education

NLP has been widely used in education to improve learning experiences and personalize instruction. The vast amount of information embedded in chatbots enables them to function as "smart teachers" that are capable of supporting personalized learning. Chatbots have a long-standing role as pedagogical agents, dating back to the development of intelligent tutoring systems (ITSs) in the 1970s. These AI-driven programs were designed to deliver individualized and adaptive instruction, acting as intelligent tutors that understand what they teach, whom they teach, and how to teach effectively [43]. AI technology allows these systems to simulate human tutors at a fraction of the cost of hiring educators, creating interactive and engaging learning experiences. Educational chatbots offer round-the-clock availability, providing students with continuous access to support and learning resources, particularly valuable for those who need to help outside regular classroom hours. Conversational ITSs enhance learning by engaging students in dialogue: posing and answering questions, identifying knowledge gaps, and addressing them using natural language [44]. Moreover, chatbots can deliver immediate feedback to students, which is important for learning progress. Timely feedback reinforces understanding and allows for immediate correction of mistakes. One example of an effective ITS is AutoTutor, which was developed by researchers at the University of Memphis. AutoTutor engages students in natural language conversations and monitors students' emotions by analyzing their dialogue patterns, facial expressions, and body posture. Based on this information, the system adapts its responses to meet each student's individual needs [45].

Several surveys conducted in 2024 and 2025 revealed that a significant majority of students use AI in their studies. For example, a global survey by the Digital Education Council found that 86% of students use AI for studying, with 54% using it weekly and nearly one in four using it daily [46]. Another survey of 11,706 undergraduate students across 15 countries found that 80% of students worldwide have used generative AI to support their university studies [47]. Surveys of faculty in higher education also indicate that AI adoption among educators is increasing globally [48]. The wide application of AI will likely promote further AI adoption in higher education.

5 Smart Automation

Further, AI is increasingly automating administrative tasks across industries, improving productivity, and reducing human errors. AI-driven chatbots use NLP to interpret the user's message and respond to inputs. The ability of chatbots to engage in natural conversation makes them effective "customer service agents." For example, AI technology is increasingly being used in healthcare. Chatbots can provide patients with around-the-clock access to health information, such as supportive information, medication reminders, symptom assessment, or appointment scheduling, allowing access to information when healthcare providers are unavailable [49]. Chatbots can ask patients about their symptoms and provide preliminary assessments based on the information. They use preprogrammed algorithms and medical knowledge to suggest potential causes or conditions [6]. Patients can receive medical information outside of normal clinic hours. They can schedule, reschedule, or cancel appointments via chatbot interactions, avoiding the need to call clinics. Further, some health organizations employ chatbots to monitor patients' health conditions or recovery from procedures. University Hospitals of Cleveland's Population Health program cares for more than 582,000 patients. University Hospitals of Cleveland care managers use chatbot technology to communicate with patients [50]. Patients use an app on their smartphones to receive messages and notifications. Through conversing with the chatbot, patients provide numeric answers (e.g., weight, heart rate, and blood glucose level) or answers to multi-choice questions about their health condition. The chatbot displays the trend of the numeric data so that the patient can see progress over time. The chatbot further educates patients in forming new health habits, such as daily weight tracking and staying hydrated throughout the day. Many patients who sign up for the chatbot engage with the app and their care managers to measure and track progress, learn more about managing their conditions, and work through challenges [50].

Universities employ chatbots to provide information to students and parents on campus life, facilities, advising, course enrollment, and so forth. The University of California at Berkeley [51] has a 24/7 AI chatbot that offers campus information in multiple languages (English, Spanish, Chinese, and Vietnamese). Georgia State University's chatbot Pounce answers thousands of questions from incoming students around the clock via text messages on their smart devices [52]. Pounce was designed to engage these students and motivate them to stay in school. In 2016, Pounce reduced "summer melt" by 22%, which meant 324 additional students showed up for the first day of fall classes [53]. Another successful example of a university AI chatbot is CSUNny of California State University at Northridge. It sends campus announcements and notifications to students' cellphones and answers basic questions. Students who were given access to CSUNny were significantly more likely to still be enrolled at the university than their control group counterparts [54].

6 Application of AI in Authentic Learning

Historically, scholars have always emphasized the importance of practicality in education. In 1798, Anglo-Irish writer Maria Edgeworth and her father advocated for practical education in their educational treatise, Practical Education. The authors highlighted the importance of teaching children in ways that prepare them for the challenges and responsibilities of adult life. Edgeworth and her father believed that education should go beyond traditional rote learning and instead focus on developing practical skills, moral reasoning, and the ability to apply knowledge in real-world situations. They [55] emphasized that learning should be aligned with the needs of society and the individual, rather than being purely theoretical. American educational reformer John Dewey believed that the relationship between theory and practice in education was vital to effective learning. In his work, *The Relation of Theory to Practice in Education*, Dewey [56] argued that education should not be confined to passive learning or the memorization of abstract concepts. Instead, he emphasized the importance of experiential learning, where students actively engage with the material and connect theoretical knowledge to real-life experiences.

Scholars from the twenty-first century [57–59] echo these voices and support the practicality of education. Education should prepare students professionally by equipping them with both the technical knowledge and the soft skills necessary to thrive in the workplace. Real-world learning in higher education is essential because it bridges the gap between academic theory and practical application, better preparing students for the workforce. Theory should guide practice, and practice in turn should inform and refine theory. This cyclical relationship allows students to make meaningful connections between what they learn in the classroom and the challenges they encounter outside it, fostering critical thinking, problem-solving, and adaptability [56].

6.1 Project-Based Learning

Over the past few decades, the world of education has seen a paradigm shift from a traditional teacher-centered and subject-centered teaching to a student-centered approach, in which the students' interests and learning outcomes become the main focal point [60]. In this new approach, students are in charge of the content, activities, and the pace of learning. Instead of passively receiving and then regurgitating information, students are encouraged to take responsibility for their learning by setting goals, asking questions, collaborating with peers, and applying knowledge to real-world situations. The educator plays the role of "guide on the side," assisting and guiding students to meet the goals that have been set by both the students and the educator. This learning method leads to better communication skills, higher confidence, and a unifying culture in our diverse society [61]. One prominent example of student-centered learning is PBL. As one of the major approaches of authentic learning [62], PBL is an inquiry-based and learner-centered instructional method where students gain knowledge and skills by completing meaningful projects and developing real-world products [63]. Krajcik and Shin [64] highlight the key elements of PBL: a driving question, the focus on learning goals, participation in educational activities, collaboration among students, the use of scaffolding technologies, and the creation of tangible artifacts. In PBL, driven by defined objectives, students begin their learning journey with a prompt. They then gather resources, participate in discussions, and work together to find resources to answer the prompt. Students take ownership of their learning by investigating the problem and practicing communication skills to settle differences, and agree on a solution. PBL differs from other pedagogical approaches in its focus on engagement and hands-on activities. The project is the vehicle for learning various knowledge and skills. The goal of PBL is to enhance learning and engagement in the learning processes by emphasizing the practical dimension and relevance to students' lives and to the society in which they live [65]. When students finish school, they will need to understand that work is not as straightforward as lectures and assignments. As employees, they will have to identify the problem, do research, talk to others, and devise a solution. PBL is well aligned with these expectations.

AI technology can be leveraged to facilitate these steps of PBL. First, AI can be used to generate prompts and project instructions. The professor simply needs to input the goal of the project, the intended audience, and the level of detail required. The AI can then create an effective project prompt and instructions. For example, the author used ChatGPT and typed in the command, "Write a prompt on how public opinions form for a college freshman class to help them understand political socialization." Within a few seconds, the chatbot produced the following output:

How do public opinions form, and what factors influence the way individuals develop their political beliefs? In your response, discuss the concept of political socialization and explore the roles of family, education, media, peers, and major life events. Use real-world examples to illustrate how these influences shape political attitudes over time. Why is understanding political socialization important in a democratic society? This prompt is designed to help students critically think about the origins of political beliefs and the social forces that shape public opinion.

Not only did ChatGPT generate a prompt that helps students understand political socialization, but also explained how the prompt would produce such an understanding.

In addition to creating prompts and questions, students and educators can rely on AI chatbots for learning materials and resources. Backed by enormous databases, AI chatbots can produce a wealth of information on just about any topic. The preprogrammed database allows chatbots to generate new text every time the user inputs a command. For example, one of the most popular chatbots, ChatGPT, contains 570 GB of data from books, web texts, and articles. The amount of data is equivalent to 300 billion words [66]. Chatbots can provide a large amount of information quickly and efficiently by using advanced NLP discussed above and access to vast knowledge repositories. A 2020 study [67] conducted at Sun Yat-Sen University in China reported that AI sufficiently supports PBL. In the study, ophthalmology students were divided into two groups, with the first group learning congenital cataracts through an AI-tutoring PBL module, whereas the second group learning the same subject through conventional lectures. Students in the first group received the material through the AI program that was trained with a large number of labeled ocular images and data, discussed within their group, and applied the knowledge to diagnose and evaluate AI-generated cases. Students in the second group attended lectures and wrote assigned papers. A T-test was performed to compare the two groups on their learning outcomes. The first group outperformed the conventionally taught group.

Third, AI can enhance student collaboration by supporting communication and organization. Intelligent platforms can continuously assess each student's needs and provide resource recommendations such as academic articles, instructional videos, and case studies accordingly [68]. The platform adapts recommendations based on each group's progress and challenges. Groups with more visual learners may receive more illustrations, images, and videos, whereas groups with more kinesthetic learners will be assigned hands-on activities. If the AI detects that one student is monopolizing the group conversation, it may suggest strategies that promote the contribution of other members. AI can match students with similar interests or complementary skills, facilitate group discussions through chatbots or virtual assistants, and help manage group tasks by tracking progress and assigning responsibilities. AI can also offer real-time feedback on group work, highlight individual contributions, and suggest ways to improve teamwork. By breaking down communication barriers and supporting equitable participation, AI empowers students to work more effectively together, both in-person and remotely.

Institutions across the globe are increasingly integrating AI as a tech tool into their PBL practices. During the COVID-19 pandemic, when schools transitioned from primarily in-person instruction to online content delivery, Kanazawa Institute of Technology (KIT) in Japan experimented with an AI-powered PBL education system. KIT applied PBL to its Design and Engineering program to produce independently minded and actively engaged future engineers [69]. A Korean university offered a 6-week AI-based PBL module in a general English course that had a total of 20 students. At the end of the program, the students reported a noticeable increase in linguistic knowledge [70].

6.2 Interdisciplinary Learning

In addition to PBL, interdisciplinarity is another key for real-world, authentic learning [3, 71]. Each discipline is an area of academic study. Disciplinary structures allow in-depth investigation of specific phenomena through sustained and systematic inquiry. They provide knowledge on specific issues in a specific field of study. Interdisciplinary learning is an approach that integrates knowledge and methods from different disciplines to address complex problems or topics that cannot be fully understood from a single perspective [72]. This learning model encourages students to make connections across subjects, fostering a more holistic understanding of the world. For example, studying environmental issues might combine biology, chemistry, economics, and social studies, allowing students to experience the interconnections between natural systems, human behavior, and policy. By engaging with multiple disciplines, students develop critical thinking, as they learn to approach challenges from diverse viewpoints [71]. Interdisciplinary learning mirrors the complexities of real-world problems, where solutions often require collaboration across fields.

Developing an interdisciplinary understanding requires learners to integrate modes of thinking from multiple disciplines. Since interdisciplinary learning demands that learners draw knowledge from two or more disciplines [73], it is critical to consistently expose students to knowledge from various academic fields. A major challenge in implementing interdisciplinary learning in higher education is that undergraduate students are novices rather than disciplinary experts [73]. Students' lack of disciplinary knowledge limits their ability to build interdisciplinary skills. Further, educators may or may not be equipped with adequate interdisciplinary knowledge to promote these skills in the classroom. To foster such interdisciplinary understanding, Kidron and Kali [73] suggest that educators harness the power of technology to address the limitations of the compartmentalization of disciplines.

AI-powered chatbots can help achieve the goal. As explained earlier, NLP deals with analyzing, understanding, and generating human language. NLP has applications in a wide range of domains, including web search, sentiment analysis, text analysis, and question answering. AI can adapt to students' needs, helping them access resources and insights from multiple disciplines. A recent empirical study indicates that chatbots can effectively enhance interdisciplinary learning. Zhong et al. [74] engaged 130 undergraduate students at a public university in Southeast Asia to study the effects of ChatGPT on interdisciplinary learning quality. The students in the experiment group used ChatGPT to brainstorm ideas to answer a complex prompt and later reflected on their experience. These students demonstrated enhanced disciplinary grounding compared to students in the control group who did not use ChatGPT to supplement their knowledge.

6.3 Personalized and Adaptive Learning

Further, personalized and adaptive learning contributes to authentic learning as well. As an approach to customize learning experiences to the unique needs, preferences, and pace of individual learners, personalized learning adjusts instructional strategies based on a student's strengths, interests, and progress [75]. Personalized learning facilitates authentic learning by aligning educational experiences with students' real-world interests, goals, and needs [76]. Although personalized learning is possible in a conventional classroom, it can be challenging when the educator has more than a few students in the class. As discussed earlier in Section 4.2, AI enables personalized learning experiences. A self-paced online class is an example of a personalized learning approach. The student controls their timing and pace of learning. Early risers can study in the morning, while working professionals and nontraditional students can work on assignments when they are available. Students can start, pause, or finish lessons at their own convenience, regulated only by the deadline at the end of the term. Self-paced online classes further allow students to choose their own learning paths based on interest or career goals. For example, a rural student with limited school resources interested in preparing for college and a retired person who wants to advance their knowledge can be in the same online calculus course. Online education offers increased flexibility to students and universities, serving as a bridge over time, space, and distance obstacles. Distance education has transformed traditional learning into an "anywhere" to an "anytime" to an "any place" delivery method [77].

Empirical evidence confirms the effectiveness of the self-paced online course format in higher education. Russell et al. [78] investigated whether online professional development courses with different levels of support improved teacher outcomes. The experiment offered variations of an online course for middle school algebra teachers: one with a high level of support (i.e., with a math education instructor, an online facilitator, and asynchronous peer interactions among participants), one self-paced, and the remaining with intermediate levels of support. All variations showed significant impact on teachers' mathematical understanding, pedagogical beliefs, and instructional practices. Another study administered by Southard et al. [79] recruited 96 students enrolled in 3 sections of Introduction to American Government courses delivered in 3 different formats: 26 students in the pure online section, 37 students in the self-paced online section, and 33 in the traditional in-class section. The authors were interested to find out whether the self-paced section would positively influence student performance as compared to its conventional online and traditional face-to-face counterparts. The findings suggest that the self-paced section outperformed the other two groups.

6.4 Authentic Assessment

To gauge the effectiveness of authentic learning, educators must employ authentic assessment, which aims to replicate the tasks and performance standards typically found in the world of work [80]. Learning assessment intends to describe what students have learned, identify challenges they face, and offer guidance to help them improve [81]. It reveals gaps between students' current understanding and the intended learning outcomes. This insight enables educators to adjust their teaching methods to better support student needs. Through assessment, the educator will know to slow down when students struggle and to speed up when students show progress. Unlike conventional assessment, which uses artifacts such as traditional tests or quizzes, authentic assessment evaluates students' skills and knowledge through real-world tasks. It emphasizes practical application, requiring students to demonstrate their learning in ways that mirror actual challenges they might face outside the classroom. Studies confirm that authentic assessment has a positive impact on student learning, motivation, and self-regulation, which are all abilities highly related to employability [80].

Authentic assessment attempts to measure learning outcomes from a holistic standpoint. Mastery involves more than being able to recall answers. Students must demonstrate a thoughtful understanding of the problem through application of their knowledge [82]. Authentic measures involve "worthy problems or questions of importance, in which students must use knowledge to fashion performances effectively and creatively. The tasks are either replicas of or analogous to the kinds of problems faced by adult citizens and consumers or professionals in the field" [83, p. 229]. An example of authentic assessment in higher education is a capstone project in a business course where students work in teams to develop a business plan for a startup idea. Instead of answering exam questions or writing an essay, students must conduct real market research, analyze competitors, create financial projections, and present their proposal. Another example of authentic assessment the author has used in her nonprofit management class is for students to create their own mock nonprofit organization, write its articles of incorporation and bylaws, design its programs, and draft a fundraising plan. These assessment projects require the application of theoretical knowledge to practical problems, thereby offering a more meaningful evaluation of student learning and readiness for professional environments.

AI provides two advantages to authentic assessment. First, AI can support educators in creating simulations or virtual environments where students engage in complex problem-solving that closely mimics the demands of professional practice. For instance, in a nursing program, a chatbot might role-play as a patient with specific symptoms, requiring students to ask questions, diagnose the issue, and suggest treatment in real time. In business or law education, chatbots can act as clients or stakeholders, allowing students to practice negotiation, consultation, or conflict resolution skills in a dynamic, responsive environment. These simulations help assess critical thinking in contexts that mirror professional challenges. Trained on large databases, chatbots can quickly generate cases and scenarios based on the educational context and purpose for students to apply their knowledge. Second, AI-powered grading tools can analyze student work not just for correctness, but for depth of analysis, creativity, and adherence to real-world criteria. These grading tools use ML algorithms and NLP to analyze and evaluate student work based on predefined criteria [84, 85]. For written responses, these tools assess grammar, coherence, organization, relevance to the prompt, and even the depth of critical thinking. The AI is trained on large datasets of previously graded assignments, allowing it to learn patterns and apply consistent scoring. Many teachers have reported positively on the efficiency of these tools [86, 87]. However, it is important to note that the use of AI grading tools raises several ethical concerns, particularly around fairness, transparency, and bias. One major issue is the potential for algorithmic bias, where the AI may favor certain writing styles or language patterns that align with the data it was trained on [88]. Another concern is the lack of transparency – students and educators often do not fully understand how grading decisions are made, making it difficult to challenge or appeal results [89]. Although these ethical challenges underscore the need for human involvement in the final assessment process, AI grading tools do offer the promise of efficiency.

7 Conclusion

British science fiction writer Arthur C. Clarke once said, "Any sufficiently advanced technology is equivalent to magic." From smart home systems and voice assistants to self-driving cars and intelligent apps, technology has advanced dramatically over the years and changed the way we live. AI has revolutionized communication, commerce, and even education. As summarized above, over the past few decades, AI has progressed from basic rule-based systems to advanced ML models that can analyze vast amounts of data and operate in complex environments. AI has transformed higher learning and university administration. NLP has been widely used in education to improve learning experiences and personalize instruction. The vast amount of information embedded in chatbots enables them to function as "smart teachers" that are capable of supporting personalized learning. Universities can rely on chatbots to communicate efficiently with students and ML technology to optimize resource utilization. This chapter intends to demonstrate how AI can play an important role in enhancing authentic learning. It summarizes the history of AI, examines its current status, and explains its applications in higher education.

To students and educators, authentic learning has always been a focal point in higher education. Authentic learning places a heavy emphasis on real-world relevance and practical application of knowledge. Different from traditional learning that focuses on memorization, authentic learning prepares students for real-life challenges. Given its versatility, AI can be integrated into college classrooms to facilitate authentic learning. Specifically, the chapter discusses how AI can assist PBL, interdisciplinary learning, personalized learning, and authentic assessment, which are key components of authentic learning. In PBL, students collaboratively work on a project that requires them to actively research, discuss, and present solutions. Interdisciplinary learning encourages students to think outside the box and integrate skills from multiple subject areas. Personalized learning enables students to pursue their own learning path and objectives. Authentic assessment measures students' ability to apply knowledge. AI facilitates authentic learning by creating opportunities for students to engage in projects that mirror the complexities of life beyond the classroom. Through adaptive learning platforms, AI flexibly adapts content to individual interests and skill levels, allowing students to pursue projects and problems that are personally relevant.

In conclusion, the integration of AI in higher education is transforming how institutions teach, support, and engage students. Historically, educators have always emphasized the importance of practicality in education. This continues to be the focal point in higher education. College students today increasingly value learning that connects directly to practical skills they can apply beyond college years. While academic knowledge remains important, many students are motivated by opportunities that will prepare them for careers. Students seek meaningful learning experiences that are relevant. With the same spirit, universities have been prioritizing practical skill development alongside academic rigor. The future of AI in higher education is promising. As AI technologies become more advanced and accessible, universities are likely to integrate them more deeply into both academic and administrative functions. However, as AI becomes more embedded in higher education, ethical concerns such as data privacy and academic integrity have surfaced. As indicated, institutions increasingly rely on digital tools that collect, process, and analyze student data. These systems often track academic performance and engagement patterns. Although the practice enhances student support and institutional decision-making, it raises important ethical and legal questions about consent, transparency, and the potential misuse of information. In addition, students are now accustomed to using AI for completing assignments, as discussed earlier. This may compromise academic integrity. Since generative AI programs create original content every time they receive a prompt, it is difficult for educators to determine if students wrote their own papers. Although AI detectors are available, they are not always reliable. Future studies are needed to explore these two subjects to safeguard the integrity of AI applications in higher education.

References

[1]   Crompton H and Burke D. Artificial intelligence in higher education: The state of the field. *International Journal of Educational Technology in Higher Education*. 2023 Apr 24;20(1):22.

[2]   Inside Higher Ed. Survey: Students considering skills-based learning [Internet]. Inside Higher Ed; 2023 [cited 2025 Jul 28]. Available from: https://www.insidehighered.com/news/student-success /life-after-college/2023/10/19/college-students-weigh-benefits-certificates

[3]   Lombardi MM and Oblinger DG. Authentic learning for the 21st century: An overview. *Educause Learning Initiative*. 2007 May 22;1(2007):1–12.

[4]   Russell SJ and Norvig P. *Artificial Intelligence: A Modern Approach*. 3rd ed. Boston: Pearson; 2016.

[5]   Research P. Artificial intelligence (AI) market size, share, and trends 2024 to 2034 [Internet]. Precedence Research; 2024 [cited 2025 Jul 28]. Available from: https://www.precedenceresearch. com/artificial-intelligence-market

[6]   Divya S, Indumathi V, Ishwarya S, Priyasankari M and Devi SK. A self-diagnosis medical chatbot using artificial intelligence. *Journal of Web Development and Web Designing*. 2018 Apr 7;3(1):1–7.

[7]   Cao L. AI in finance: Challenges, techniques, and opportunities. *ACM Computing Surveys*. 2022 Feb 3;55(3):1–38.

[8]   Pillai R, Sivathanu B and Dwivedi YK. Shopping intention at AI-powered automated retail stores (AIPARS). *Journal of Retailing and Consumer Services*. 2020 Nov 1;57:102–207.

[9]   Goertzel B. Artificial general intelligence: Concept, state of the art, and future prospects. *Journal of Artificial General Intelligence*. 2014;5(1):1–46.

[10] Kurzweil R. The singularity is near. In: *Ethics and Emerging Technologies*. London: Palgrave Macmillan UK; 2005. pp. 393–406.

[11] Tyson LD and Automation ZJ. AI & work. *Daedalus*. 2022 May 1;151(2):256–271.

[12] Goertzel B. Artificial general intelligence: Concept, state of the art, and future prospects. *Journal of Artificial General Intelligence*. 2014;5(1):1–46.

[13]   Kuusi O and Heinonen S. Scenarios from artificial narrow intelligence to artificial general intelligence – Reviewing the results of the international work/technology 2050 study. *World Futures Review*. 2022 Mar;14(1):65–79.

[14]   McCarthy J, Minsky ML, Rochester N and Shannon CE. A proposal for the Dartmouth summer research project on artificial intelligence, august 31, 1955. *AI Magazine*. 2006 Dec 15;27(4):12.

[15]   New Jersey Institute of Technology. ELIZA: A very basic Rogerian psychotherapist chatbot [Internet]. New Jersey Institute of Technology; 2025 [cited 2025 Jul 28]. Available from: https://web.njit.edu/~ronkowit/eliza.html

[16]   IBM. What is NLP? [Internet]. IBM; 2025a [cited 2025 Jul 28]. Available from: https://www.ibm.com/topics/natural-language-processing

[17]   Colby KM. PARRYing. *Behavioral and Brain Sciences*. 1981 Dec;4(4):550–560.

[18]   BBC News. Chatbot bids to fool humans [Internet]. BBC News; 2003 Sep 16 [cited 2025 Jul 28]. Available from: http://news.bbc.co.uk/2/hi/technology/3116780.stm

[19]   IBM. What is machine learning? [Internet]. IBM; 2025b [cited 2025 Jul 28]. Available from: https://www.ibm.com/topics/machine-learning

[20]   NPR News. It has to have a soul: How chatbots get their personalities [Internet]. NPR News; 2017 [cited 2025 Jul 28]. Available from: https://www.npr.org/sections/alltechconsidered/2017/03/10/519002884/it-has-to-have-a-soul-how-chatbots-get-their-personalities

[21]   Watson IBM, 'Jeopardy!' champion [Internet]. IBM; 2025c [cited 2025 Jul 28]. Available from: https://www.ibm.com/history/watson-jeopardy

[22]   Progressive Insurance. Chatting with Flo FAQs [Internet]. Progressive Insurance; 2025 [cited 2025 Jul 28]. Available from: https://www.progressive.com/contact-us/chatbot-faqs/

[23]   Bank of America. Bank of America's Erica tops 1 billion client interactions, now nearly 1.5 million per day [Internet]. Bank of America Newsroom; 2022 [cited 2025 Jul 28]. Available from: https://newsroom.bankofamerica.com/content/newsroom/press-releases/2022/10/bank-of-america-s-erica-tops-1-billion-client-interactions–now-.html

[24]   Allen G. Understanding AI technology. *Joint Artificial Intelligence Center (JAIC) The Pentagon US*. 2020;2(1):24–32.

[25]   Defense Advanced Research Projects Agency. Summary [Internet]. AI Next Campaign; 2025 [cited 2025 Jul 28]. Available from: https://www.darpa.mil/research/programs/ai-next

[26]   Jordan MI and Mitchell TM. Machine learning: Trends, perspectives, and prospects. *Science*. 2015 Jul 17;349(6245):255–260.

[27]   Mahesh B. Machine learning algorithms – A review. *International Journal of Science and Research*. 2020;9(1):381–386.

[28]   Nasteski V. An overview of the supervised machine learning methods. *Horizons. B*. 2017 Dec 15;4(51–62):56–66.

[29]   Jung A. *Machine Learning: The Basics*. Cham: Springer Nature; 2022.

[30]   Kaelbling LP, Littman ML and Moore AW. Reinforcement learning: A survey. *Journal of Artificial Intelligence Research*. 1996 May 1;4:237–285.

[31]   Google for education. Ivy tech develops machine learning algorithm to identify at-risk students and provide early intervention [Internet]. Google for Education; 2025 [cited 2025 Jul 28]. Available from: https://edu.google.com/resources/customer-stories/ivytech-gcp/

[32]   McKinsey & Company. Using machine learning to improve student success in higher education [Internet]. McKinsey & Company; 2022 [cited 2025 Jul 28]. Available from: https://www.mckinsey.com/industries/education/our-insights/using-machine-learning-to-improve-student-success-in-higher-education

[33]   Khan T, Tian W, Zhou G, Ilager S, Gong M and Buyya R. Machine learning (ML)-centric resource management in cloud computing: A review and future directions. *Journal of Network and Computer Applications*. 2022 Aug 1;204:103–405.

[34] Osypanka P and Nawrocki P. Resource usage cost optimization in cloud computing using machine learning. *IEEE Transactions on Cloud Computing*. 2020 Aug 11;10(3):2079–2089.

[35] Inside Higher Ed. Chicago Art School deploys machine learning in admissions [Internet]. Inside Higher Ed; 2024 Apr 29 [cited 2025 Jul 28]. Available from: https://www.insidehighered.com/news/admissions/2024/04/29/art-focused-university-using-ai-admissions

[36] Chowdhary K. Natural language processing. *Fundamentals of Artificial Intelligence*. 2020 Apr;5:603–649.

[37] Liddy ED. *Natural Language Processing*. Syracuse (NY): Syracuse University; 2001.

[38] Rahman L, Mohammed N and Al Azad AK. A new LSTM model by introducing biological cell state. In: *2016 3rd International Conference on Electrical Engineering and Information Communication Technology (ICEEICT); 2016 Sep 22–24*. Dhaka, Bangladesh. Piscataway (NJ): IEEE; 2016. pp. 1–6.

[39] Abbott LF and Kepler TB. Model neurons: From Hodgkin Huxley to Hopfield. In: *Statistical Mechanics of Neural Networks: Proceedings of the XIth Sitges Conference Sitges, Barcelona, Spain, 3–7 June 1990*. Berlin, Heidelberg:: Springer; 2005 Jul 10. pp. 5–18.

[40] Calitoiu D, Oommen JB and Nussbaum D. Analytic Results on the Hodgkin-Huxley neural network: Spikes annihilation. In: *Conference of the Canadian Society for Computational Studies of Intelligence 2007 May 28*. Berlin, Heidelberg:: Springer. pp.320–331

[41] Chomsky N. *Syntactic Structures*. Mouton de Gruyter; 1957.

[42] Fanni SC, Febi M, Aghakhanyan G and Neri E. Natural language processing. In: *Introduction to Artificial Intelligence*. Cham: Springer International Publishing; 2023 Sep 16. pp. 87–99.

[43] Alkhatlan A and Kalita J. Intelligent tutoring systems: A comprehensive historical survey with recent developments. *arXiv Preprint arXiv:1812.09628*. 2018 Dec 23.

[44] Ilagan JBR and Ilagan JR. *A Prototype of A Chatbot for Evaluating and Refining Student Startup Ideas Using A Large Language Model*. 2023.

[45] University of Memphis. AutoTutor [Internet]. University of Memphis; 2025 [cited 2025 Jul 28]. Available from: https://www.memphis.edu/iis/projects/autotutor.php

[46] Technology C. Survey: 86% of students already use AI in their studies [Internet]. Campus Technology; 2024 Aug 28 [cited 2025 Jul 28]. Available from: https://campustechnology.com/articles/2024/08/28/survey-86-of-students-already-use-ai-in-their-studies.aspx

[47] Chegg. Chegg Global Student Survey 2025: 80% of undergraduates worldwide have used GenAI to support their studies – But accuracy a top concern [Internet]. Chegg; 2025 Jan 28 [cited 2025 Jul 28]. Available from: https://investor.chegg.com/Press-Releases/press-release-details/2025/Chegg-Global-Student-Survey-2025-80-of-Undergraduates-Worldwide-Have-Used-GenAI-to-Support-their-Studies–But-Accuracy-a-Top-Concern/default.aspx

[48] University C. AI in higher education: A meta summary of recent surveys of students and faculty [Internet]. Campbell Academic Technology Services; 2025 Mar 6 [cited 2025 Jul 28]. Available from: https://sites.campbell.edu/academictechnology/2025/03/06/ai-in-higher-education-a-summary-of-recent-surveys-of-students-and-faculty/

[49] Clark M and Bailey S. Chatbots in health care: Connecting patients to information. *Canadian Journal of Health Technologies*. 2024 Jan 22;4(1):1–22.

[50] Schario ME, Bahner CA, Widenhofer TV, Rajaballey JI and Thatcher EJ. Chatbot-assisted care management. *Professional Case Management*. 2022 Jan 1;27(1):19–25.

[51] The University of California at Berkeley. Chatbot [Internet]. University of California at Berkeley; 2025 [cited 2025 Jul 28]. Available from: https://studentcentral.berkeley.edu/chatbot/

[52] Georgia State University. Reduction of summer melt: A strategic approach [Internet]. Georgia State University; 2025 [cited 2025 Jul 28]. Available from: https://success.gsu.edu/initiatives/reduction-of-summer-melt/

[53] EdTech. Successful AI examples in higher education that can inspire our future [Internet]. EdTech Magazine; 2020 [cited 2025 Jul 28]. Available from: https://edtechmagazine.com/higher/article/2020/01/successful-ai-examples-higher-education-can-inspire-our-future

[54] California State University at Northridge. CSUNny program summary and outcomes [Internet]. California State University at Northridge; 2023 [cited 2025 Jul 28]. Available from: https://edtechmagazine.com/higher/sites/edtechmagazine.com.higher/files/CSUNny%20data.pdf

[55] Edgeworth M and Edgeworth RL. *Practical Education*. New York: Harper & Brothers; 1798.

[56] Dewey J. The relation of theory to practice in education. *Teachers College Record*. 1904 Nov;5(6):9–30.

[57] Lopes PN and Salovey P. Toward a broader education: Social, emotional, and practical skills. *Building Academic Success on Social and Emotional Learning: What Does the Research Say*. 2004 Apr;15:76–93.

[58] Millar R Practical work. Good practice in science teaching: What research has to say. 2010 May 1;2:108–134.

[59] Wellington J. *Practical Work in School Science: Which Way Now*. London: Routledge; 1998.

[60] Lathika K. Student centered learning. *International Journal of Current Research in Modern Education*. 2016;1(1):677–680.

[61] Overby K. Student-centered learning. *Essai*. 2011;9(1):1–4.

[62] Rule AC. The components of authentic learning. *Journal of Authentic Learning*. 2006;3(1):1–10.

[63] Brundiers K and Wiek A. Do we teach what we preach? An international comparison of problem-and project-based learning courses in sustainability. *Sustainability*. 2013 Apr;5(4):1725–1746.

[64] Krajcik J and Shin N. Project-based learning. In: Sawyer RK (ed.). *The Cambridge Handbook of the Learning Sciences*. 2nd ed. New York: Cambridge University Press; 2014. pp. 275–297.

[65] Shpeizer R. Towards a successful integration of project-based learning in higher education: Challenges, technologies and methods of implementation. *Universal Journal of Educational Research*. 2019 Aug 1;7(8):1765–1771.

[66] BBC News. ChatGPT: Everything you need to know about OpenAI's GPT-4 tool [Internet]. BBC Science Focus; 2023 [cited 2025 Jul 28]. Available from: https://www.sciencefocus.com/future-technology/gpt-3

[67] Wu D, Xiang Y, Wu X, Yu T, Huang X, Zou Y, Liu Z and Lin H. Artificial intelligence-tutoring problem-based learning in ophthalmology clerkship. *Annals of Translational Medicine*. 2020 Jun;8(11):700.

[68] Herrington J, Reeves TC and Oliver R. *A Guide to Authentic E-learning*. New York: Routledge; 2010.

[69] Ito T, Sode Tanaka M, Shin M and Miyazaki K. The online PBL (project-based learning) education system using AI (artificial intelligence). In: *DS 110: Proceedings of the 23rd International Conference on Engineering and Product Design Education (E&PDE 2021)* 2021 Sep 9–10. Herning, Denmark: VIA Design, VIA University; 2021.

[70] Kim MK. PBL using AI technology-based learning tools in a college English class. *Korean Journal of General Education*. 2023 Apr 30;17(2):169–183.

[71] Dalrymple J and Miller W. Interdisciplinarity: A key for real-world learning. *Planet*. 2006 Dec 1;17(1):29–31.

[72] Ivanitskaya L, Clark D, Montgomery G and Primeau R. Interdisciplinary learning: Process and outcomes. *Innovative Higher Education*. 2002 Dec;27(2):95–111.

[73] Kidron A and Kali Y. Boundary breaking for interdisciplinary learning. *Research in Learning Technology*. 2015 Oct 28;23:1–17.

[74] Zhong T, Zhu G, Hou C, Wang Y and Fan X. The influences of ChatGPT on undergraduate students' demonstrated and perceived interdisciplinary learning. *Education and Information Technologies*. 2024 Dec;29(17):23577–23603.

[75] Taylor DL, Yeung M and Bashet AZ. Personalized and adaptive learning. In: *Innovative Learning Environments in STEM Higher Education: Opportunities, Challenges, and Looking Forward 2021 Mar 12*. Cham: Springer International Publishing. pp.17–34

[76] Abedi R, Nili Ahmadabadi MR, Taghiyareh F, Aliabadi K and Pourroustaei Ardakani S. The effects of personalized learning on achieving meaningful learning outcomes. *Interdisciplinary Journal of Virtual Learning in Medical Sciences*. 2021 Sep 1;12(3):177–187.

[77] Shachar M and Neumann Y. Twenty years of research on the academic performance differences between traditional and distance learning: Summative meta-analysis and trend examination. *MERLOT Journal of Online Learning and Teaching*. 2010 Jun;6(2):318–334.

[78] Russell M, Kleiman G, Carey R and Douglas J. Comparing self-paced and cohort-based online courses for teachers. *Journal of Research on Technology in Education*. 2009 Jun 1;41(4):443–466.

[79] Southard S, Meddaugh J and France-Harris A. Can SPOC (self-paced online course) live long and prosper? A comparison study of a new species of online course delivery. *Online Journal of Distance Learning Administration*. 2015 Jun 1;18(2):8–16.

[80] Villarroel V, Bloxham S, Bruna D, Bruna C and Herrera-Seda C. Authentic assessment: Creating a blueprint for course design. *Assessment and Evaluation in Higher Education*. 2018 Jul 4;43(5):840–854.

[81] Berry R. *Assessment for Learning*. vol. 1. Hong Kong: Hong Kong University Press; 2008.

[82] Wiggins GP. *Assessing Student Performance: Exploring the Purpose and Limits of Testing*. San Francisco: Jossey-Bass/Wiley; 1993.

[83] Wiggins G. A true test: Toward more authentic and equitable assessment. *Phi Delta Kappan*. 2011 Apr;92(7):81–93.

[84] Li Y, Raković M, Srivastava N, Li X, Guan Q, Gašević D and Chen G. Can AI support human grading? Examining machine attention and confidence in short answer scoring. *Computers and Education*. 2025 Apr 1;228:1–18.

[85] Yeung WE, Qi C, Xiao JL and Wong FR. *Evaluating the Effectiveness of AI-based Essay Grading Tools in the Summative Assessment of Higher Education [Internet]*. Hong Kong: The Hong Kong Polytechnic University; 2023 [cited 2025 Jul 28] Available from: https://ira.lib.polyu.edu.hk/handle/10397/107742.

[86] CNN. Teachers are using AI to grade essays. But some experts are raising ethical concerns [Internet]. CNN; 2024 [cited 2025 Jul 28]. Available from: https://www.cnn.com/2024/04/06/tech/teachers-grading-ai

[87] Education week. This AI tool cut one teacher's grading time in half. How it works [Internet]. Education Week; 2024 [cited 2025 Jul 28]. Available from: https://www.edweek.org/technology/this-ai-tool-cut-one-teachers-grading-time-in-half-how-it-works/2024/04

[88] Chinta SV, Wang Z, Yin Z, Hoang N, Gonzalez M, Quy TL and Zhang W. FairAIED: Navigating fairness, bias, and ethics in educational AI applications. *arXiv Preprint arXiv:2407.18745*. 2024 Jul 26.

[89] Hofman J. *Transparency in AI-driven Grading Tools for Open-ended Questions in Higher Education* [bachelor's thesis]. Enschede (NL): University of Twente; 2023.

Luis Manuel Cerdá-Suárez*

Shaping Higher Education: Student-Teacher Interactions and Artificial Intelligence Tool Integration in Latin America

Abstract: In higher education, positive perceptions often highlight artificial intelligence (AI) as a supportive tool akin to mentorship, enhancing connectivity with instructors. However, a recent literature review evidences that the integration of AI in educational settings presents several challenges and limitations that impact both teaching and learning experiences. Furthermore, negative perceptions raise alarm over potential issues such as increased surveillance and a homogenized learning experience, leading some students to feel constrained by technology, and making it essential to evaluate how AI influences these dynamics. This work explores the transformative role of AI in enhancing student-teacher interactions within higher education, focusing on evidence from Latin America. Its purpose is to show how an AI tool can enhance the understanding of student's learning experiences in higher education by analyzing their interactions with teachers and providing personalized feedback. Grounded in a comparative case analysis conducted across universities in Chile and Spain, a structured questionnaire was administered to 385 participants. The findings reveal how AI systems can both facilitate and complicate communication, support, and engagement between students and teachers in several learning environments. This chapter provides actionable strategies to mitigate these challenges, ensuring a balanced integration of AI that enhances both learning outcomes and teacher-student interactions.

Keywords: Artificial intelligence (AI), personalized learning, teaching, higher education, student-teacher interactions

1 Introduction

In recent years, artificial intelligence (AI) has become central to innovation across sectors, including higher education. It shapes content delivery, performance assessment, and student support [1–3]. Tools such as personalized algorithms, automated assessments, and adaptive systems are now integrated into teaching and governance [1, 4].

---

*Corresponding author: Luis Manuel Cerdá-Suárez**, Faculty of Economics and Business, Universidad Internacional de La Rioja (UNIR), Logroño, La Rioja, Spain, e-mail: luis.cerda@unir.net

https://doi.org/10.1515/9783112206393-007

However, while much attention focuses on AI's benefits for learning and efficiency, less is said about how it transforms the relational aspects of teaching – such as emotional dynamics and student-teacher connections [5, 6].

In Latin America, the impact of AI is complicated by structural inequality, diverse pedagogical traditions, and varying institutional capacities. Evaluating AI in this region requires more than a technical perspective [7]. A relational and culturally grounded approach is essential to understand how AI influences engagement and educational quality [2, 8, 9].

Although literature on AI in education has grown, major gaps persist: emotional dynamics are underexplored; pedagogical models like the TTI framework are seldom applied; Latin American case studies are rare; and cross-cultural comparisons are limited. These oversights risk promoting a narrow, decontextualized narrative of AI in education. In general terms, several critical gaps remain as follows [1, 10, 11]:

– Neglect of Relational Dynamics: While AI has been widely studied for its cognitive and operational contributions [2, 5], its implications for emotional and relational aspects of teaching are under-theorized [9, 12]. Few studies explore how AI affects trust, empathy, or classroom connectedness [13–15].
– Limited Application of Frameworks: Pedagogical frameworks, such as the teaching through interactions (TTI) model, which focuses on emotional support, classroom organization, and instructional support [10, 11], are rarely used to evaluate AI's influence, despite their utility for examining interpersonal processes in education.
– Scarcity of Latin American Case Studies: Most AI in education research draws from North America, Europe, or East Asia [3, 14, 16], while the sociotechnical realities of Latin American universities remain comparatively underrepresented considering their distinct cultural and infrastructural conditions [4–7, 9], particularly in studies focusing on student-teacher dynamics.
– Lack of Cross-Cultural Comparisons: There is a lack of cross-national studies that examine how similar AI tools function differently across varying educational ecosystems – particularly between Global North and Global South contexts [16, 17].

Addressing these gaps is critical to avoid a decontextualized narrative about AI in education, ensuring that AI does not simply replicate existing inequalities or reduce teaching to a set of quantifiable metrics, but instead contributes meaningfully to inclusive, empathetic, and context-sensitive educational transformation. Particularly in Latin America, within higher education institutions, the integration of AI tools offers promising avenues to transform student-teacher interactions, enhance personalized learning, and address long-standing educational inequalities. However, these opportunities also bring challenges that necessitate a deeper understanding of how relational, organizational, and emotional dimensions of education are affected by technology. For this reason, while AI offers transformative potential, its implementation must be carefully balanced with frameworks like the TTI, that emphasize emotional, organizational, and instructional interconnectedness contexts [6, 18, 19].

The TTI framework – originally designed to study the quality of classroom interactions – focuses on three core domains: emotional support, classroom organization, and instructional support. When applied to the integration of AI tools in higher education, this framework allows for a nuanced exploration of how digital interventions affect the socio-emotional fabric of teaching and learning. This is particularly relevant in a region such as Latin America, characterized by profound social and economic diversity, where education systems often grapple with limited resources, uneven digital access, and diverse pedagogical traditions contexts [1, 2, 20].

The TTI framework provides a structured approach to bridge the gap between AI-driven innovations and the relational dynamics essential for effective education. Recognizing that classrooms function as complex social systems, this framework facilitates a more comprehensive understanding of the dynamics that incorporates AI-driven solutions to support teacher development and provide real-time feedback on student engagement. This feedback enables teachers to refine their practices, adapt to diverse student needs, and has the potential to foster meaningful classroom interactions and enhance educational quality across various contexts [6, 21, 22].

While the integration of AI in education offers significant opportunities for enhancing learning experiences, its successful implementation requires sensitivity to cultural and contextual factors unique to each educational system contexts [16, 23]. In Latin America, for instance, diverse socio-economic contexts and institutional structures demand tailored approaches that align technological innovations with the realities of educators and students. This necessity underscores the importance of adapting AI strategies to the specific cultural and local dynamics of each region to maximize their effectiveness and sustainability contexts [1, 17, 21].

While AI systems offer advantages, they also pose challenges that are addressed in this work through recommendations for educators and institutions contexts [1, 4, 5, 22, 23]. This chapter underscores the potential of AI not only to enhance learning experiences but also to shape the future of education in ways that prioritize both technological innovation and human connection [1, 4, 11].

Furthermore, this chapter explores intersections and strategies using the TTI framework to analyze how AI mediates and reshapes human interaction in educational spaces, that is, how AI technologies influence the nature and quality of student-teacher interactions in higher education based on primary data, collected through a comparative study of higher education institutions in Chile and Spain. While Spain provides a European benchmark for comparison, Chile offers insights into the challenges and innovations specific to Latin America contexts [6, 7]. Using the TTI framework [10] as an analytical lens to examine how AI tools mediate several processes in digital and hybrid classrooms – which emphasizes emotional support, classroom organization, and instructional support – this framework analyzes the perceptions and experiences of both students and faculty regarding AI's role in shaping educational relationships.

The research is based on a case study design. Through mixed-method research involving structured surveys and contextual analysis across both national contexts, this chapter examines how AI tools mediate communication, support, and engagement in both digital and hybrid classroom environments. The goal is not only to evaluate technological effectiveness but also to reflect critically on how AI alters the fundamental nature of teaching and learning relationships in different cultural and institutional settings. The main specific objectives of this research are the following:

1. To identify the contextual, institutional, and cultural factors that influence the effectiveness and limitations of AI-mediated interactions in Latin American higher education [5, 7, 20, 24].
2. To analyze the perceptions of students and faculty in Chile and Spain regarding the use of AI tools in academic communication, emotional support, and instructional guidance [6, 10, 25].
3. To propose evidence-based strategies for enhancing student-teacher interactions through AI, drawing on the TTI framework to ensure relational integrity and contextual relevance [9, 10, 26].

This chapter argues that understanding the impact of AI in higher education requires a shift in analytical perspective: from a focus solely on efficiency and outcomes to a broader, relational lens that considers how AI mediates the quality of engagement between educators and learners. This is especially pressing in Latin America, where structural disparities, digital divides, and diverse sociocultural contexts pose unique challenges to the implementation of educational technologies. The region's experiences with digital transformation are shaped not only by technological capacity but also by pedagogical traditions, relational norms, and institutional inequalities contexts [1, 16, 27].

Ultimately, the chapter argues that successful AI integration in higher education depends not solely on the technological sophistication of the tools deployed, but also on how they are embedded within pedagogical frameworks that prioritize meaningful interaction, cultural sensitivity, and equitable access. By aligning AI innovation with the TTI framework, educators and policymakers can make informed decisions that enhance – not hinder – the relational essence of teaching and learning.

2 Theoretical Framework: Teaching Through Interactions (TTI) and AI Integration

Certainly, the integration of AI in higher education has garnered widespread attention due to its potential to personalize learning, automate feedback, and improve educational outcomes. However, the relational and human dimensions of teaching and learning are often sidelined in technological discourses contexts [1, 23]. To address this gap, this chapter adopts the TTI framework, originally proposed by contexts [9], which conceptualizes teaching as a set of dynamic interactions between students and teachers that unfold in emotional, organizational, and instructional domains. These domains offer a robust lens for evaluating the effectiveness and appropriateness of AI integration in higher education contexts [4, 5, 24, 25].

To adequately analyze the relational implications of AI in higher education, this study is guided by four main pillars: (1) an overview of the TTI framework, which offers a structured understanding of student-teacher relational dynamics; (2) a socio-technical perspective on AI integration, which contextualizes technology adoption within broader institutional and cultural ecosystems; (3) a description of the cultural dimensions and localized adaptations in Latin America; and finally, (4) a contextualization of AI in Latin American higher education in the recent years.

2.1 Overview of the TTI Framework

Originally developed by Pianta et al. [9–11], the TTI framework conceptualizes classroom interactions as comprising three key interrelated domains: (1) Emotional support: the degree to which teachers foster a positive learning environment through warmth, respect, and responsiveness in teacher-student relationships. It includes sensitivity, encouragement, and the creation of a positive climate into the classroom. (2) Classroom organization: the ways in which teachers manage student behavior, time, and attention to create a productive learning environment. It refers to the structuring of learning environments through clear expectations, efficient routines, and proactive behavior management. (3) Instructional support: the strategies used by teachers to promote cognitive engagement, critical thinking, and conceptual understanding. It encompasses the quality of feedback, scaffolding, and cognitive stimulation provided to students.

These domains are particularly relevant when evaluating the role of AI tools in classrooms. For instance, an AI chatbot might provide timely feedback (instructional support), but could also diminish emotional support if it replaces human interaction contexts [26, 27]. In this sense, the TTI framework becomes an essential analytical tool for understanding both the advantages and unintended consequences of AI integration in education contexts [1, 17]. Thus, in terms of mapping AI applications onto the TTI domains, AI tools can be evaluated for their influence across the TTI domains:

– In the domain of emotional support, AI applications such as AI-driven sentiment analysis or emotion-detection algorithms can detect student frustration or disengagement, promise to help teachers identify students' affective states in real time, and offer teachers real-time emotional diagnostics. However, such tools risk over-reliance on quantifiable emotion proxies, potentially undermining nuanced human understanding contexts [1, 12]. While such tools can alert educators to signs of disengagement or stress, they also risk reducing complex emotional expressions to quantifiable signals. Furthermore, excessive reliance on automated affective monitoring may erode trust and authenticity in teacher-student relationships. Emotional support, as envisioned in TTI, requires not just recognition of emotional cues but genuine, responsive human interaction – something that AI alone cannot replicate contexts [5, 15].

– In the domain of classroom organization, AI systems often offer clear benefits by streamlining administrative tasks, automating attendance, managing assignment deadlines, and even optimizing group formation through data-driven algorithms contexts [16, 17]. Thus, AI can help manage large classes by automating attendance, participation tracking, or content delivery pacing. These capabilities can free teachers to focus more on pedagogy and relationships. However, while these features offer efficiency, they might also constrain spontaneity and flexibility, qualities often essential for inclusive teaching. There is also a risk that overly rigid algorithmic structures can undermine teacher autonomy and fail to adapt to the fluid dynamics of actual classrooms. A system designed to optimize efficiency may inadvertently prioritize compliance over engagement, creating environments that are procedurally effective but pedagogically sterile contexts [18].

– In the domain of instructional support, AI tools such as adaptive learning platforms and intelligent tutoring systems are often promoted for their ability to personalize learning experiences, offer real-time feedback, and scaffold student understanding contexts [13, 28]. In this sense, AI-based adaptive learning systems can tailor content difficulty to individual learners and fostering differentiated instruction contexts [10]. Within the TTI framework, these tools align most directly with the instructional support domain. However, concerns arise when such systems substitute for teacher guidance rather than supplement it contexts [9, 17, 29]. The cognitive dimension of instruction is deeply intertwined with dialogic exchange, contextual understanding, and formative assessment – elements that are often diminished in standardized AI-generated feedback, but their effectiveness hinges on culturally and contextually relevant design, especially in diverse regions like Latin America contexts [10, 11, 30].

Moreover, the TTI framework emphasizes the interconnectedness of these three domains contexts [3, 6, 17]. Changes in one area inevitably affect the others. For example, a system that improves instructional feedback may still lead to poorer overall classroom interactions if it diminishes emotional support or imposes rigid organizational structures. Therefore, evaluating AI's role through TTI requires not only assessing discrete functions but understanding how technology reshapes the relational ecology of the classroom.

The TTI model has been widely validated across educational levels and international contexts [11, 13], but its application in higher education – particularly in relation to AI-mediated learning environments – is still nascent. By adopting this framework, the current study bridges pedagogical research with technological innovation, emphasizing that the quality of classroom interactions remains a core determinant of educational outcomes, regardless of technological mediation. In this chapter, we argue that AI tools should be critically evaluated not just for their instructional outcomes or technological sophistication, but for their impact on these three core dimensions of classroom life. By using the TTI framework, we are better positioned to assess whether AI enhances or undermines the relational quality of teaching – a factor that remains central to student engagement, persistence, and learning outcomes.

2.2 Conceptual Integration: AI and Relational Pedagogy

One criticism of AI in education is that it promotes a transactional rather than relational model of learning. The TTI framework helps counterbalance this by emphasizing the humanistic dimensions of education. By examining AI applications through this lens, we move beyond metrics and performance to consider empathy, connection, and mutual respect as educational outcomes contexts [12, 31].

Moreover, adopting TTI in AI-enhanced environments supports the development of hybrid pedagogies where technological tools are not replacements but amplifiers of pedagogical intentions contexts [10, 32]. This is especially pertinent in culturally diverse settings where relational dynamics are deeply influenced by social norms, power structures, and institutional constraints. For this reason, a second strand of the theoretical framework draws on sociotechnical theories, which stress the interplay between technological tools and the social, cultural, and institutional settings in which they are deployed [14, 16, 22]. From this perspective, AI is not a neutral or universally applicable solution but a system that reflects and amplifies existing organizational values, biases, and power structures contexts [1, 17].

In the context of Latin American higher education, such an approach is essential contexts [33, 34]. The uneven availability of technological infrastructure, the diversity of teaching cultures, and the legacy of hierarchical institutional arrangements influence how AI tools are interpreted and used by educators and learners [6, 7, 35]. Furthermore, critical scholars have warned that AI may exacerbate surveillance, standardization, and depersonalization in education if not grounded in participatory, ethical design [20, 21, 36].

Integrating the TTI and sociotechnical perspectives allows us to interrogate how AI systems influence classroom dynamics on multiple levels. While TTI provides pedagogical vocabulary for understanding micro-level interactions – emotional support, classroom organization, and instructional feedback – sociotechnical theory reveals how macro-level structures, such as institutional priorities or national policy, shape the deployment and impact of AI in education. For example, in universities where performance metrics dominate quality assurance systems, AI tools may be configured primarily to optimize student throughput, regardless of relational contexts [36–38]. Conversely, in institutions where pedagogical autonomy and interpersonal engagement are prioritized, the same technologies may be appropriated in more relationally meaningful ways. This dual lens also helps surface hidden tensions – such as when a learning analytics dashboard intended to support engagement is perceived by students as a surveillance mechanism.

In Latin America, the adoption of AI tools in higher education takes place in settings shaped by deep-rooted inequalities, limited digital infrastructure, and a wide range of teaching philosophies. Across the region, universities continue to grapple with persistent challenges such as inadequate funding, unreliable connectivity, low levels of digital literacy, and complex governance structures [2, 38]. These constraints strongly affect how AI is implemented, adapted to local needs, and perceived by both faculty and students. For instance, institutions with limited digital capacity often lack the infrastructure to effectively use adaptive learning technologies or intelligent tutoring systems tailored to students' learning profiles. Additionally, the legacy of top-down policy reforms in the region can result in the adoption of AI tools that conflict with the pedagogical values and culturally responsive practices embraced by educators [18, 39].

Chile provides a clear example of this uneven landscape. Although national efforts like the "Digital Talent for Chile" strategy and broader investments in educational technologies have driven digital transformation, the outcomes have been distributed unequally across the country's higher education sector [12, 40]. Prestigious urban universities typically have better access to technology and administrative support to deploy AI effectively. In contrast, regional and rural institutions often face challenges such as limited bandwidth, outdated infrastructure, and insufficient training for faculty. These gaps contribute to unequal opportunities and a fragmented AI implementation across the system. Moreover, the country's highly stratified and market-driven higher education model often prioritizes efficiency metrics over meaningful teaching practices, further amplifying these disparities [6, 17].

Furthermore, relational pedagogy in Latin American settings, including Chile, is rooted in traditions that emphasize community, dialogue, affective connection, horizontal communication, social justice, and collective learning. These values often stand in tension with the logics of algorithmic personalization, standardization, and automation that underlie many AI systems developed in the Global North contexts [7, 25]. For instance, while AI-powered tutoring systems may be effective in delivering customized content, they may also limit opportunities for spontaneous, dialogic engagement that many Latin American educators view as central to meaningful learning contexts [4, 14].

If implemented without thoughtful cultural adaptation, AI may reinforce educational models misaligned with local pedagogical values [11, 15]. As such, relational frameworks like TTI should be understood not only as analytical instruments but also as ethical blueprints that guide the equitable and inclusive application of AI in education. This is especially critical in Latin America, where longstanding inequalities, institutional disparities, and diverse sociocultural traditions shape the educational context. In these settings, AI adoption must go beyond technical adequacy to include pedagogical sensitivity and ethical responsibility [5, 8, 17]. The integration of TTI with sociotechnical perspectives provides a robust lens for the design, assessment, and adaptation of AI systems that respect both relational dynamics and cultural specificities in education [8, 17].

This chapter therefore positions the integration of AI and relational pedagogy as a critical juncture in the evolution of higher education. By articulating how AI systems intersect with the emotional, organizational, and instructional domains of the TTI framework, we aim to provide both a diagnostic and developmental lens – highlighting not only what is at stake, but also how educators, institutions, and developers can respond.

2.3 Cultural Dimensions and Localized Adaptations

In Latin America, the teacher-student relationship is often shaped by hierarchies, affective bonds, and collective identities. These cultural features are rooted in historical, social, and educational traditions that value interpersonal proximity, emotional expressiveness, and mutual dependence between teachers and students. Unlike in some Anglo-European models that emphasize individual autonomy and academic detachment, the Latin American pedagogical ethos tends to foster strong relational ties as integral to the learning process. AI tools, if uncritically adopted, may ignore these relational norms, leading to mismatched expectations and suboptimal outcomes contexts [1, 7, 41]. The TTI framework offers a culturally adaptable model that enables educators and designers to align AI functionalities with local pedagogical realities. By emphasizing emotional support, classroom organization, and instructional responsiveness, TTI can serve as a lens for adapting AI systems to enhance – rather than displace – human connection in diverse educational contexts [16, 17].

In Chile, for instance, where educational inequalities are often intertwined with geographical and socioeconomic divides, AI tools must be adapted not only to the infrastructure of each institution but also to the pedagogical relationships that define learning environments [2, 12, 21]. A practical example includes the deployment of AI-driven adaptive learning platforms in several Chilean universities. While these platforms can provide personalized content, they must be integrated in ways that do not undermine the affective mediation typically provided by instructors. In rural or marginalized areas, where face-to-face interactions carry additional symbolic and social weight, the substitution of human tutoring with algorithmic feedback can erode trust and disengage students [16, 17]. Therefore, the design and implementation of AI must be context-sensitive, involving local educators in co-design processes to ensure that tools resonate with students' experiences and cultural expectations [3, 17].

Furthermore, relational frameworks such as TTI can help mitigate the risk of technocentric determinism by prioritizing the values of equity, inclusivity, and dialogical engagement. In the face of growing interest in automating educational processes, a culturally grounded, relational approach can serve as a counterbalance, reminding institutions that meaningful education is not only about performance metrics but also about fostering humane, empathetic, and reciprocal relationships between learners and educators [2, 42].

Understanding these dynamics is crucial for the effective deployment of AI. In Latin America – and Chile in particular – student-teacher relationships are often described using terms such as affection, respect, and horizontality. These are closely aligned with the emotional and organizational domains in TTI. The TTI framework thus offers a culturally meaningful analytical lens for understanding how AI can amplify – or undermine – traditional pedagogical values in the region [9, 17, 21].

In this chapter, we argue that the successful integration of AI in higher education in Latin America must not only consider technical capabilities but also be anchored in relational theories like TTI that account for the emotional, organizational, and instructional dimensions of learning. This means viewing AI not as a universal solution, but as a set of tools that must be tailored to the social, cultural, and pedagogical ecosystems of each institution and country.

2.4 Contextualizing AI in Latin American Higher Education

Latin American universities face unique structural, cultural, and pedagogical challenges in adopting AI technologies. These include wide disparities in technological infrastructure, socio-economic inequalities, diverse linguistic, and cultural contexts, and often rigid institutional policies. While AI offers solutions to many of these issues, its application must be tailored to these realities' contexts [43, 44]. While AI offers solutions to many of these issues, its application must be tailored to these realities. Unlike more homogeneous education systems, Latin America's diversity – ranging from centralized systems in countries like Chile to more decentralized models in Brazil or Mexico – demands flexible, localized approaches. Failure to do so may result in technology-led reforms that fail to address the core pedagogical and relational needs of learners.

2.4.1 The Digital Divide and Technological Infrastructure

One of the most significant barriers to AI integration in Latin America is the digital divide. Although internet penetration has improved, access to stable and affordable connectivity remains uneven across urban and rural areas. Furthermore, many universities lack the necessary hardware, software, or IT support to implement AI-based solutions comprehensively [11]. This creates a risk of deepening existing educational inequalities.

In Chile, rural regions in both the south and north often face serious limitations in digital infrastructure, including slow internet connections, outdated technology, and minimal IT support. These constraints significantly hinder the implementation of AI-driven educational tools, particularly those that rely on cloud computing, real-time analytics, or synchronous interactions. As a result, the digital divide deepens existing inequalities within higher education: students with reliable digital access gain the benefits of personalized AI-enhanced learning, while those without fall further behind – exacerbating socio-economic disparities. Moreover, institutions in resource-scarce areas often lack the hardware, licensed software, and cybersecurity infrastructure necessary for safe and effective AI deployment. Many universities rely heavily on inconsistent government funding or short-term development grants, making it difficult to implement sustainable, long-term AI strategies. These challenges point to the urgent need for digital transformation plans that are context-sensitive – emphasizing not only technical access, but also faculty training, platform usability, and pedagogical alignment [1, 5, 23].

2.4.2 Institutional and Pedagogical Readiness

Latin American institutions are often characterized by hierarchical administrative structures and resistance to change. These systems can delay or obstruct the adoption of innovative technologies, particularly those requiring cross-departmental collaboration or pedagogical experimentation. Many faculty members frequently have limited digital literacy, and professional development opportunities for AI use in pedagogy are scarce. As a result, AI tools are frequently introduced without a pedagogical framework, leading to underutilization or misapplication [45–47].

In Chile, although national policies promote digital education, many universities lack formal training programs to help faculty integrate AI into their teaching practices. This gap results in two main consequences: (1) AI tools are underutilized or applied in ways that conflict with relational teaching values and (2) decisions about technology adoption are often made by administrators without meaningful input from teaching staff. As a result, AI is sometimes seen as a top-down imposition rather than a bottom-up innovation, leading to skepticism, resistance, or disengagement among educators.

Pedagogical readiness also involves curriculum design, student assessment, and course delivery models. In settings where lecture-based teaching remains dominant, the pedagogical integration of AI tools – such as adaptive learning platforms or automated feedback systems – may be limited unless accompanied by broader instructional reforms [10, 17, 23].

2.4.3 Cultural Attitudes Toward AI and Technology

Attitudes toward AI in education vary significantly across the region. While some educators view it as a valuable ally for workload management and student support, others are skeptical, fearing surveillance, loss of autonomy, or depersonalization [1, 17].

Concerns about AI implementation in education are well justified. In Chile and other Latin American countries, faculty have voiced apprehension about AI being used to monitor their behavior, assess teaching effectiveness without adequate context, or replace genuine human interactions with automated processes. Students, likewise, express mixed reactions. While they recognize the benefits of AI – such as faster feedback, greater efficiency, and personalization, especially in large classes – they also worry about the erosion of authentic human connection. This concern is particularly acute in sensitive scenarios like academic counseling or disciplinary actions [6, 7, 48]. The key challenge, then, is to ensure that AI tools reinforce, rather than erode, the trust, empathy, and mutual recognition that underpin effective student-teacher relationships [6, 17, 47, 48].

By comparison, Spain illustrates a more advanced phase of AI integration in higher education. National digital strategies like the "Plan de Digitalización y Competencias Digitales" and institutional bodies such as "Comisión Sectorial de Tecnologías de la Información y las Comunicaciones (CRUE-TIC)" have supported broader technological infrastructure and a more robust educational technology ecosystem [6, 20, 23, 49]. Spanish universities also benefit from public funding that supports digital transformation efforts [1, 8, 17, 50]. However, even with this progress, ethical and pedagogical challenges remain. Faculty frequently cite concerns around data privacy, algorithmic bias, and opaque decision-making processes. Moreover, many institutions are still working to align AI systems with pedagogical priorities, as opposed to allowing them to serve primarily administrative or efficiency-focused purposes [2, 3, 7, 51]. This reinforces the idea that technological advancement alone does not guarantee improved relational or educational outcomes [16, 17, 52].

This comparison serves as a useful reference point for identifying which AI practices could be adapted to Latin American contexts and which require significant modification [3, 6]. Importantly, it reiterates that technological readiness does not inherently lead to more meaningful or relationally effective teaching. Rather, it highlights the necessity of viewing AI not as a one-size-fits-all solution, but as a set of adaptable tools and practices that must be thoughtfully aligned with institutional values, cultural contexts, and relational pedagogical goals [6, 53, 54].

This chapter proposes a novel analytical synthesis of the TTI and sociotechnical frameworks. While TTI enables a relational evaluation of teaching quality, sociotechnical perspectives offer the cultural and institutional lenses necessary to understand the complexity of AI integration. Together, they allow for a nuanced, context-sensitive examination of how AI mediates emotional, organizational, and instructional aspects of student-teacher interaction. This combined approach is particularly valuable for Latin America, where relational pedagogy and institutional constraints coexist. It moves beyond technologically deterministic models to instead foreground how AI can be used to support, rather than replace, meaningful human connections in the classroom [4, 6, 9, 17].

The next section presents the study's design and empirical basis, grounded in a comparative approach involving Chile and Spain, and anchored in the TTI framework to evaluate how AI affects student-teacher interactions in diverse socio-educational contexts.

3 Methodology

This study employs a comparative case study methodology grounded in the TTI framework to explore how AI influences student-teacher interactions in higher education and investigate the impact of AI on student-teacher interactions in higher education. By juxtaposing experiences from Chile and Spain, the research aims to uncover both context-specific insights and generalizable patterns related to AI integration. Furthermore, the research captures regional specificities while contextualizing Latin America's challenges and innovations within a broader educational framework. The methodological design combines quantitative and qualitative elements to capture the complexity of human-technology interactions in educational settings.

Although originally developed for primary and secondary education, recent literature supports the extension of TTI to higher education, particularly when the focus is on relational pedagogy, not just content delivery. Emotional, organizational, and instructional aspects of interaction are seen as essential for student engagement and learning outcomes [4, 14, 18]. For example, several studies and adaptations in adult education, medical education, and university teacher training have extended TTI constructs to higher education with success. In this sense, focusing on student-teacher interaction aligns with AI research in higher education, current AI and EdTech literature increasingly emphasizes the importance of human-AI collaboration in learning environments and the relational impacts of automation and personalization, avoiding depersonalization, surveillance, or teacher deskilling [16, 17, 55].

3.1 Methodological Design

The primary objective of this study is to analyze the influence of AI tools on the emotional, organizational, and instructional dimensions of student-teacher interactions as defined by the TTI framework in different cultural and institutional settings, in terms of the nature of teaching and learning relationships. In empirical terms, the main specific objectives were to identify how AI tools are currently used in higher education institutions in Chile and Spain, analyze the perceived impact of these tools on each of the TTI domains, and compare the cultural and institutional factors and common challenges that mediate the effectiveness of AI-supported pedagogy to ensure relational integrity and contextual relevance, as evidence-based strategies for enhancing student-teacher interactions.

A mixed-methods design was used, combining survey-based quantitative data with qualitative interviews to capture the depth and variation of participants' experiences [3, 14]. This triangulated approach enables a more comprehensive analysis of how AI mediates relational dynamics in university settings, consistent with the TTI framework, and enables a multi-dimensional analysis of how AI technologies affect the relational, organizational, and instructional fabric of university teaching. By grounding the empirical design in the TTI framework and validated instruments tailored for higher education, the study produces robust, context-sensitive insights that are both theoretically informed and practically relevant.

3.2 Participants and Sampling

The study involved 385 participants, including undergraduate and graduate students, as well as faculty members from major universities in Chile and Spain between January and March 2025. A stratified sampling strategy was adopted to ensure representation across disciplines, institutional types, and levels of digital adoption. Sampling was purposive and stratified to ensure representation across various institutional types (public and private), academic disciplines (STEM (science, technology, engineering, and mathematics) and non-STEM), and geographic regions (urban and nonurban campuses). This diversity was designed to reflect the heterogeneous nature of AI implementation and relational culture within Latin American and Spanish higher education systems. Ethical clearance was obtained from the relevant ethics committees at participating institutions. All participants provided informed consent, and strict confidentiality protocols were observed during data collection, analysis, and reporting.

The study was conducted across four higher education institutions – two in Chile and two in Spain – selected for their diversity in institutional type, technological infrastructure, and the degree to which they have implemented hybrid or virtual learning models enhanced by AI. In Chile, the research involved "Pontificia Universidad Católica de Valparaíso" (PUCV), a traditional public university with a growing commitment to digital transformation, particularly in hybrid learning and virtual classroom interaction tools; and "Universidad Técnica Federico Santa María" (USM), a leading technological institution recognized for its engineering programs and for integrating AI-supported systems in both synchronous and asynchronous learning environments. In Spain, data were collected from "Universidad Internacional de La Rioja" (UNIR), a fully online university with extensive experience in digital learning, AI-enabled tutoring systems, and real-time academic analytics; and "Universidad de Valladolid" (UVa), a historic public university that has gradually adopted hybrid teaching models, incorporating AI-driven platforms for formative assessment and instructional personalization. These universities represent diverse institutional profiles and pedagogical models, enabling a robust comparative analysis of how AI tools impact student-teacher interactions within virtual and blended higher education settings. Their participation provides insight into how AI is adapted to relational, cultural, and technological contexts aligned with the emotional, organizational, and instructional dimensions of the TTI framework.

3.3 Instruments and Data Collection

The quantitative component included a structured questionnaire administered to 385 participants (students and faculty) across public and private universities in both countries and was developed based on the TTI framework. The survey instrument was adapted from two validated tools for higher education: the Teaching Styles Inventory for Higher Education (TSIH) and the Clinical Teaching Behavior Inventory (CTBI) contexts [36, 37, 56]. These instruments were selected for their theoretical alignment with the TTI domains and their prior validation in diverse higher education contexts while accounting for the complexity of adult learning environments. In contrast, the CLASS-S instrument, although aligned with TTI domains, was developed for secondary education and is primarily observational rather than self-reported. Its application in university-level, AI-mediated settings would risk construct misalignment and reduce the validity of findings. By using these higher education-specific instruments, this study ensures theoretical consistency, psychometric robustness, and cultural relevance in the analysis of student-teacher interactions within AI-enhanced classrooms.

The primary instrument was a structured questionnaire aligned with the TTI domains, student satisfaction, and relational quality as two separate dependent variables. In the study, both variables were measured using quantitative self-report items embedded in the questionnaire after the main TTI items. Thus, student satisfaction was measured using two Likert-type items adapted from validated higher education satisfaction scales university course evaluations): "Overall, I am satisfied with how this course integrated AI tools" and "This course met my expectations in terms of learning and support," using the mean score of these items (5-point scale: 1 = strongly disagree to 5 = strongly agree, and reliability: Cronbach's alpha > 0.80). Relational quality was measured using two Likert-type items adapted from student-teacher relationship literature (Student-Teacher Relationship Scale)) contexts [38]: "I felt a strong sense of connection with my instructor, even when AI tools were used" and "AI-mediated teaching preserved the quality of human interaction in this course," scoring the same 5-point Likert scale (reliability: Cronbach's alpha > 0.82).

The instrument included Likert-scale items on the perceived impact of AI on emotional support, classroom organization, instructional quality, and open-ended questions to allow narrative responses regarding experiences and expectations of AI use in learning environments. The survey was administered online, and interviews were conducted with selected participants to gain deeper qualitative insights.

Moreover, items were designed and adapted to assess perceived changes in the three core dimensions of the TTI framework – emotional support, classroom organization, and instructional support – specifically in AI-mediated educational settings. To ensure contextual validity and cultural relevance, the initial phase of this study incorporated a qualitative exploratory design to inform the development and adaptation of the questionnaire. Semi-structured interviews with university faculty and students in both Chile and Spain were conducted prior to finalizing the survey instrument. This step allowed to identify key relational dynamics, terminologies, emotional responses to AI integration, perceived shifts in classroom management, changes in the quality of instructional interactions, and AI usage patterns specific to each educational context, ensuring alignment with the TTI framework's domains (emotional support, classroom organization, and instructional support).

The qualitative data were thematically analyzed and used to adapt existing validated items from the Teaching Styles Inventory Questionnaire for Higher Education (TSIQ-HE) and Clinical Teaching Behavior Inventory (CTBI) instruments, refine their wording, and eliminate culturally ambiguous or irrelevant items. Primary categories derived deductively from the TTI framework and supplemented by inductive subcategories captured emergent themes, especially those reflecting sociotechnical tensions and cultural adaptation. This participatory process also enhanced the instrument's face and content validity, addressing cross-cultural variations in how AI is perceived and experienced in pedagogical settings. By integrating practitioner and student perspectives early in the research design, the study strengthened the reliability and contextual sensitivity of the final questionnaire.

A novel, context-specific instrument – TTI-AI-HE scale – was created, consisting of 15 Likert-type items (5-point scale), with 5 items under each TTI domain (see Appendix). Following instrument development, a pilot test was conducted with a subsample of 42 participants from both countries. Results from the pilot informed final refinements, improving item clarity and cultural appropriateness.

Informed consent was obtained from all participants in accordance with ethical guidelines. In Chile, approval was granted by the ethics committees of PUCV and USM, while in Spain, ethical review and clearance were provided by UVa and UNIR. Participants were informed of the voluntary nature of the study, data confidentiality, and their right to withdraw at any time. Consent forms were distributed digitally and included prior to survey access.

3.4 Data Analysis

Qualitative responses were thematically coded according to the TTI domains. Triangulation of survey and interview data enhanced the validity of the findings and allowed for a comprehensive portrayal of the relational dynamics shaped by AI. Quantitative data were analyzed using descriptive, inferential statistics, and cross-tabulation to identify trends and correlations by country across demographic variables and national contexts.

The validated scale was finally distributed to participants across universities in Chile and Spain. Prior to inferential analyses, an exploratory factor analysis (EFA) was conducted to confirm the scale's dimensional structure and to test its alignment with the TTI framework. The Kaiser-Meyer-Olkin (KMO) measure and Bartlett's test of sphericity were applied to verify sampling adequacy and the suitability of factor analysis. Three distinct factors corresponding to the emotional, organizational, and instructional domains were extracted using principal axis factoring with Promax rotation.

Reliability was assessed through Cronbach's alpha, yielding values above 0.80 for each subscale, confirming internal consistency. Descriptive statistics (means, standard deviations, and frequencies) were calculated for each item. To assess cross-country differences, independent samples t-tests and multivariate analysis of variance (MANOVA) were employed. Further, regression models were applied to examine how perceived AI effectiveness predicted student's satisfaction and perceived relational quality. This analytical strategy ensured that the instrument not only reflected the theoretical constructs of the TTI framework but also provided valid and reliable data for comparing AI-mediated relational dynamics in diverse higher education systems.

The following section presents the empirical findings, structured by TTI domain and analyzed comparatively between the Chilean and Spanish cases.

4 Results and Discussion

4.1 Profile of the Sample

The sample consisted of participants engaged in AI-mediated higher education environments, whether hybrid or fully virtual, drawn from the four universities here described. In this research, the stratified purposive sampling strategy was employed to ensure diversity across institutional type, gender, academic role, disciplinary field, and technological exposure. Table 1 summarizes the distribution of participant characteristics. Students comprised entire sample, ensuring a strong representation of learner perspectives in the analysis (in this research, faculty participants provided complementary insights into the design and instructional use of AI tools).

Analysis of the descriptive data revealed several notable patterns. As it is shown, gender distribution was balanced, with a slight majority of female respondents (54%). The disciplinary composition of the sample reflected a mix of STEM (47.8%), social sciences (34%), and humanities (18.2%), allowing for cross-disciplinary comparisons. Regarding mode of study, 62% of respondents reported engaging in hybrid learning environments, while 38% were in fully virtual programs.

A final noteworthy finding was the variation in participants' self-reported familiarity with AI tools: 21.8% indicated low familiarity, 49.1% medium, and 2.19% high. This distribution reflects the uneven diffusion of AI in higher education and justifies the need to examine how perceptions and usage differ across these levels.

Table 1: Sample profile by country (count and percentage).

Characteristics, variables	Chile			Spain			Total Sample ($n = 385$)
	PUCV ($n = 95$)	USM ($n = 100$)	Total, Chile ($n = 195$)	UNIR ($n = 95$)	UVa ($n = 95$)	Total, Spain ($n = 190$)	
Gender:							
–Female	50 (25.6%)	55 (28.2%)	105 (53.8%)	50 (26.3%)	52 (27.4%)	102 (53.7%)	207 (54%)
–Male	44 (22.6%)	45 (23.1)	89 (45.6%)	43 (22.6%)	43 (22.6%)	86 (45.3%)	175 (45%)
–Other/no resp.	1 (0.5%)	–	1 (0.5%)	2 (1.0%)	–	2 (1.0%)	3 (0.7%)
Area:							
–STEM	45 (23.01%)	50 (25.6%)	95 (48.7%)	49 (25.8%)	40 (21.1%)	89 (46.8%)	184 (47.8%)
–Social sciences	35 (17.9%)	33 (16.9%)	68 (34.9%)	33 (17.4%)	30 (15.8%)	63 (33.2%)	131 (34%)
–Humanities	15 (7.7%)	17 (8.7%)	32 (6.4%)	13 (6.8%)	25 (13.16%)	38 (20.0%)	70 (18.2%)
Mode:							
–Hybrid	60 (30.8%)	64 (32.8%)	124 (63.6%)	20 (10.5%)	95 (50%)	115 (60.5%)	239 (62%)
–Fully virtual	35 (17.9%)	36 (18.5)	71 (36.4%)	75 (39.5%)	–	75 (39.5%)	146 (38%)
AI familiarity:							
–Low	28 (14.4%)	20 (10.26%	48 (24.6%)	1 (0.5%)	35 (18.4%)	36 (18.9%)	84 (21.8%)
–Medium	54 (27.7%)	40 (20.51%)	94 (48.2%)	55 (28.9%)	40 (21%)	95 (50.0%)	189 (49.1%)
–High	13 (6.6%)	40 (20.51%)	53 (27.2%)	39 (20.5%)	20 (10.5%)	59 (31.1%)	112 (29.1%)

4.2 Contextual, Institutional, and Cultural Factors Influencing AI-Mediated Interactions

To complement the empirical results and the quantitative and qualitative findings of this chapter, it is essential to explore the contextual, institutional, and cultural dimensions that shape AI-mediated student-teacher interactions in higher education. From a methodological perspective, this subsection builds on the comparative case study approach and mixed-methods design regarding the sociotechnical systems theory, which emphasizes the interdependence between technological tools and the social environments in which they operate, allowing for an in-depth interpretation of how macro-level structures influence micro-level relational dynamics captured through the TTI framework. In line with contextualist approaches to educational research (e.g., Bronfenbrenner's ecological systems theory and sociocultural theory), this study recognizes that teaching and learning processes are embedded within multilayered systems – spanning the micro (classroom), meso (institutional), and macro (sociocultural) levels [4, 16, 47, 48].

The TTI framework provides a robust lens for capturing interactional dynamics at the classroom level, while the inclusion of contextual analysis ensures that variations in AI adoption and impact are interpreted considering institutional cultures, governance models, and regional educational norms [57, 58].

A critical dimension in understanding the dynamics of AI-mediated interactions in Latin American higher education involves these factors are not peripheral but foundational, influencing how students and faculty perceive, adopt, and engage with technological tools within relational pedagogical environments [3, 6, 7].

From a contextual standpoint, countries like Chile face persistent disparities in technological infrastructure, particularly between urban and rural regions. Despite increased investments in digital transformation, access to high-speed internet, reliable hardware, and faculty training remains uneven [6, 8]. This digital divide introduces structural asymmetries in how AI-supported platforms are implemented across universities, such as PUCV and USM, affecting the consistency of AI tools in fostering emotional and instructional support [3]. Moreover, participants from these institutions reported inconsistent experiences with digital environments, particularly in hybrid settings, where bandwidth and digital fluency influenced their perceived quality of student-teacher interactions [48, 49, 59].

At the institutional level, governance models in Latin America often follow hierarchical structures, with limited decision-making autonomy for individual instructors [6, 8]. In this research, faculty members from both Chilean and Spanish universities reported that AI adoption was often led by administrative or IT units, with minimal input from educators regarding pedagogical alignment. This reduced teacher engagement and hindered effective use of AI for enhancing emotional and instructional support in two of the three core domains of the TTI framework [1, 2]. Furthermore, universities often lacked robust professional development programs to train staff on integrating AI tools into relational teaching practices [10, 60].

From a cultural perspective, the affective dimension of education is a defining feature of teacher-student interactions in Latin America [6, 20]. Empathy, relational proximity, and personalized mentorship are deeply rooted in pedagogical traditions and remain highly valued by both students and teachers. However, many AI tools are designed with efficiency, standardization, and data extraction as core principles, which may conflict with these culturally embedded values [11, 14]. For instance, while students at PUCV appreciated AI feedback for its immediacy, several expressed discomfort with the lack of emotional nuance or the "coldness" of the AI-mediated communication. Similar sentiments were echoed by students at UVa and UNIR, where the widespread use of automated learning management systems (LMS) was sometimes perceived as "mechanical" or "detached."

Moreover, concerns regarding surveillance and autonomy were recurrent among both student and faculty respondents [13, 15]. These concerns were particularly pronounced in Chilean universities, where some students viewed data and several tools as mechanisms of control rather than support. This aligns with broader critiques in the literature, which warn that uncritical AI adoption may exacerbate educational inequalities and reduce pedagogical freedom if not embedded within a culturally responsive and participatory framework [14, 15, 61].

Taken together, these findings highlight the necessity of designing AI implementations that are culturally sensitive, institutionally grounded, and contextually adaptive. Effective deployment must not rely solely on the technological sophistication of AI tools but must also be rooted in the everyday practices and values of university classrooms. Integrating frameworks like TTI allows institutions to evaluate AI not only in terms of efficiency but also through its capacity to support relational teaching, emotional connection, and pedagogical integrity in diverse cultural settings [1, 4, 6, 46].

4.3 Results on Perceptions, Experiences, and Outcomes

This section presents the findings from both quantitative and qualitative data. The analysis is structured around the three core TTI domains: emotional support, classroom organization, and instructional support. The comparative perspective between Chile and Spain reveals how national context shapes the integration and reception of AI in higher education.

The results of the quantitative analysis were derived from the responses to the adapted TTI-based questionnaire, which is the TTI-AI-HE scale, which comprised 15 items across three core domains: emotional support, classroom organization, and instructional support. Preliminary data checks confirmed the suitability of the dataset for factor analysis. Skewness and kurtosis values for all items fell within acceptable ranges (−1 to +1), indicating a normal distribution of responses across the sample [36, 37].

As described, an EFA using SPSS was first conducted using principal axis factoring with Promax rotation to uncover the underlying structure of the questionnaire. The choice of Promax (oblique) rotation in the EFA was theoretically and empirically justified because the three core domains of the TTI framework are conceptually interrelated in real educational settings (e.g., strong emotional support can facilitate better classroom organization, and instructional strategies are often influenced by both the emotional climate and classroom management) [38, 48]. Given this expected correlation among latent factors, an oblique rotation method such as Promax was more appropriate than orthogonal alternatives like Varimax: Promax allows the factors to correlate, producing a more realistic and interpretable factor structure aligned with the complex relational dynamics in higher education teaching. This choice ensured that the EFA reflected the interconnected nature of student-teacher interactions, particularly in culturally nuanced contexts like Latin America and Spain.

Furthermore, the KMO measure of sampling adequacy was 0.86, and Bartlett's test of sphericity was significant (χ^2 = 1384.2, p < 0.001), confirming that the data was factorable. Three factors emerged with eigenvalues above 1, explaining a cumulative variance of 67.3%, which aligns with the original conceptualization of the TTI framework. Items loaded cleanly onto their intended factors, with all loadings above 0.60, supporting the construct validity of the adapted scale.

To ensure both theoretical alignment and empirical robustness, a final nine-item scale was derived from the original 15-item pool. To further validate the measurement model, a confirmatory factor analysis (CFA) was performed using AMOS software and the model demonstrated good fit indices: χ^2/df = 2.04, CFI = 0.95, Tucker-Lewis index (TLI) = 0.94, root mean square error of approximation (RMSEA) = 0.048, and standardized root mean square residual (SRMR) = 0.045.These results confirmed a three-factor structure and indicated an acceptable model fit: thus, the CFA confirmed that the nine-item, three-factor model had superior fit than the 15-item version. These values meet recommended thresholds for model adequacy in educational research and indicate good construct validity. Cronbach's alpha coefficients for the emotional support (α = 0.87), classroom organization (α = 0.83), and instructional support (α = 0.85) dimensions further confirmed the internal consistency of the scale. Based on the results of the EFA, only three items per domain were retained for the final model. The decision to reduce from five to three items per factor was guided by both statistical and theoretical considerations: (1) Factor loadings: only items with loadings ≥0.60 were retained to ensure strong representation of each latent construct. Several items showed cross-loadings or failed to meet the threshold, particularly in the emotional and organizational domains, justifying their removal. (2) Internal consistency: Cronbach's alpha remained high for each of the reduced domains (α > 0.80), indicating that scale reliability was not compromised by item reduction. (3) Theoretical clarity: retaining only the items most closely aligned with the TTI domains improved construct interpretability and minimized redundancy, especially important in comparative cross-cultural studies with language and contextual sensitivity.

Items in Table 2 reflect the core aspects of the TTI framework, while ensuring high validity, reliability, and cultural adaptability for use in Chilean and Spanish higher education contexts. Their performance in both the EFA and CFA processes supports their inclusion in the final model.

Table 2: Exploratory factor analysis (EFA): factor loadings (Promax rotation).

Item	Emotional support	Classroom organization	Instructional support
Item 1: ES1: My instructor shows genuine interest in students' well-being.	0.78		
Item 2: ES2: I feel emotionally supported during AI-mediated classes.	0.81		
Item 3: ES3: The use of AI allows instructors to be more responsive to students' emotional needs.	0.75		
Item 4: CO1: AI tools help instructors manage class time and activities efficiently.		0.69	
Item 5: CO2: AI-based systems make the learning environment more structured.		0.73	
Item 6: CO3: My instructor uses AI to ensure smooth transitions and task flow.		0.71	
Item 7: IS1: AI tools provide personalized feedback that helps me understand difficult concepts.			0.84
Item 8: IS2: My instructor uses AI to promote higher-order thinking and problem-solving.			0.80
Item 9: IS3: AI integration improves the clarity and coherence of instructional content.			0.77

Items ES1 and ES3 were synthesized based on both statistical results from factor analysis and thematic patterns identified in qualitative interviews. ES1 combines relational aspects from Q1 and Q2, focusing on the instructor's genuine care, a key component of the TTI emotional support domain. ES3 draws from Q2 and Q5, highlighting the mediating role of AI in enabling instructors to detect and respond to student emotional needs. Both items showed strong factor loadings (≥ 0.70), were psychometrically validated through EFA and CFA, and enhanced conceptual coherence across culturally sensitive settings like Chile and Spain.

To ensure analytical robustness, descriptive statistics (means, standard deviations, and frequencies) were calculated for each item, followed by inferential analyses using independent-samples t-tests, MANOVA, and regression modeling (see Table 3). These methods ensured that the adapted nine-item TI-AI-HE scale provided valid, reliable, and comparative data reflecting the theoretical constructs of the TTI framework.

Preliminary tests confirmed normal distribution of item responses (skewness and kurtosis within ±1 range), supporting the use of parametric techniques. Each factor was constructed by aggregating the scores of its corresponding three items, typically by computing the mean or sum of the three item scores (Likert-type scale: e.g., 1–5), ensuring internal consistency (Cronbach's alpha > 0.80 for each factor), and validating factor structure via CFA with good model fit indicators (CFI, TLI, RMSEA, and SRMR). All three-item constructs were shown to be reliable ($\alpha > 0.80$), valid (via CFA and theoretical coherence), and non-redundant, covering the breadth of each domain without overlap.

Independent-sample t-tests showed statistically significant differences between Chilean and Spanish participants across all three domains: emotional support: t (383) = −3.42, $p < 0.01$; classroom organization: t (383) = −4.11, $p < 0.001$; and instructional support: t (383) = −3.13, $p < .01$. The MANOVA confirmed that country of origin had a multivariate effect on the combined TTI domain scores (Wilks' lambda = 0.937, F (3, 381) = 8.48, $p < 0.001$). Thus, this suggested that national context significantly influences perceptions of AI-mediated interactions.

Subsequent regression models showed that perceived AI effectiveness (measured through composite TTI scores) significantly predicted student satisfaction ($\beta = 0.51$, $p < 0.001$) and relational quality ($\beta = 0.48$, $p < 0.001$), with moderate to strong effect sizes ($R^2 = 0.26$ and 0.23, respectively). These models underscore the relational importance of AI implementation, particularly in educational cultures valuing affectivity and personalization.

Table 3: Country's descriptive statistics for TTI domains.

Domain	Country	Mean	SD	N
Emotional support	Chile	3.45	0.87	195
	Spain	3.72	0.78	190
Classroom organization	Chile	3.58	0.81	195
	Spain	3.89	0.69	190
Instructional support	Chile	3.67	0.76	195
	Spain	3.94	0.73	190

The empirical findings revealed both convergences and divergences between Chilean and Spanish participants in terms of how AI impacts student-teacher interactions, indicating that participants perceive AI tools as having differential impacts across the TTI domains. Faculty across both countries agreed that while AI could augment feedback and early alerts (e.g., identifying at-risk students), it could not substitute authentic teacher-student relationships. Students expressed concern that AI sometimes made them feel monitored rather than supported, reinforcing surveillance cultures rather than empathy. Moreover, in Chile, students highlighted emotional distance and reduced interpersonal engagement in AI-mediated interactions. However, they acknowledged improved feedback and instructional clarity. Spanish participants emphasized gains in classroom organization, particularly in hybrid environments, although emotional support was perceived as highly dependent on the instructor's digital literacy and pedagogical style.

Qualitative findings from the pilot study reinforced these perceptions. Interview participants described AI tools as useful for scaffolding instruction and organizing content delivery but often struggled with the absence of emotional nuance. Participants from PUCV and USM mentioned the challenge of maintaining student motivation and empathy when mediated by automated systems, especially in asynchronous or automated settings. Students from UNIR and UVa generally reported greater acceptance of AI tools, although concerns regarding depersonalization remained recurrent. These insights support the thesis that while AI can enhance certain instructional and organizational aspects, its integration into higher education must be balanced by strategies that preserve relational and affective components of teaching, particularly in the Latin American context where such dimensions are pedagogically significant.

Related to the emotional support domain, participants in both countries acknowledged that AI tools such as learning analytics dashboards and emotion recognition systems helped teachers become more aware of student moods and stress levels. However, participants in both Chile and Spain reported mixed experiences regarding AI's impact on emotional support. For example, Chilean respondents were more skeptical of the impersonality of AI tools and expressed concern that AI might erode traditional affective bonds between teachers and learners, only 47% agreed, with many emphasizing that AI often lacked "human warmth," particularly in under-resourced institutions where face-to-face support was already limited. By contrast, Spanish respondents, while also cautious, were generally more positive about the augmentative role of AI in detecting emotional needs. In Spain, 62% of respondents felt that AI tools (e.g., adaptive platforms and automated feedback) contributed to a sense of personalized engagement with instructors, even in asynchronous settings, especially in hybrid models at UNIR and UVa.

Qualitative interviews reinforced this trend. Spanish educators viewed AI tools as freeing up time for higher-quality interactions ("AI gives me more time to focus on students' emotional needs"), while Chilean faculty highlighted challenges in emotional connection, particularly in fully online courses with limited AI customization. Faculty members across both countries emphasized that AI can be useful for identifying at-risk students but warned that automated emotional cues should never replace authentic human interaction. Students often reported feeling "observed" rather than "supported," highlighting the risk of AI reinforcing surveillance rather than empathy.

Regarding the classroom organization domain, AI tools were perceived as beneficial for managing large groups, streamlining administrative tasks, and supporting blended learning environments. AI tools improved task tracking, resource allocation, and pacing. Both Chilean and Spanish respondents reported that AI tools enhanced classroom management by streamlining administrative tasks such as attendance, assignment tracking, and content delivery. Chilean teachers noted improvements in time management and workload distribution, particularly in large classes. However, several respondents in Chile expressed concerns over the rigidity introduced by algorithmic pacing and AI scheduling tools, which were sometimes poorly adapted to the local curriculum and student habits. Chilean participants also acknowledged these benefits (66%) but pointed to infrastructural limitations and bandwidth issues that often impeded full utilization. That is, rigidity in Chilean platforms led to friction with local teaching styles. Spanish educators, benefiting from institutional support and training, used AI tools more flexibly and reported fewer constraints: they tended to report greater institutional support and customization options for these systems. In Spain, 71% of respondents noted improvements in scheduling, task tracking, and resource distribution due to AI-based LMS.

Moreover, qualitative insights suggested that in Chile, AI-driven organization tends to reinforce centralized control in classrooms, sometimes reducing flexibility for teacher-led adaptation. In contrast, Spanish universities exhibited more institutional support and training for faculty to adjust AI tools to their teaching styles.

In terms of the instructional support domain, instructional support was the area where AI received the most positive feedback. AI-based personalized learning systems were widely used in both countries. In both cases, participants observed that AI freed up time for more meaningful interactions, such as mentoring and dialogic teaching. Students in both countries appreciated automated feedback mechanisms, personalized learning paths, and AI-based tutoring systems. Seventy-eight percent of Spanish students and 69% of Chilean students agreed that AI helped clarify difficult concepts, particularly in STEM disciplines. Thus, on one hand, Chilean students noted the lack of culturally adapted content, and some faculty resisted AI integration due to limited training and pedagogical autonomy concerns: some students reported similar benefits but pointed to issues of content relevance and lack of localization. Furthermore, the digital divide among instructors led to uneven integration, with some relying heavily on AI and others avoiding it due to lack of training. On the other hand, Spanish educators demonstrated greater digital literacy and institutional support. Thus, Spanish institutions appeared to be more advanced in the deployment of adaptive platforms, with students appreciating how these systems aligned content with their individual learning pace.

A key emergent theme was the need for AI tools that respect pedagogical autonomy [3, 19]. Some Chilean professors reported discomfort with "algorithmic rigidity" that clashed with their instructional approaches, and many faculty voiced concerns over over-reliance on technology and the risk of deskilling the teaching profession [1, 17, 62].

4.4 Discussion: Strategies for Enhancing Interactions Through AI

In terms of discussion on AI, engagement, and TTI dimensions, the findings from Chile and Spain offer valuable insights into the multifaceted impact of AI on student-teacher interactions in higher education, particularly when interpreted through the TTI framework. This discussion synthesizes the results across the emotional, organizational, and instructional domains, highlighting the affordances and tensions that AI introduces in relational pedagogies.

The findings highlight how AI integration in higher education can enhance or hinder student-teacher interactions depending on how well it aligns with the emotional, organizational, and instructional dynamics described in the TTI framework [21, 25, 37]. While both Chile and Spain demonstrate increasing adoption of AI tools, the cultural, infrastructural, and institutional contexts significantly influence outcomes. That is, while Spain demonstrated higher levels of digital maturity, the relational dynamics emphasized by the TTI framework were similarly affected in both contexts. Notably, cultural perceptions of teacher authority and affective engagement were more pronounced in Chile, leading to more resistance to AI replacing human roles [7, 63]. Spanish respondents showed more willingness to integrate AI as a pedagogical partner [16, 17, 50, 51]. These results underscore the importance of contextual adaptation and the need for professional development strategies that align AI tools with relational pedagogies grounded in emotional and instructional responsiveness [52, 53].

4.4.1 Emotional Support: Augmentation or Alienation? Balancing Automation and Connection

AI tools designed to monitor and respond to student emotions hold promise for early intervention and personalized care. However, their use must be critically evaluated to ensure they augment rather than alienate [5, 32]. The perception among Chilean students that AI-mediated feedback can feel impersonal signals a potential erosion of the affective bonds that are central to Latin American pedagogical cultures. In contrast, Spanish participants, accustomed to digital mediation in education, were more accepting of these tools – indicating that digital literacy and familiarity mediate how emotional AI features are received. This finding reinforces the importance of aligning emotional AI with cultural expectations and ensuring that human interaction remains central to learning environments [50, 51].

Furthermore, AI tools offer efficiency but can create relational distance. This tension is particularly evident in Chile, where emotional support is often mediated through less-resourced systems. The TTI framework helps foreground the importance of affective dimensions in learning, reminding institutions that AI must be implemented in ways that support – not replace – human connection. For example, while Spanish universities have deployed chatbots with affect-sensitive algorithms, Chilean institutions often lack the resources to do the same, leading to one-size-fits-all feedback that students find cold or impersonal.

4.4.2 Classroom Organization: Efficiency/Centralization Versus Flexibility

AI's contribution to organizing learning environments – through automating schedules, tracking participation, and managing content delivery – has clear administrative benefits. Both Chilean and Spanish educators reported improvements in workflow and time management. However, AI systems that lack cultural and institutional flexibility can impose rigid structures that conflict with localized pedagogical practices. This is particularly relevant in Latin America, where relational flexibility and dialogical engagement are often essential to inclusive education. The challenge is to design AI systems that support, rather than supplant, context-sensitive organization strategies.

The classroom organization domain revealed that AI tools can both facilitate and constrain teaching practices. While streamlining administrative work and fostering blended learning, AI systems may also impose rigid structures on learning environments. The comparative analysis shows how institutional readiness matters: Spanish universities empower teachers to customize LMS platforms, whereas Chilean counterparts tend to impose top-down adoption. This suggests that AI tools need to be designed with flexible interfaces and accompanied by training that enhances – not dictates – teacher agency.

4.4.3 Instructional Support: Personalization with Pedagogical Intent

In this research, the strongest consensus between Chile and Spain was that AI can enhance instructional support. However, this benefit is unevenly distributed. Spanish institutions often offer teacher development programs to accompany AI adoption; Chilean institutions, by contrast, show fragmented support structures.

A recurring concern across both countries was the fear of "algorithmic pedagogy" – where teaching becomes overly dependent on machine-generated decisions. While adaptive learning can personalize content, it may also create homogeneous pathways that neglect critical thinking or reduce space for improvisation in teaching. Adaptive learning technologies – highlighted by both Chilean and Spanish students – demonstrate AI's power to personalize education. These tools cater to individual learning paths, offer differentiated feedback, and promote learner autonomy. Yet, their success depends on their integration into broader pedagogical goals [16, 17, 53, 54].

Many participants expressed concern about the potential deskilling of teaching and the risk that AI could reduce complex instructional tasks to algorithmic sequences. This suggests that while AI can enhance instructional support, its implementation must be guided by intentional pedagogical design grounded in the TTI domain of instructional quality [16, 17, 55].

4.4.4 Relational Lens for Technological Integration, Tensions, and Trade-Offs

The TTI framework proves to be a powerful heuristic for assessing the relational quality of AI-supported education. Rather than focusing solely on technological efficiency, this framework encourages evaluation based on the human-centered values of respect, engagement, and responsiveness. This is particularly pertinent in Latin American contexts, where educational relationships are often grounded in strong interpersonal connections. Moreover, the contrast with Spain underscores the need for culturally responsive AI implementation. While Spain's digital readiness facilitates broader AI adoption, Chile's more relationally oriented pedagogical culture requires tools that prioritize connection over automation [55, 64].

Applying the TTI framework in a cross-cultural context reveals how AI affects not only instructional content but also the relational ecosystem of classrooms. In Latin America, where interpersonal relationships and affective presence are central to pedagogy, AI needs to be context-sensitive and relationally informed. The findings also challenge the universalist assumptions of many AI tools developed in Euro-American contexts, urging for regional customization and participatory design involving local educators [1, 56, 57].

Several tensions emerge from this comparative analysis are as follows:
- Automation Versus Connection/Empathy: Tools that streamline teaching may simultaneously depersonalize it.
- Efficiency/Scalability Versus Flexibility: AI enables efficiency and scalability in teaching support but may overlook the socio-emotional nuances that underlie learning.
- Data-Driven Insight Versus Privacy: Increased reliance on learning analytics may compromise student's trust and autonomy.

Navigating these tensions requires institutional policies that frame AI not as a replacement for teachers to enhance their relational capacities. This entails ongoing professional development, ethical guidelines, and participatory design involving both educators and students [16, 17, 58].

5 Conclusions, Recommendations, and Future Research

Through this cross-national comparison, this work aimed to assess not only how AI tools influence learning outcomes and instructional strategies, but also how they mediate affective and interpersonal dynamics between students and educators. Thus, this chapter examined the tension between personalization and standardization, autonomy and surveillance, innovation and inclusion, and how these tensions manifest differently in varied cultural contexts. Our empirical approach was grounded in a structured survey administered to 385 participants – students and faculty alike – who shared their experiences and perceptions related to AI-supported learning environments. Based on a comparative analysis between Chile and Spain, the study shows that while AI presents valuable opportunities for improving instructional support and organizational efficiency, its emotional and relational implications remain deeply complex and context-dependent [50–52].

To sum up, this study contributes to this literature by analyzing how AI mediates those same relational domains in culturally distinct higher education contexts [51–53]. Using diverse adapted, validated higher education instruments (TSIH and CTBI), this chapter alignment and validate measurement of constructs like emotional support, instructional quality, and classroom management in university classrooms: it is crucial for grounding the study in transferable educational research to other contexts [53, 54].

5.1 Conclusions and Limitations

One of the most significant conclusions drawn is that AI's success in educational contexts is not solely a matter of technical capacity but of cultural alignment, ethical sensitivity, and pedagogical coherence. AI can personalize learning and support at scale, but if implemented without regard for local pedagogical traditions or relational dynamics – as is often the case in Latin America – it risks reinforcing inequalities and reducing meaningful human interaction [65, 66]. The TTI framework serves as a valuable tool to rebalance these considerations, ensuring that technology is assessed not only for its efficiency but also for its capacity to support emotional connection and instructional responsiveness [67, 68]. The cross-cultural comparison with Spain further illustrates that higher technological readiness does not necessarily equate to better relational outcomes, reinforcing the need for context-sensitive strategies [10, 55, 56].

Furthermore, this chapter explores how AI technologies are reshaping the nature of student-teacher interactions in higher education, particularly in Latin America. Drawing on comparative data from Chile and Spain, the study highlights that while AI offers promising tools for personalized learning and instructional support, it also introduces significant challenges, especially concerning emotional connection, organizational flexibility, and pedagogical agency.

A key insight is that the effectiveness of AI integration depends not only on the sophistication of technology but also on the sociocultural and institutional contexts in which it is embedded. In Latin America, where educational practices are deeply relational and often constrained by structural inequalities, AI must be thoughtfully adapted to support rather than replace meaningful interactions [3, 5, 17, 67].

This alignment enables the chapter to transcend technological determinism by offering contextually grounded insights vital for ethical and effective AI integration in Latin American higher education. Given the layered and situated nature of AI adoption – particularly within Chilean universities – this analysis deepens our understanding of how both students and faculty interpret, navigate, and respond to AI technologies. Such context-sensitive perspectives are crucial to anchoring empirical findings in the sociocultural fabric of the studied regions, informing AI strategies that are relationally attuned and locally meaningful. Drawing on sociotechnical systems theory, the study affirms that AI tools are not neutral or universally applicable but are shaped by, and embedded within, specific institutional and social ecologies. Additionally, the research incorporates a contextualist lens derived from Bronfenbrenner's ecological systems theory, emphasizing the nested dimensions of educational environments [68]. This methodological approach underscores the importance of interpreting TTI-informed interactional patterns within broader systems of institutional governance, technological infrastructure, and pedagogical values [69, 70]. By integrating critical-theoretical traditions from Latin American education, the study further explores how enduring inequalities and structural asymmetries shape AI's reception, appropriation, and resistance [57, 58].

Together, these methodological perspectives offer a robust framework to investigate how macro-level forces impact the equity, efficacy, and relational dynamics of AI-mediated interactions in Latin American higher education. The study also engages critical pedagogy and Latin American educational sociology to foreground cultural and historical principles – such as emotional connectedness, horizontal relationships, and social justice – that may either challenge or reshape how AI tools are introduced. This lens allows for deeper interpretation of participant narratives, revealing not only patterns of adoption but also instances of resistance, negotiation, and agency in the face of technological change.

5.2 Recommendations for Implementation

Drawing from empirical findings and theoretical insights, this section outlines actionable recommendations for educators, institutions, and policymakers seeking to integrate AI tools in higher education while preserving the relational integrity of teaching and learning.

For educators, it is essential to:
- Leverage AI for Relational Awareness: Use sentiment analysis and engagement metrics not as ends in themselves but as prompts for deepening human connection.
- Balance Automation with Presence: Delegate routine tasks to AI to free up time for mentoring and dialogic teaching.
- Participate in Tool Design: Provide feedback to developers to ensure tools align with real classroom needs and relational dynamics.

Some of the main recommendations for institutions can be detailed as follows:
- To Develop Context-Sensitive Policies: Avoid one-size-fits-all technology mandates and allow for institutional and disciplinary variation in AI use.
- To Promote Digital and Pedagogical Literacy: Offer training that combines technical skills with relational pedagogy, anchored in frameworks like TTI.
- To Foster Participatory Governance: Involve faculty and students in decisions about which AI tools to adopt and how they are used.

For policymakers and technology developers, the following guidelines are essential:
- Embed Cultural Awareness in AI Design: Ensure that AI tools align with the relational and cultural dynamics of Latin American education systems.
- Enforce Ethical Standards: Prioritize data protection, algorithmic fairness, and transparency as non-negotiable principles in AI deployment.
- Invest in Collaborative Research: Support comparative, cross-country studies that examine how AI influences student-teacher relationships in varied regional contexts.

Together, these recommendations support an integrated approach that views AI not as a substitute for human interaction but as a tool to enrich it, especially when evaluated through the holistic lens of the TTI framework [15, 21, 58].

5.3 Limitations and Future Research

Despite these insights, several limitations should be acknowledged as follows:
- Sample Limitations: The study focused on universities in Chile and Spain, which, while informative, may not fully represent the diversity of Latin American or global higher education contexts. Further studies involving countries like Brazil, Colombia, or Argentina are needed.

- Temporal Constraints: The data were collected over a relatively short time frame (January–March 2025), which may not capture evolving attitudes toward rapidly changing AI technologies.
- Self-Reported Data: Much of the evidence relies on self-reported perceptions and may be subject to response bias, particularly concerning sensitive topics like surveillance or autonomy.
- Technology Access Variability: Differences in infrastructure and institutional capacity across sampled universities may have skewed comparative insights, particularly when evaluating organizational efficiency.
- Tool-Specific Impact: The study did not disaggregate findings based on specific AI tools or platforms, which could reveal variation in how different systems affect TTI domains.

These limitations point to a rich agenda for future research, which should include the following:
- Longitudinal studies to assess the sustained impact of AI on relational teaching practices, learning relationships, and student outcomes over time.
- Tool-specific case studies to evaluate how particular AI applications (e.g., adaptive learning systems, chatbots, and analytics dashboards) align with TTI principles. It is relevant to collaborate on participatory design initiatives that involve educators and students in the development of AI tools, and facilitate ethnographic and participatory research to gain deeper, contextualized insights into how AI mediates human relationships in diverse learning cultures.
- Comparative regional analyses within Latin America to identify best practices and scalable innovations that respect local pedagogical identities and expand the geographic scope of comparative studies across Latin America, incorporating diverse institutional types, languages, and cultural traditions.
- Exploration of faculty development models that integrate AI training with relational pedagogy, ensuring that educators are empowered – not replaced – by technological tools.

In conclusion, this chapter affirms that the promise of AI in higher education lies not in replacing teachers or its ability to automate teaching but in enabling more intentional, inclusive, relationship-rich teaching environments, and its capacity to enhance the human connections at the heart of learning. The thoughtful application of AI – guided by culturally grounded frameworks like TTI – can help ensure that the future of education remains deeply human, even as it becomes increasingly digital.

Appendix

TTI-AI-HE Scale: Questionnaire on Student-Teacher Interactions in AI-Enhanced University Courses.
Response options are as follows:
(1) Strongly disagree
(2) Disagree
(3) Neutral
(4) Agree
(5) Strongly agree

Domain 1: Emotional Support: Assesses perceptions of warmth, respect, empathy, and positive climate influenced by AI use.
1. My instructor uses AI tools in a way that makes me feel supported and respected.
2. AI-based systems used in this course have enhanced my sense of connection with the instructor.
3. The integration of AI in this class contributes to a positive emotional climate.
4. I feel more comfortable asking questions or participating in class when AI tools are used.
5. AI applications have helped humanize the learning experience rather than depersonalize it.

Domain 2: Classroom Organization: Assesses how AI influences the structuring of the class, time management, and behavioral expectations.
6. The use of AI has made classroom tasks and expectations more structured and transparent.
7. AI tools have helped the instructor manage time and resources more efficiently.
8. I can follow the course more easily thanks to how AI supports the organization of class activities.
9. AI-based systems contribute to smoother transitions and better pacing during lessons.
10. AI helps the teacher maintain an organized, well-managed learning environment.

Domain 3: Instructional Support: Assesses feedback, cognitive challenge, scaffolding, and support for deeper learning.
11. AI tools provide feedback that improves my understanding of complex concepts.
12. The AI technologies used in this course stimulate critical thinking and reflection.
13. I receive more timely and personalized instructional support because of AI integration.
14. The instructor uses AI tools to explain content more clearly and effectively.

AI has helped me engage more deeply with the academic material in this course.

References

[1] Hofkens T, Pianta RC and Hamre B. Teacher-student interactions: Theory,measurement, and evidence for universal properties that support students' learning across countries and cultures. In: Maulana R, Helms-Lorenz M and Klassen RM (eds.). *Effective Teaching around the World*. Cham: Springer; 2023. https://doi.org/10.1007/978-3-031-31678-4_18.

[2] Holmes W, Bialik M and Fadel C. *Artificial Intelligence in Education: Promises and Implications for Teaching and Learning*. Boston: Center for Curriculum Redesign; 2019.

[3] Sangwa S, Ngobi D, Ekosse E and Mutabazi P. AI governance in African higher education: Status, challenges, and a future-proof policy framework. *Artificial Intelligence and Education*. 2025;1(1):2054. https://doi.org/10.62617/aie2054.

[4] Zawacki-Richter O, Marín VI, Bond M and Gouverneur F. Systematic review of research on artificial intelligence applications in higher education – Where are the educators?. *International Journal of Educational Technology in Higher Education*. 2019;16(1):39. https://doi.org/10.1186/s41239-019-0171-0.

[5] Ismail IA and Aloshi JM. Data privacy in AI-driven education: An in-depth exploration into the data privacy concerns and potential solutions. In: Keeley K (ed.). *AI Applications and Strategies in Teacher Education. IGI Global Scientific Publishing*. 2025. pp. 223–252. https://doi.org/10.4018/979-8-3693-5443-8.ch008.

[6] Langa VG, Balkaran S, Mahlala S and Muronda B. Promoting social inclusion through AI integration in higher tertiary institutions: A South African perspective. *International Journal of Business Ecosystem and Strategy*. 2025;7(4):299–308. https://bussecon.com/ojs/index.php/ijbes/article/view/848.

[7] Mehak F and Jafree S. R. Bridging the digital divide: Predictors of positive attitudes and functional use of AI among university students in Pakistan. *Social Sciences Spectrum*. 2025;4(1):617–632. https://doi.org/10.71085/sss.04.01.237.

[8] Edam-Agbor IB, Orim FS, Ofem UJ, Ekpang P, Echu A, Okim TO, Undie MA, Ogunjimi B, Egbe IM, Akin-Fakorede OO, Gombe AB, Angrey CU, Abua D and Enidiok MS. Librarians' awareness, acceptability, and application of artificial intelligence in academic research libraries. Multigroup analysis via PLS–SEM. *Social Sciences & Humanities Open*. 2025;11:101333. https://doi.org/10.1016/j.ssaho.2025.101333.

[9] Roll I and Wylie R. Evolution and revolution in artificial intelligence in education. *International Journal of Artificial Intelligence in Education*. 2016;26(2):582–599. https://doi.org/10.1007/s40593-016-0110-3.

[10] Luckin R, Holmes W, Griffiths M and Forcier LB. *Intelligence Unleashed: An Argument for AI in Education*. London: Pearson; 2016.

[11] Chen L, Chen P and Lin Z. Artificial intelligence in education: A review. *IEEE Access*. 2020;8:75264–75278. doi: 10.1109/ACCESS.2020.2988510.

[12] Xia L, Cao Z and Bilawal Khaskheli M. How digital technology and business innovation enhance economic–environmental sustainability in legal organizations. *Sustainability*. 2025;17(14):6532. https://doi.org/10.3390/su17146532.

[13] Ramabina M. Exploring the utilisation of chatGPT in academic libraries: A self-reflection perspective. *Regional Journal of Information and Knowledge Management*. 2024;9(1):47–60. https://rjikm.org/index.php/rjikm/article/view/105.

[14] Ayyoub AM, Khlaif ZN, Shamali M, Abu Eideh B, Assali A, Hattab MK, Barham KA and Bsharat TRK. Advancing higher education with GenAI: Factors influencing educator AI literacy. *Frontiers in Education*. 2025;10:1530721. doi: 10.3389/feduc.2025.1530721.

[15] Ramirez-Montoya MS. Challenges for open education with educational innovation: A systematic literature review. *Sustainability*. 2020;12(17):7053. https://doi.org/10.3390/su12177053.

[16] García ZA, Jorge S, Dalio M, Makwakwa O and Albornoz BN. *Digital Inclusion Strategies: A Primer for Latin American Policymakers*. NY: Inter-American Development Bank; 2024. https://doi.org/10.18235/0013168.

[17] Esteban AJ, Park I, Nga NT, Perunovic S, Park SE and Shehzadi YJI. Undergraduate students' perceptions on the use of chatGPT for English learning at a Korean university. *rEFLections*. 2025;32 (2):994–1016. https://so05.tci-thaijo.org/index.php/reflections/article/view/282889.

[18] MdM R, Siddiqee MS, MdN S and Ahamed R. Assessing AI adoption in developing country academia: A trust and privacy-augmented UTAUT framework. *Heliyon*. 2024;10(18):e37569. https://doi.org/10. 1016/j.heliyon.2024.e37569.

[19] Patel S and Ragolane M. The implementation of artificial intelligence in South African higher education institutions: Opportunities and challenges. *Technium Education and Humanities*. 2024;9:51–65. https://techniumscience.com/index.php/education/article/view/11452.

[20] Pianta RC, Hamre BK and Allen JP. Teacher-student relationships and engagement: Conceptualizing, measuring, and improving the capacity of classroom interactions. In: Christenson SL, Reschly AL and Wylie C (eds.). *Handbook of Research on Student Engagement*. Boston: Springer; 2012. pp. 365–386. https://doi.org/10.1007/978-1-4614-2018-7_17.

[21] Hamre BK, Pianta RC, Mashburn AJ and Downer JT. Building a science of classrooms: Application of the CLASS framework in over 4,000 U.S. *Early Childhood and Elementary Classrooms*. 2007. https://www.researchgate.net/publication/237728991_Building_a_Science_of_Classrooms_Applica tion_of_the_CLASS_Framework_in_over_4000_US_Early_Childhood_and_Elementary_Classrooms/cita tion/download.

[22] Allen JP, Pianta RC, Gregory A, Mikami AY and Lun J. An interaction-based approach to enhancing secondary school instruction and student achievement. *Science*. 2011;333(6045):1034–1037. doi: 10.1126/science.1207998.

[23] Oberle E, Domitrovich CE, Meyers DC and Weissberg RP. Establishing systemic social and emotional learning approaches in schools: A framework for schoolwide implementation. *Cambridge Journal of Education*. 2016;46(3):277–297. doi: 10.1080/0305764X.2015.1125450.

[24] Williamson B, Eynon R and Potter J. Pandemic politics, pedagogies and practices: Digital technologies and distance education during the coronavirus emergency. *Learning, Media and Technology*. 2020;45(2):107–114. https://doi.org/10.1080/17439884.2020.1761641.

[25] Tsai YS and Gasevic D. Learning analytics in higher education – Challenges and policies: A review of eight learning analytics policies. In: *Proceedings of the Seventh International Learning Analytics & Knowledge Conference (LAK '17)*. New York, NY, USA: Association for Computing Machinery; 2017. pp. 233–242. https://doi.org/10.1145/3027385.3027400.

[26] Selwyn N. *Should Robots Replace Teachers? AI and the Future of Education*. 1st ed. Cambridge UK: Polity Press; 2019. p. 160.

[27] Redecker C. *European Framework for the Digital Competence of Educators (Digcompedu)*. Luxembourg: Publications Office of the EU; 2017. doi: 10.2760/159770.

[28] Mora-Cantallops M, Inamorato Dos Santos A, Villalonga-Gómez C, Lacalle Remigio JR, Camarillo Casado J, Sota Eguzábal JM, Velasco JR and Ruiz Martínez PM. *Competencias Digitales Del Profesorado Universitario En España. Un Estudio Basado En Los Marcos Europeos DigCompEdu Y OpenEdu, EUR 31127 ES*. Publications Office of the European Union: Luxembourg; 2022. doi: 10.2760/448078.

[29] Instefjord EJ and Munthe E. Educating digitally competent teachers: A study of integration of professional digital competence in teacher education. *Teaching and Teacher Education*. 2017;67:37–45. https://doi.org/10.1016/j.tate.2017.05.016.

[30] Riordan A, Echeverria V, Jin Y, Yan L, Swiecki Z, Gašević D and Martinez-Maldonado R. Human-centred learning analytics and AI in education: A systematic literature review. *Computers and Education: Artificial Intelligence*. 2024;6:100–215. https://doi.org/10.1016/j.caeai.2024.100215.

[31] Williamson B, Bayne S and Shay S. The datafication of teaching in higher education: Critical issues and perspectives. *Teaching in Higher Education*. 2020;25(4):351–365. doi: 10.1080/ 13562517.2020.1748811.

[32] Green E, Divya Singh D and Chia R (eds.). *AI Ethics and Higher Education Good Practice and Guidance for Educators, Learners, and Institutions*. Geneva: Globethics.net; 2022.

[33] Ifenthaler D and Yau JYK. Utilising learning analytics to support study success in higher education: A systematic review. *Educational Technology Research and Development*. 2020;68:1961–1990. https://doi.org/10.1007/s11423-020-09788-z.

[34] Viberg O, Hatakka M, Bälter O and Mavroudi A. The current landscape of learning analytics in higher education. *Computers in Human Behavior*. 2018;89:98–110. https://doi.org/10.1016/j.chb.2018.07.027.

[35] Ferguson R and Clow D. Where is the evidence? a call to action for learning analytics. In: *Proceedings of the Seventh International Learning Analytics & Knowledge Conference (LAK '17). Association for Computing Machinery*. New York, NY, USA; 2017. pp. 56–65. https://doi.org/10.1145/3027385.3027396.

[36] Bond M, Marín VI, Dolch C, Bedenlier S and Zawacki-Richter O. Digital transformation in German higher education: Student and teacher perceptions and usage of digital media. *International Journal of Educational Technology in Higher Education*. 2018;15(48). doi: 10.1186/S41239-018-0130-1.

[37] Lustosa R, Carolina A, Yaacov BB, Franco Segura C, Arias Ortiz E, Heredero E, Botero J, Brothers P, Payva T and Spies M. *Higher Education Digital Transformation in Latin America and the Caribbean*. NY: Inter-American Development Bank; 2021. https://doi.org/10.18235/0003829.

[38] OECD. *Digital Education Outlook 2021: Pushing the Frontiers with Artificial Intelligence, Blockchain and Robots*. Paris: OECD Publishing; 2021. https://doi.org/10.1787/589b283f-en.

[39] Dritsas E and Trigka M. Exploring the intersection of machine learning and big data: A survey. *Machine Learning and Knowledge Extraction*. 2025;7(1):13. https://doi.org/10.3390/make7010013.

[40] Gordon M, Daniel M, Ajiboye A, Uraiby H, Xu NY, Bartlett R and Hanson J. A scoping review of artificial intelligence in medical education: BEME guide no. 84. *Medical Teacher*. 2024;46(4):446–470. doi: 10.1080/0142159X.2024.2314198.

[41] Pérez-López E and Alzás T. Marco analítico para la educación remota de emergencia en las universidades en tiempos de confinamiento. *Revista Electrónica de Investigación Educativa*. 2023;25 (e12):1–15. https://doi.org/10.24320/redie.2023.25.e12.4965.

[42] Ogunleye B, Zakariyyah KI, Ajao O, Olayinka O and Sharma H. A systematic review of generative AI for teaching and learning practice. *Education Sciences*. 2024;14(6):636. https://doi.org/10.3390/educsci14060636.

[43] Hinojo-Lucena F-J, Aznar-Díaz I, Cáceres-Reche M-P and Romero-Rodríguez J-M. Artificial intelligence in higher education: A bibliometric study on its impact in the scientific Literature. *Education Sciences*. 2019;9(1):51. https://doi.org/10.3390/educsci9010051.

[44] Demaidi MN. Artificial intelligence national strategy in a developing country. *AI & Society*. 2025;40:423–435. https://doi.org/10.1007/s00146-023-01779-x.

[45] Bahroun Z, Anane C, Ahmed V and Zacca A. Transforming education: A comprehensive review of generative artificial intelligence in educational settings through bibliometric and content analysis. *Sustainability*. 2023;15(17):12983. https://doi.org/10.3390/su151712983.

[46] Duarte N, Montoya Y and Beltran A, Bolaño Garcia M use of artificial intelligence in education: A systematic review. In *4th South American International Conference on Industrial Engineering and Operations Management* 2023; May 9, https://doi.org/10.46254/SA04.20230169.

[47] Creswell JW and Plano Clark VL. *Designing and Conducting Mixed Methods Research*. 3rd ed. Thousand Oaks, CA: SAGE; 2018.

[48] Abello DM, Alonso Tapia J and Panadero Calderón E. Development and validation of the teaching styles inventory for higher education (TSIH). *Anales de Psicología*. 2020;36(1):143–154. Available from https://revistas.um.es/analesps/article/view/370661.

[49] Lee-Hsieh J, O'Brien A, Liu CY, Cheng SF, Lee YW and Kao YH. The development and validation of the clinical teaching behavior inventory (CTBI-23): Nurse preceptors' and new graduate nurses'

perceptions of precepting. *Nurse Education Today*. 2016;38:107–114. https://doi.org/10.1016/j.nedt.2015.12.005.

[50] Moreno García R and Martínez-Arias R. Adaptación española de la escala de relación profesor-alumno (STRS) de Pianta. *Psicología Educativa. Revista de Los Psicólogos de la Educación*. 2008;14(1): 11–27. https://www.redalyc.org/articulo.oa?id=613765491002.

[51] Cohen L, Manion L and Morrison K. *Research Methods in Education*. 8th ed. New York: Routledge; 2018. https://doi.org/10.4324/9781315456539.

[52] Floridi L, Cowls J, Beltrametti M, et al. AI4People – An ethical framework for a good AI society: Opportunities, risks, principles, and recommendations. *Minds & Machines*. 2018;28:689–707. https://doi.org/10.1007/s11023-018-9482-5.

[53] European Commission. Directorate-general for education, youth, sport and culture, ethical guidelines on the use of artificial intelligence (AI) and data in teaching and learning for educators. Publications Office of the European Union; 2022. Available from: doi:10.2766/153756.

[54] UNESCO. *Recommendation on the Ethics of Artificial Intelligence*. Paris: UNESCO; 2022. https://digitallibrary.un.org/record/4062376?v=pdf&ln=es#files.

[55] Kroff FJ, Coria DF and Ferrada CA. Inteligencia Artificial en la educación universitaria: Innovaciones, desafíos y oportunidades. *Espacios*. online. 2024;45(5):120–135. https://doi.org/10.48082/espacios-a24v45n05p09.

[56] González -Campos J, López - Núñez J and Araya - Pérez C. Educación superior E Inteligencia artificial: Desafíos Para La Universidad Del Siglo XXI». *Aloma: Revista De Psicologia, Ciències De l'Educació I De l'Esport*. 2024;42(1):79–90. https://doi.org/10.51698/aloma.2024.42.1.79-90.

[57] Satama Pereira WI and Sánchez Ramírez L Del C. Integración de la Inteligencia Artificial en el Contexto Educativo Latinoamericano: Una Exploración a las Perspectivas Emergentes y los Desafíos Futuros. *Revista de Ciencias Multidisciplinarias*. SAGA [Internet]. 2024;1(3):1–13. https://revistasaga.org/index.php/saga/article/view/1.

[58] Salas-Pilco SZ and Yang Y. Artificial intelligence applications in Latin American higher education: A systematic review. *International Journal of Educational Technology in Higher Education*. 2022;19(21): 1–20. https://doi.org/10.1186/s41239-022-00326-w.

[59] Niño-Carrasco SA, Castellanos-Ramírez JC, Perezchica Vega JE and Sepúlveda Rodríguez JA. Percepciones de estudiantes universitarios sobre los usos de inteligencia artificial en educación. *Revista Fuentes REFU*. 2025;27(1):94–106. https://revistascientificas.us.es/index.php/fuentes/article/view/26356.

[60] Moradi H. The role of language teachers' perceptions and attitudes in ICT integration in higher education EFL classes in China. *Humanities and Social Sciences Communications*. 2025;208:1–12. https://doi.org/10.1057/s41599-025-04524-5.

[61] Akram H, Abdelrady AH, Al-Adwan AS and Ramzan M. Teachers' perceptions of technology integration in teaching-learning practices: A systematic review. *Frontiers in Psychology*. 2022;1. doi: 10.3389/fpsyg.2022.920317.

[62] Haleem A, Javaid M, Asim Qadri M and Suman R. Understanding the role of digital technologies in education: A review. *Sustainable Operations and Computers*. 2022;13:275–285. https://doi.org/10.1016/j.susoc.2022.05.004.

[63] Romero M, Romeu T, Guitert M and Baztán P. La transformación digital en la educación superior: El caso de la UOC. *Revista RIED*. 2023;26(1):163–179. https://doi.org/10.5944/ried.26.1.33998.

[64] Timmis S and Valladares-Celis MC. Digital inequalities and the COVID legacy in higher education in the global South and North: Intersecting inaccessibilities and institutional assumptions. *Compare: A Journal of Comparative and International Education*. 2025 April;1–19. doi: 10.1080/03057925.2025.2483691.

[65] Marimon F, Arias Valle MB, Coria Augusto CJ and Larrea Arnau CM. Del optimismo a la confianza: El impacto de ChatGPT en la confianza de los estudiantes en el aprendizaje asistido por IA. *Revista RIED*. 2025;28(2):131–153. https://revistas.uned.es/index.php/ried/article/view/43238.

[66] Akinlar A and Küçüksüleymanoğlu R. Leveraging technology to Enhance educational equity and diversity introduction. In: Alabay GG, et al. (ed.). *Creating Positive and Inclusive Change in Educational Environments*. IGI Global Scientific Publishing; vol. 2025. pp. 1–22. https://doi.org/10.4018/979-8-3693-5782-8.ch001.

[67] Aljemely Y. Challenges and best practices in training teachers to utilize artificial intelligence: A systematic review. *Frontiers in Education*. 2024;9:1470853. doi: 10.3389/feduc.2024.1470853.

[68] Romero Rodríguez JM, Ramírez-Montoya MS, Buenestado Fernández M and Lara Lara F. Use of ChatGPT at university as a tool for complex thinking: Students' perceived usefulness. *NAER*. 2023;12:323–339. doi: 10.7821/naer.2023.7.1458

[69] Tan X, Cheng G and Man Ling MH. Artificial intelligence in teaching and teacher professional development: A systematic review. *Computers and Education: Artificial Intelligence*. 2025;8(100355). https://doi.org/10.1016/j.caeai.2024.100355.

[70] Zhang K and Aslan AB. AI technologies for education: Recent research & future directions. *Computers and Education: Artificial Intelligence*. 2021;2:Article 100025. https://doi.org/10.1016/j.caeai.2021.100025.

Rita Cersosimo

Rethinking Inclusive Design in Higher Education: Integrating Generative AI and UDL Practices Through the UDL-AI-TPACK Model

Abstract: Universal Design for Learning (UDL) offers a flexible and inclusive framework for designing learning environments that address learner variability. While widely discussed in K-12 settings, UDL remains underused in higher education, partly due to limited faculty training and the perceived complexity of its implementation. This chapter explores the potential of generative artificial intelligence (AI) as a driving force for UDL adoption in university teaching and faculty development. Drawing from a training experience with preservice teachers, who used AI tools to support inclusive lesson planning, we propose a model for extending such practices to professors in higher education. The integration of UDL principles, AI literacy, and the Technological Pedagogical Content Knowledge (TPACK) framework can foster reflective and accessible course design among university faculty, resulting in what we will define as UDL-AI-TPACK model. Generative AI is examined not as a replacement for pedagogical intention, but as a design companion that prompts inclusive thinking and adaptive planning. The chapter discusses the theoretical underpinnings of this integration, presents insights from the initial case study, and outlines strategies for using AI to scaffold UDL-oriented faculty training. In doing so, it contributes to reimagining digital transformation in higher education through a lens of equity, flexibility, and pedagogical depth.

Keywords: Universal Design for Learning, artificial intelligence, preservice teachers

1 Introduction

In an era marked by accelerating digital transformation and increasing student diversity, higher education institutions are required to adopt teaching practices that are both technologically current and pedagogically inclusive. While many universities have expanded access and invested in digital infrastructure, the implementation of inclusive design principles in everyday teaching remains inconsistent. Faculty members are often expected to meet the needs of increasingly diverse learners without sufficient training, institutional guidance, or practical tools [1, 2].

Rita Cersosimo, Department of Education, University of Genoa, Italy, e-mail: rita.cersosimo@unige.it

https://doi.org/10.1515/9783112206393-008

Universal Design for Learning (UDL) offers a promising response to these challenges. Developed by the CAST (Center for Applied Special Technology) [3], UDL is a framework for designing flexible learning environments that anticipate and address learner variability. Its core principles aim to reduce barriers and increase accessibility for all students and focus on providing multiple means of engagement, representation, and action/expression. Although UDL has gained traction in K-12 and special education [4], its uptake in higher education remains limited, often hindered by misconceptions that associate it exclusively with disability accommodations or by structural resistance to pedagogical change [2, 5].

At the same time, the rapid evolution of generative artificial intelligence (AI) is transforming the educational landscape. AI tools are increasingly being used to support student learning, assessment, and content creation. However, the potential of these technologies to support faculty learning and inclusive instructional design is still partly unexplored. While the literature on AI in education tends to focus on student-facing applications [6, 7], less attention has been given to how instructors might use AI as a reflective tool and pedagogical partner in their own professional development.

This chapter seeks to address that gap. We explore how integrating UDL principles with AI literacy and pedagogical knowledge can support more inclusive and adaptive teaching in higher education. Drawing from a pilot study conducted in a university course on inclusive pedagogy, we propose a conceptual and practical model, namely UDL-AI-TPACK (Technological Pedagogical Content Knowledge), that helps instructors dealing with the intersection of technology, pedagogy, and content with a focus on equity and accessibility.

The contribution of this chapter to the existing literature is threefold: first, it sheds light on the underexplored role of AI in supporting faculty training for inclusive course design; then, it introduces an original integrative model that aligns UDL, AI literacy, and the TPACK framework [8, 9]; third, it offers practical insights from an applied case study that can inform faculty development strategies and promote systemic change in higher education.

The chapter is structured as follows. We begin by outlining the conceptual foundations of UDL in higher education, with a particular focus on faculty training and the current gaps in the literature. Next, we introduce the concept of AI literacy and connect it to teacher professional knowledge through the TPACK framework. Building on this, we propose an integrated model (UDL-AI-TPACK) to guide inclusive instructional design using generative AI. We then describe a pilot activity conducted with undergraduate education students, illustrating how the model can be operationalized in practice. The final sections discuss key findings, explore implications for faculty development, and propose recommendations for implementing the framework at institutional level.

2 Conceptual Framework

This section lays the theoretical groundwork for the chapter by exploring the key conceptual frameworks that guide our reflection on inclusive instructional design in higher education. The aim is to contextualize the proposed integration of UDL and AI within established educational theories and current scholarly debates, thus providing both depth and coherence to the model we introduce later.

Section 2.1 examines UDL as a foundational pedagogical framework, grounded in cognitive neuroscience. We trace its core principles and review its application in higher education settings. Particular attention is given to recent literature on the impact of UDL on student learning and the role of faculty training in enabling its adoption. This section also highlights persistent challenges in institutionalizing UDL and identifies a growing need for approaches that support reflective and scalable implementation across disciplines.

Section 2.2 shifts focus to the concept of AI literacy, an emerging area of inquiry that intersects with educational technology, ethics, and teacher professional development. Building on recent theoretical work, we present AI literacy as a multidimensional construct encompassing conceptual knowledge of AI systems, practical competencies for interacting with generative tools, and ethical-pedagogical judgment in educational contexts. We link this perspective to broader understandings of teacher expertise, emphasizing the need for professional learning models that go beyond technical training and foster critical engagement with digital technologies.

Taken together, these sections form the basis for the integrated framework we propose in Section 3: a model that combines UDL, the TPACK framework, and AI literacy to support inclusive and intentional course design in higher education. By situating this model within robust theoretical traditions, we aim to discuss its conceptual validity and its practical relevance in current academic contexts.

2.1 Universal Design for Learning in Higher Education

UDL is a research-based educational framework aimed at making learning environments more accessible, inclusive, and effective for all students by proactively addressing the variability in how individuals engage with, perceive, and express their knowledge. It takes inspiration from cognitive neuroscience and the understanding that three broad brain networks, that is, affective, recognition, and strategic, play distinct roles in learning [3, 4]. These networks provide a conceptual foundation for UDL's three core principles: providing multiple means of engagement (the "why" of learning), multiple means of representation (the "what"), and multiple means of action and expression (the "how").

Practically speaking, these principles encourage instructors to design learning experiences that offer various ways to spark student interest (e.g., choice in topics or formats), multiple formats for presenting content (e.g., text, video, and audio), and flexible options for demonstrating understanding (e.g., essays, presentations, and multimedia). For example, in a science class, an instructor might use an interactive simulation to represent a concept visually, accompany it with a narrated explanation for auditory learners, and provide hands-on experiments or real-world applications to support engagement and action. UDL thus promotes intentional flexibility, with the goal of reducing barriers to learning while maintaining high expectations for all.

Although UDL has become increasingly influential in K-12 and special education contexts, its adoption in higher education has been more gradual and uneven. Many university faculty members are unaware of the framework or regard it as a set of accommodations specific to students with disabilities, rather than as a holistic pedagogical approach applicable to all learners [1]. Moreover, academic structures and disciplinary traditions in higher education often discourage instructional experimentation, particularly when inclusive practices are perceived as peripheral to the core mission of knowledge transmission [5].

Nonetheless, a growing body of research highlights the potential of UDL to transform university teaching. Systematic reviews indicate that UDL can support diverse student populations, including those from diverse backgrounds, international students, and learners with varying levels of academic preparedness [10, 11]. When effectively implemented, UDL has been associated with increased student engagement, improved accessibility, and more inclusive classroom climates. However, as Fovet [2] points out, many UDL initiatives in higher education remain fragmented, short-lived, or overly reliant on disability service units. Without institutional commitment, faculty development, and integrated policy frameworks, UDL risks being reduced to a technical fix rather than embraced as a transformative pedagogical orientation. Some contributions also emphasize the need for a shift from individual faculty efforts to systemic adoption, supported by leadership, cross-departmental collaboration, and sustained investment in instructional design and professional development [1]. At the same time, some scholars advocate for a more critical and context-sensitive interpretation of UDL, that is, one which also considers the sociocultural dynamics of higher education and resists overly normative or standardized applications [2].

Recent studies have further expanded our understanding of UDL in higher education, particularly by examining its dual impact on students and faculty. Moriña et al. [12], for instance, offer a comprehensive systematic review that highlights two primary areas of focus: the benefits of UDL for students and the role of faculty training in effective implementation. While much of the existing literature has concentrated on student outcomes, these studies consistently report positive effects on learning processes and academic performance. Evidence suggests that UDL can lead to improved exam and assessment results [13–17], higher engagement, motivation, and participation [13, 18, 19], as well as reduced student anxiety [15, 20]. These findings support the framework's potential to foster more inclusive and effective learning environments for diverse student populations [21–23].

Nonetheless, in this chapter we focus on the second, and comparatively underexplored, dimension: faculty training. While demonstrating UDL's value for students is a necessary first step, its successful and sustained application in higher education hinges on instructors' understanding, buy-in, and capacity to translate principles into practice. Addressing this area is essential for moving beyond isolated interventions and toward a systemic adoption of UDL that is embedded in institutional culture and pedagogy.

In the systematic review by Moriña et al. [12], 7 of the 20 analyzed studies explicitly addressed this dimension, describing a range of professional development initiatives aimed at equipping instructors with the necessary competencies to apply UDL principles in their teaching. These initiatives varied widely in format and delivery. Some involved traditional in-person formats such as workshops, guided tutorials, and departmental meetings, while others employed online resources and asynchronous materials to enhance accessibility and scalability. For instance, Olivier and Potvin [24] describe a virtual training model that combined live sessions with online discussions, whereas Kim et al. [25] adopted a blended approach that included panel discussions, asynchronous workshops, and tailored departmental interventions.

Innovative formats also emerged, such as the self-study group documented by Azam et al. [26], in which faculty collaboratively explored UDL principles and reflected on their own pedagogical practices. This emphasis on reflection was echoed in other studies, where participants reported gaining not only practical strategies for inclusive course design, such as accessible materials [25, 27], flexible instructional approaches [24], and inclusive assessment methods [28], but also a deeper awareness of their teaching habits and assumptions. Several studies highlighted the importance of opportunities for self-assessment and collegial dialogue during training, which fostered a more critical and intentional engagement with UDL [24, 26].

Regarding training effectiveness, instructors generally reported improved understanding and confidence in applying UDL principles following participation. In some cases, this growth in knowledge was measured using pre- and post-intervention assessments [29]. In addition to acquiring concrete tools, many faculty members expressed an increased motivation to incorporate UDL strategies after gaining insight into the barriers their students might face [24]. This suggests that meaningful training can serve not only to disseminate technical knowledge but also to shift mindsets, an essential step toward broader institutional adoption of UDL.

These findings underscore the effectiveness of UDL-focused faculty training in enhancing instructors' capacity to create more inclusive learning environments. Yet, it is striking that, despite the recency of these studies, none of them consider the potential role of AI in supporting faculty development in UDL competence. While the current literature has begun to explore how students interact with AI tools in learning contexts, it largely overlooks the faculty perspective – particularly the opportunities AI presents as a dynamic resource for professional growth. As a result, a significant gap remains in understanding how AI might serve as a pedagogical partner in helping educators design more inclusive, responsive, and flexible teaching practices.

In this chapter, we adopt a developmental and reflective view of UDL in higher education. Rather than treating UDL as a fixed protocol to be implemented, we frame it as a pedagogical lens through which educators can rethink course design, assessment, and classroom interaction in light of learner variability. In the sections that follow, we explore how the integration of generative AI and UDL principles can support this reimagining, particularly when situated within a framework that fosters both technological fluency and inclusive intentionality.

2.2 AI Literacy and Teacher Professional Knowledge

As AI continues to expand its presence across educational domains, the need to cultivate AI literacy among educators has become increasingly urgent. AI literacy encompasses more than a technical understanding of algorithms or tools; it involves the ability to critically engage with AI systems, make informed pedagogical choices, and reflect on the ethical, social, and epistemological implications of their use. This competence is particularly relevant in higher education, where instructors are often disciplinary experts with limited exposure to educational technologies or inclusive design frameworks.

In the context of teacher education, AI literacy can be conceptualized as a multidimensional construct that integrates three core domains of professional [9]:
1. Conceptual understanding (*episteme*): Knowing what AI is, how it works, and how it affects decision-making processes. This includes awareness of machine learning models, algorithmic bias, data provenance, and the sociotechnical systems in which AI operates.
2. Practical competence (*techne*): The ability to interact effectively with AI tools, such as through prompt engineering, selecting appropriate platforms, and adapting AI-generated content for pedagogical purposes. This dimension involves not just technical fluency but also the capacity to critically evaluate the affordances and limitations of AI tools in instructional design.
3. Ethical and pedagogical judgment (*phronesis*): The capacity to evaluate when, why, and how to use AI in a manner that aligns with educational values, inclusivity, and student agency. This includes considerations of data privacy, transparency, fairness, and the potential impact of AI on the relational and human dimensions of teaching.

This tripartite articulation, rooted in ancient philosophy, is consistent with broader models of teacher professionalism that emphasize the integration of different forms of knowledge. Gustavsson [30] and Kinsella and Pitman [31], for instance, advocate for a vision of teacher expertise that goes beyond instrumental skills to include reflective and value-based reasoning.

In practical terms, developing AI literacy in teacher education means offering learning opportunities that scaffold the exploration, critical use, and pedagogical appropriation of AI tools. This includes guided practice with generative models (e.g., large language models like ChatGPT), discussions around ethical dilemmas, collaborative design tasks, and structured reflection. These activities can help future educators not only to become competent users of AI but also to question its assumptions and articulate their own pedagogical positions in relation to it.

Moreover, AI literacy intersects meaningfully with the goals of inclusive education. Teachers equipped with a critical understanding of AI are better positioned to evaluate whether these technologies reinforce or challenge educational inequities. As highlighted in recent work on AI and special education [6, 7], uncritical adoption of AI may risk reproducing biases or oversimplifying complex pedagogical decisions. Conversely, when situated within a framework like UDL, AI can become a tool for expanding access, supporting differentiation, and amplifying student voice.

In sum, AI literacy is not an optional add-on but a foundational component of contemporary teacher professionalism, particularly in higher education, where the pressure to adopt new technologies often outpaces the pedagogical support provided. Integrating AI literacy into faculty development and teacher preparation programs is therefore essential not only for technological fluency, but for fostering critical, inclusive, and ethically grounded teaching practices.

2.3 Toward an Integrated UDL-TPACK-AI Framework

To support the intentional and inclusive use of AI in educational design, this chapter proposes an integrated framework that combines the UDL model with the TPACK framework, expanded to explicitly incorporate dimensions of AI literacy. This conceptual synthesis aims to scaffold the use of digital tools with the pedagogical reasoning and ethical discernment required for inclusive teaching in higher education.

Originally formulated by Mishra and Koehler [8], TPACK identifies three core domains of teacher knowledge: Content knowledge (CK), referring to subject-matter expertise; pedagogical knowledge (PK), encompassing understanding of teaching strategies and learning theories; and technological knowledge (TK), relating to the use and evaluation of educational technologies. The model emphasizes the dynamic interplay among these domains and their intersections: Pedagogical CK (PCK), technological CK (TCK), technological PK (TPK), and the central overlap, that is, TPACK itself, which represents a situated, integrative understanding of teaching with technology.

In recent years, scholars have called for an expansion of the TPACK model to account for the emergence of AI in education. As Goldman et al. [6] suggest, AI should not be treated as a mere technological add-on, but as a transformative element that intersects with all dimensions of teacher knowledge. In this perspective, AI becomes embedded within TK (as knowledge of how AI tools work), PK (as strategies for using AI to support inclusive and adaptive pedagogy), and CK (as understanding how disciplinary content can be generated, adapted, or mediated by AI systems).

This extended model, which several scholars refer to as AI-TPACK (see Karataş et al. [32] for a comparative overview), acknowledges that educators need to develop the capacity to integrate AI tools into their practice from the early stages of instructional design. As Goldman et al. [6] argue, teachers should be supported in designing learning experiences that incorporate AI from the outset, through guided examples, practical exercises, and reflective activities. This approach aligns closely with the principles of UDL, which call for proactive and flexible planning to accommodate learner variability from the beginning of lesson design.

Our integrated UDL-TPACK-AI framework thus positions AI as a mediating tool that can support inclusive instructional design when used intentionally and reflectively. UDL provides the normative foundation, which focuses on accessibility, flexibility, and equity, while TPACK offers a model for diagnosing and developing the types of knowledge needed to implement those principles in specific disciplinary and technological contexts. The integration of AI adds a new layer of complexity and opportunity: AI systems can assist in the personalization of learning pathways, generation of multimodal resources, and formative feedback, but they also demand heightened awareness of issues such as bias, transparency, and pedagogical coherence.

This model was operationalized in the pilot activity described in the following section, in which university students were invited to design a UDL-based lesson with the support of a generative AI tool. Through this experience, they engaged with each component of the integrated framework: applying disciplinary knowledge (CK), selecting inclusive strategies (PK), experimenting with AI-enabled tools, and reflecting on their pedagogical choices at the intersection of all three. The experience illustrates how AI can serve not as a substitute for teacher expertise, but as a scaffold for developing inclusive, situated, and critically informed professional practice.

3 Case Study: A Pilot Activity in Higher Education

This section presents a pilot training activity conducted with postgraduate students enrolled in an education program at the University of Genoa (Italy). The activity was designed as part of a broader course on inclusive pedagogy and aimed to explore how generative AI tools could support the application of UDL principles in instructional planning. While originally developed for students preparing for careers in school education, the structure and findings of the activity are relevant to broader discussions on inclusive teaching practices and the integration of AI in higher education.

The training was conducted in several phases. The first phase, spanning 6 h within a course on special pedagogy, introduced students to the theoretical foundations of UDL, the rationale for inclusive design, and the practical application of UDL principles in lesson planning. This phase included collaborative analysis of classroom scenarios, discussion of student variability, and the exploration of UDL checkpoints as a dynamic planning tool.

Another phase was a 4-h hands-on lab focused on applying this knowledge in a concrete instructional design task. Students were organized into two cohorts (approximately 35 students each) and tasked with designing a lesson on the water cycle using UDL principles. To support this process, they engaged with a generative AI tool, namely *UDL Lesson Planner*, developed using a customized interface based on ChatGPT by the teacher Roberto Castaldo. The tool was selected for its ability to support exploratory design thinking by prompting students to reflect on engagement, representation, and expression in their planning.

Before beginning the task, students received a structured planning template to guide their design. The template included fields for specifying lesson objectives, activity phases, resources, timing, and assessment strategies, with dedicated prompts for indicating which UDL principles were being addressed at each stage. This format was intended to scaffold the design process and to encourage metacognitive awareness of how inclusive strategies were being operationalized.

To enrich the quality and relevance of the lesson content, students were invited to consult curated digital resources, including a website with educational materials and multimedia content related to the water cycle. In addition, the instructional goals for the activity were explicitly framed in terms of both CK (e.g., understanding the states of matter, the phases of the water cycle, and sustainable water use) and competencies (e.g., observing natural phenomena scientifically, interpreting environmental changes, and recognizing the role of water in ecosystems).

Throughout the lab, the instructor played an active facilitative role, modeling how to formulate effective prompts for the AI tool and guiding students in critically evaluating the outputs. This included discussions on the relevance and clarity of AI-generated suggestions, the alignment between content and student needs, and the feasibility of proposed activities within real classroom contexts.

While the primary aim of the activity was to promote familiarity with inclusive design frameworks, it also fostered important academic skills such as critical use of digital technologies and reflective thinking. Following the design task, students engaged in structured peer review and received formative feedback from the instructor. Although the evaluation phase is not the focus of this chapter, it provided additional opportunities for students to consolidate their understanding and reflect on the pedagogical implications of integrating UDL and AI in learning design.

3.1 Student Interaction with AI: A Scaffolded Exploration of Inclusive Design

Under the guidance of the instructor, students were introduced to basic strategies for interacting effectively with generative AI systems, with particular emphasis on how to formulate pedagogically meaningful prompts. Initial modeling by the instructor showed how different types of questions could elicit more or less relevant, actionable, or creative suggestions from the AI. Students were then encouraged to experiment iteratively with their own prompts, testing how small changes in wording, focus, or specificity would shape the AI's responses. This phase of the activity aimed to develop not just technical fluency, but a deeper awareness of how language mediates human-machine interaction and instructional design.

Rather than using AI as a shortcut or solution-provider, students were invited to view it as a "cognitive partner" that could extend their planning capacity, offer alternative perspectives, and scaffold inclusive thinking [7]. The AI was not presented as authoritative, but as a conversational agent whose outputs required interpretation, adaptation, and critical filtering. In this way, students learned to position themselves not as passive recipients of technological suggestions, but as active designers capable of leveraging AI within a pedagogical frame grounded in UDL.

As participants oriented themselves between human judgment and machine-generated suggestions, they gradually developed a more nuanced understanding of the opportunities and limits of AI in inclusive education. This included recognizing moments when AI proposals were generic, insufficiently contextualized, or even misaligned with pedagogical goals, prompting valuable discussions about educational intentionality, transparency, and the critical role of teacher agency. Thus, beyond operational proficiency, the activity cultivated a reflective stance toward the integration of AI, positioning it not as a neutral tool but as a site of pedagogical choice and ethical responsibility.

The five steps (Figure 1) of the pilot training experience offer a situated illustration of how the UDL-AI-TPACK framework was operationalized. Each phase supports different dimensions of the model:

–   Step 1 (theoretical introduction) focuses primarily on PK and the foundational principles of UDL, helping students conceptualize variability in learning and recognize the rationale for inclusive design. This phase establishes the normative dimension of the model.
–   Step 2 (practical lab) brings in TK through the use of an AI tool, while reinforcing CK. This intersection activates early forms of TPK and TCK, now guided by UDL principles.
–   Step 3 (AI interaction) explicitly cultivates AI literacy. Students develop *techne* through prompting, and also begin to exercise *phronesis* by evaluating AI outputs in light of UDL goals. This phase bridges TK, PK, and the ethical component of AI use.

- Step 4 (peer review and feedback) reinforces PCK and invites reflective practice, a key aspect of both UDL implementation and teacher professional growth. AI remains in the background, but its earlier role becomes part of the design conversation.
- Step 5 (evaluation and reflection) consolidates the model. Students reflect on their own development across all three domains (content, pedagogy, and technology) while critically assessing their inclusive design choices. The dimension of UDL intentionality is reactivated here, now informed by direct experience with AI tools and collaborative design.

In this way, the pilot training does not simply present UDL, AI, and TPACK as separate knowledge areas, but guides students through an experience where these dimensions are dynamically integrated and reflected upon.

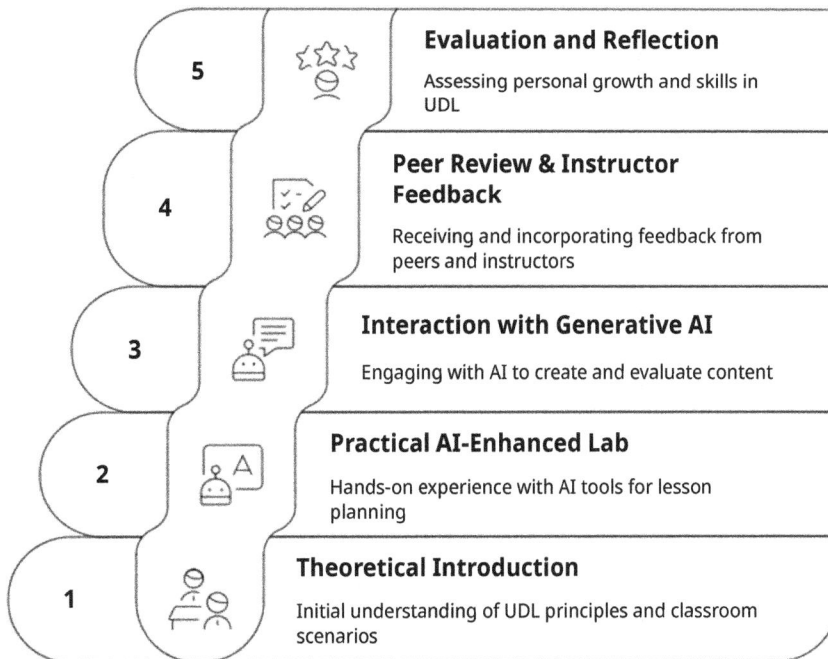

Figure 1: Overview of the pilot activity steps followed in the case study, from theoretical introduction to AI-supported design, peer feedback, and reflective evaluation.

3.2 Findings

To assess the perceived impact of the training, participants completed a self-assessment questionnaire before and after the lab activity. The questionnaire included Likert-scale items (1–5) related to their understanding of UDL principles, perceived ability to apply them in practice, and confidence in designing inclusive activities. Quantitative data were analyzed using paired-sample t-tests to determine statistically significant changes between pre- and post-intervention responses.

The results indicated a statistically significant increase in students' confidence in designing lessons based on UDL principles, particularly in areas related to engagement (+0.46, $p < 0.001$), use of multiple representations (+0.39, $p = 0.007$), and diverse forms of expression (+0.37, $p = 0.008$). Students also reported improved clarity regarding UDL concepts (+0.33, $p = 0.02$) and a greater ability to identify the barriers that UDL aims to address (+0.25, $p = 0.04$).

Interestingly, there was also a significant increase in students' recognition of their need for further training in applying UDL in practice (−0.51, $p = 0.01$), suggesting a more realistic and reflective awareness of the complexity involved in inclusive design. Other items, such as belief in UDL's effectiveness for inclusion (−0.17, $p = 0.21$) and its general applicability to school settings (+0.005, $p = 0.83$), showed no statistically significant change, indicating that students likely held positive views on these aspects prior to the activity.

Overall, the data suggest that the intervention had a meaningful impact on students' perceived competence and reflective understanding, particularly in operational aspects of inclusive planning. These findings provide a preliminary but encouraging indication of the potential of AI-supported training for fostering UDL-aligned design thinking in higher education. However, while the quantitative data provide encouraging evidence of increased student confidence and conceptual clarity, they capture only part of the learning that occurred. The following reflections are based on systematic observation during the training sessions and aim to shed light on the more nuanced, process-oriented dimensions of how students engaged with AI as a tool for inclusive instructional design.

Under the guidance of the instructor, students were introduced to basic strategies for interacting effectively with generative AI systems, with a particular emphasis on how to formulate pedagogically meaningful prompts. Initial modeling showed how different types of questions could elicit more or less relevant, actionable, or creative suggestions from the AI. Students were then encouraged to experiment iteratively with their own prompts, testing how changes in wording or specificity shaped the system's responses. This phase revealed a steep but productive learning curve. Students quickly realized that AI was not a "magic solution" for lesson design but a tool that responded in highly variable ways depending on how it was used. This aligns with current discussions in the literature, suggesting that effective use of AI in education requires prompt literacy and critical awareness [9]. Rather than accepting outputs at face value, students learned to evaluate the relevance, coherence, and inclusiveness of AI-generated suggestions.

One of the most important pedagogical shifts observed was the move from passive to active interaction with technology. While at the start some students used the AI tool to simply "fill in" lesson plans, by the end of the activity many were leveraging it to test alternative instructional strategies, reflect on accessibility barriers, and generate multimodal resources, behaviors that map directly onto UDL checkpoints and TPACK intersections. In this sense, the tool functioned not as a replacement for pedagogical reasoning, but as a scaffold for it.

This approach contrasts with more instrumentalist uses of AI reported in some studies, where automation is primarily framed as a way to save time or personalize content [6]. While personalization is certainly relevant, our findings suggest that the greater educational value lies in co-design with AI, where the human user maintains epistemic, ethical, and creative control [33]. This resonates with the idea of "AI as cognitive partner" discussed in contemporary scholarship, particularly in the context of teacher education [7].

Furthermore, student interaction with the AI surfaced important tensions. In some cases, outputs were overly generic, culturally decontextualized, or distant from the needs of students with specific disabilities; this prompted valuable classroom discussions about bias, representation, and educational intentionality that offered a natural entry point into the phronetic dimension of AI literacy [30], helping students reflect not only on *how* to use AI, but *why, when,* and *with what implications*. Such reflection is rare in traditional instructional design tasks, and was one of the most significant added values of this AI-integrated activity.

From the perspective of TPACK, this phase strengthened TPK by requiring students to make instructional decisions that took into account both content and technological affordances. Moreover, when students tailored AI-generated materials to align with UDL principles, such as offering audio-visual alternatives, varied task formats, or differentiated scaffolds, they engaged in complex decision-making at the intersection of TK, PK, and CK.

4 Discussion

The results of this pilot study invite several levels of reflection on the pedagogical and conceptual implications of integrating UDL and generative AI in higher education. First, the statistically significant gains in students' self-perceived competence, particularly in areas related to multiple means of engagement, representation, and expression, suggest that the training succeeded in making UDL not only understandable, but actionable. This aligns with broader findings in the literature, indicating that hands-on, design-based learning is essential for the internalization of inclusive frameworks [10, 11].

At the same time, the marked increase in students' recognition of their own need for further training may reflect a metacognitive shift: having engaged in a realistic design task, students became more aware of the complexities and uncertainties involved in applying UDL principles to real-world teaching scenarios. This result resonates with the tripartite model of AI literacy discussed earlier, particularly the dimension of *phronesis*, or ethical-practical reasoning. Rather than simply increasing their confidence, the activity appears to have fostered a deeper, more reflective stance toward the responsibilities of inclusive design.

One of the most promising aspects of the activity was the role of generative AI not as a source of definitive answers, but as a catalyst for pedagogical exploration. The AI tool encouraged students to experiment with different instructional strategies and to test their ideas against the criteria of accessibility and variability. This aligns with recent scholarship arguing that AI can serve as a mediating artifact in instructional design – one that stimulates critical inquiry, supports differentiation, and scaffolds novice designers' decision-making processes [6, 7]. These elements speak directly to the competencies identified by Trust et al. [34] in their framework of teacher educator technology competencies (TETCs). In particular, competencies such as "modeling effective use of digital tools to support diverse learners," "engaging in critical reflection," and "collaborating with others to improve practice" are clearly reflected in the structure and outcomes of the pilot activity. Throughout the training, students were consistently encouraged to evaluate the functionality of AI tools along with the inclusivity and accessibility of the outputs they produced. This mirrors what Trust et al. emphasize as the need for technology integration grounded in values of equity and learner diversity. For example, students used the AI to generate multiple representations of content and then discussed whether these formats met the needs of hypothetical learners with different abilities or backgrounds. This type of reasoning reveals an emergent awareness of what Trust and colleagues define as the intersection between digital fluency and social responsibility. Moreover, the design of the training supported peer learning and collaborative refinement, which the literature increasingly recognizes as crucial for sustainable professional growth. As students reviewed each other's lesson plans and AI interactions, they began to co-construct shared standards for inclusive design, moving beyond individual intuition toward more intentional, criteria-driven decisions. This dynamic supports the notion – found in both Trust et al. [34] and Fovet [2] – that professional learning environments must be dialogic and community-oriented, particularly when dealing with complex innovations like AI. From a methodological perspective, the integration of observation and qualitative prompts within the training allowed us to document these shifts in thinking, which are often overlooked in quantitatively driven studies. While survey data can capture perceived confidence or awareness, it is through discourse, negotiation, and reflection that pedagogical dispositions are made visible and transformed. In this regard, our study contributes to ongoing efforts to bridge the gap between technical skill acquisition and pedagogical transformation, showing that when AI tools are framed within inclusive design challenges and supported by collaborative structures, they can foster deeper learning about teaching itself. This approach aligns with Trust et al.'s call to recenter educational technology around meaningful, equity-driven practices, rather than treating AI integration as a purely instrumental goal.

Importantly, the findings suggest that the UDL-TPACK-AI framework can function as more than a theoretical abstraction. It provides a practical structure through which students can articulate and connect different forms of knowledge: content expertise (CK), pedagogical reasoning (PK), technological fluency (TK), and inclusive intentionality (UDL). The integration of these domains helped learners understand that teaching is not a linear application of knowledge, but a situated and iterative design process. A recent contribution by Cun and Huang [35] outlines a four-pathway model for integrating generative AI into teacher education through the lens of TPACK. Their framework describes a progression from basic technical exposure (exposure pathway), through efficiency-oriented use (efficiency pathway), to pedagogical enrichment (enrichment pathway), and finally to critical transformation (transformation pathway). Each stage represents a deepening integration of technological, pedagogical, and CK, with increasing emphasis on reflection and intentionality. The training model presented in our study aligns most closely with the enrichment and transformation pathways, particularly in its later phases. In fact, participants were encouraged to use AI to support lesson planning and also to interrogate themselves on the quality, accessibility, and ethical implications of the content generated. However, a key distinction lies in the incorporation of UDL as a foundational principle rather than an optional lens. Whereas Cun and Huang's [35] model focuses primarily on how teachers interact with AI to support learning goals, our model embeds inclusion and learner variability at the core of the design logic. Moreover, the UDL-AI-TPACK framework proposed here adds a normative layer; by situating accessibility, flexibility, and representation as nonnegotiable criteria for lesson quality, our model positions AI not simply as a cognitive amplifier but as a tool to help educators meet diverse learner needs. In this sense, the model contributes to the ethical transformation of teacher practice – not only in response to technological change, but in proactive pursuit of equity and inclusive education. This distinction is not merely theoretical. In our pilot, students reported that AI prompted them to consider previously unacknowledged learning barriers, and to articulate more flexible, multimodal instructional strategies. This level of critical design awareness, fostered through meta-reflection, suggests that models integrating both UDL and AI literacy may offer a more holistic approach to preparing future educators in digitally mediated learning environments.

The context of higher education adds another layer of significance. Many university programs still treat educational technology and inclusion as separate or secondary concerns, often relegated to specialized courses or support units. This activity, by contrast, embedded AI-supported inclusive design within the core of the curriculum, modeling an approach that treats accessibility, flexibility, and critical technological engagement as central to professional formation. A key implication of this study concerns the central role of AI literacy in initial teacher education. As Langran et al. [36] argue, the rapid emergence of generative AI demands a shift in how educators are prepared, not only to use AI tools effectively, but to understand, critique, and model their responsible integration in the classroom. The authors emphasize that teacher educators must actively shape future teachers' perceptions of AI by incorporating both technical competencies and ethical reflection into training experiences. This dual perspective aligns closely with the design of our pilot activity, in which students were introduced to AI not merely as a time-saving instrument, but as a pedagogical partner capable of supporting inclusive design when used with intentionality. Our findings support this position. While initial student interactions with the AI tool revealed a tendency toward passive use, the scaffolded design of the training encouraged more strategic and critical engagement. Over time, students moved from asking the AI for lesson content to using it as a collaborative ideation tool – evaluating the appropriateness, inclusiveness, and multimodality of its suggestions in light of UDL principles. This evolution reflects the kind of AI-supported pedagogical reasoning envisioned by Langran et al. [36], where technology enhances – not replaces – professional judgment. Moreover, the activity created space for what Langran and colleagues describe as "critical conversations" around AI's risks and limitations. For example, several students questioned the neutrality of AI-generated materials or identified accessibility gaps in suggested strategies. These discussions, although emergent, mirror the reflective stance that scholars argue is essential for preparing teachers to make responsible decisions about technology use in complex, diverse learning environments.

In sum, the discussion points to a set of pedagogical dynamics activated by the intervention: the shift from declarative to procedural and reflective knowledge; the transition from AI as content generator to AI as design interlocutor; and the emergence of design as a site of ethical and epistemic negotiation. These dynamics underscore the value of integrated frameworks in preparing educators – at all levels – for the complexity and promise of teaching in a digitally mediated, inclusivity-oriented landscape.

4.1 Translating the UDL-AI-TPACK Model into Faculty Development

While the pilot activity described in this chapter focused on students of education, the principles and structure of the UDL-AI-TPACK model can be fruitfully extended to faculty development in higher education. As documented in the literature, inclusive pedagogies are often adopted by individual instructors on a voluntary basis, with limited institutional support [1, 2]. Yet for UDL to become a truly transformative force in higher education, it must be embedded into faculty training initiatives that explicitly address the challenges and affordances of teaching in increasingly diverse and digitally mediated environments.

One promising direction involves the use of generative AI tools as scaffolding mechanisms in faculty learning. Just as students in our case study engaged with AI to design lessons aligned with UDL principles, university instructors can be introduced to structured, low-stakes opportunities to explore AI-assisted course design. These experiences can be framed within a reflective pedagogy, in which faculty are encouraged to analyze how generative outputs align with the UDL checkpoints, challenge their assumptions about learner variability, and revise instructional materials to enhance accessibility and engagement.

Drawing on ecological models of institutional change [2], we propose a multi-level approach that integrates our framework into faculty development through three interconnected dimensions:

1. Individual Level: Workshops and lab-based sessions where instructors learn to use AI tools (e.g., ChatGPT and Microsoft Copilot) for syllabus revision, resource creation, or assessment design, while engaging with UDL principles in practice.
2. Departmental Level: Communities of practice that encourage cross-disciplinary dialogue on inclusive design and AI literacy, fostering a shared language and gradual normalization of UDL-informed strategies across programs.
3. Institutional Level: Strategic alignment with teaching and learning centers, accessibility services, and academic leadership to embed the model in faculty onboarding, course evaluation policies, and quality assurance metrics.

To facilitate this transition, institutions should prioritize the development of dedicated resources that concretely support inclusive instructional design. Guided templates, annotated lesson exemplars, and curated libraries of AI prompts aligned with UDL checkpoints can offer faculty both inspiration and structure, enabling them to apply UDL principles in technologically mediated settings. These tools should not be prescriptive, but rather act as scaffolds that support creative and contextualized pedagogical thinking. As the integration of AI introduces new layers of complexity into the design process, the role of instructional designers becomes particularly strategic: they can co-develop resources in collaboration with faculty, ensuring alignment between technological tools and inclusive pedagogical intentions.

Equally important is the integration of sustained professional development that does not treat AI integration as a technical upgrade, but as a catalyst for deeper pedagogical reflection. Workshops and faculty learning communities should address core ethical concerns such as data privacy, algorithmic bias, and the potential reproduction of educational inequalities. Discussions should invite educators to interrogate how AI tools might reinforce normative assumptions about learning, and how they can instead be mobilized to dismantle systemic barriers. As Sperling et al. [9] and

Miao and Cukurova [37] emphasize, AI literacy entails operational competence and also a critical and ethical stance toward the role of automation in education.

Ultimately, the proposed UDL-AI-TPACK framework encourages a shift from accommodation to anticipation, from technical compliance to pedagogical creativity. Rather than positioning AI as a neutral enhancer of instruction, this perspective foregrounds its potential to support justice-oriented teaching, provided it is embedded within a reflective, value-driven approach to curriculum design. In this view, inclusive pedagogy is not a fixed set of practices but a continuous negotiation between learner diversity, institutional constraints, and technological affordances. By equipping faculty with both the mindset and the means to design for variability, higher education can move toward a model of teaching that is not only more accessible but more responsive, dynamic, and human-centered.

5 Conclusions

This chapter has explored the integration of UDL and AI within teacher education, proposing an original conceptual and practical framework – the UDL-AI-TPACK model. Grounded in cognitive, pedagogical, and ethical dimensions, the model was tested through a pilot training activity that guided students in inclusive lesson design with generative AI tools. Drawing on both quantitative and observational data, the study highlighted how future teachers can develop technological fluency while maintaining a strong commitment to learner variability and accessibility.

The findings suggest that generative AI, when situated within a structured pedagogical process, can support critical reflection and inclusive design thinking rather than fostering superficial or instrumental use. Participants demonstrated increased awareness of UDL principles, growing capacity to engage with AI tools, and a more realistic understanding of the complexities involved in designing for diverse learners. These outcomes align with recent calls in the literature [34, 36] to reframe AI literacy in teacher education as a multidimensional construct that includes technical, pedagogical, and ethical components.

A key contribution of this study lies in the structured, replicable model it offers. As highlighted by Cun and Huang [35], many current initiatives to introduce AI into teacher education are fragmented and lack theoretical coherence. Our model responds to this gap by explicitly connecting AI use to established frameworks (TPACK and UDL), and by proposing a scaffolded training process that combines modeling, experimentation, peer feedback, and critical reflection. This structure can be adapted across different disciplines and institutional contexts, and could serve as a basis for sustained faculty development programs.

At the same time, several limitations must be acknowledged. The pilot involved a small cohort of undergraduate students enrolled in a specific course within education studies, which may limit the generalizability of the findings to other populations, particularly in-service faculty across different disciplines. Additionally, the study focused on short-term outcomes; it did not assess the long-term retention or transferability of the skills acquired, nor did it explore how students might apply inclusive design in real teaching contexts. Finally, while AI use was scaffolded, the evolving nature of generative tools means that students' experiences may vary significantly depending on future changes in access, features, and institutional policies.

These limitations open up promising directions for future research. Longitudinal studies could investigate how sustained engagement with AI and UDL principles affects instructional practices over time. Comparative research across disciplinary contexts could illuminate how different epistemic cultures mediate the adoption of inclusive design tools. Further development of AI-powered platforms tailored to UDL criteria – e.g., including automatic accessibility audits, culturally responsive suggestions, or embedded scaffolding – could support more effective and intentional use.

In conclusion, the integration of AI in teacher education cannot be reduced to technical training or policy mandates. It must be accompanied by pedagogical vision, ethical reflection, and inclusive values. The UDL-AI-TPACK model represents an initial attempt to articulate such a vision. By foregrounding learner variability, collaboration, and reflective judgment, it offers a roadmap for shaping technological change in the service of equity and educational transformation.

References

[1]   Fornauf BS and Erickson JD. Toward an inclusive pedagogy through universal design for learning in higher education: A review of the literature. *Journal of Postsecondary Education and Disability*. 2020;33 (2):183–199.

[2]   Fovet F. Developing an ecological approach to the strategic implementation of UDL in higher education. *Journal of Education and Learning*. 2021;10(4):27–39. doi: 10.5539/jel.v10n4p27.

[3]   CAST. *Universal Design for Learning Guidelines Version 2.2*. Wakefield (MA): CAST; 2018 [cited 2025 Jul 21]. Available from: https://udlguidelines.cast.org.

[4]   Rose DH and Meyer A. *Teaching Every Student in the Digital Age: Universal Design for Learning*. Alexandria (VA): Association for Supervision and Curriculum Development (ASCD); 2002.

[5]   Capp MJ. The effectiveness of universal design for learning: A meta-analysis of literature between 2013 and 2016. *International Journal of Inclusive Education*. 2017;21(8):791–807. doi.org/10.1080/ 13603116.2017.1325074

[6]   Goldman SR, Carreon A and Smith SJ. Exploring the integration of artificial intelligence into special education teacher preparation through the TPACK Framework. *Journal of Special Education Preparation*. 2024;4(2):52–64. doi.org/10.33043/6zx26bb2

[7]   Holman K, et al. Navigating AI-powered personalized learning in special education: A guide for preservice teacher faculty. *Journal of Special Education Preparation*. 2024;4(2):90–95. doi.org/ 10.33043/5b2x-qcb3

[8]   Mishra P and Koehler MJ. Technological pedagogical content knowledge: A framework for teacher knowledge. *Teachers College Record*. 2006;108(6):1017–1054. doi.org/10.1111/j.1467-9620.2006.00684.x

[9] Sperling K, et al. In search of artificial intelligence (AI) literacy in teacher education: A scoping review. *Computers and Education Open.* 2024;6:100–169. doi.org/10.1016/j.caeo.2024.100169.

[10] Griful-Freixenet J, Struyven K and Vantieghem W. Exploring pre-service teachers' beliefs and practices about two inclusive frameworks: Universal design for learning and differentiated instruction. *Teaching and Teacher Education.* 2021;107:103–503. doi.org/10.1016/j.tate.2021.103503.

[11] Rao K, Torres C and Smith SJ. Digital tools and UDL-based instructional strategies to support students with disabilities online. *Journal of Special Education Technology.* 2021;36(2):105–112. doi.org/10.1177/0162643421998327.

[12] Moriña A, Carballo R and Doménech A. Transforming higher education: A systematic review of faculty training in UDL and its benefits. *Teaching in Higher Education.* 2025;1–18. doi.org/10.1080/13562517.2025.2465994.

[13] Casebolt T and Humphrey K. Use of universal design for learning principles in a public health course. *Annals of Global Health.* 2023;89(1):48. doi.org/10.5334/aogh.4045.

[14] Espada-Chavarria R, et al. Universal design for learning and instruction: Effective strategies for inclusive higher education. *Education Sciences.* 2023;13(6):620. doi.org/10.3390/educsci13060620.

[15] Fleming EC. UDL for inclusive teaching: Offering choice to increase belonging through technology. *Journal of Teaching and Learning with Technology.* 2023;12(1). doi.org/10.14434/jotlt.v12i1.36327.

[16] Manly CA. A panel data analysis of using multiple content modalities during adaptive learning activities. *Research in Higher Education.* 2024;65(6):1112–1136. doi.org/10.1007/s11162024-09784-9.

[17] Nieminen JH and Pesonen HV. Taking universal design back to its roots: Perspectives on accessibility and identity in undergraduate mathematics. *Education Sciences.* 2019;10(1):12. doi.org/10.3390/educsci10010012.

[18] Ismailov M and Chiu TKF. Catering to inclusion and diversity with universal design for learning in asynchronous online education: A self-determination theory perspective. *Frontiers in Psychology.* 2022;13:819–884. doi.org/10.3389/fpsyg.2022.819884.

[19] Lohmann MJ, et al. Engaging graduate students in the online learning environment: A universal design for learning (UDL) approach to teacher preparation. *Networks: An Online Journal for Teacher Research.* 2018;20(2):n2. doi.org/10.4148/2470-6353.1264.

[20] Owenz M and Cruz L. Addressing student test anxiety through universal design for learning alternative assessments. *College Teaching.* 2025;73(3):134–144. doi.org/10.1080/87567555.2023.2245945.

[21] Davies PL, Schelly CL and Spooner CL. Measuring the effectiveness of universal design for learning intervention in postsecondary education. *Journal of Postsecondary Education and Disability.* 2013;26 (3):195–220.

[22] Dean T, Lee-Post A and Hapke H. Universal design for learning in teaching large lecture classes. *Journal of Marketing Education.* 2017;39(1):5–16. doi.org/10.1177/0273475316662104.

[23] Reyes CT, et al. "Every little thing that could possibly be provided helps": Analysis of online first-year chemistry resources using the universal design for learning framework. *Chemistry Education Research and Practice.* 2022;23(2):385–407. doi.org/10.1039/D1RP00171J.

[24] Olivier E and Potvin M-C. Faculty development: Reaching every college student with universal design for learning. *Journal of Formative Design in Learning.* 2021;5(2):106–115. doi.org/10.1007/s41686-021-00061-x.

[25] Kim HJ, Kong Y and Tirotta-Esposito R. Promoting diversity, equity, and inclusion: An examination of diversity-infused faculty professional development programs. *Journal of Higher Education Theory & Practice.* 2023;23(11). doi.org/10.33423/jhetp.v23i11.6224.

[26] Azam S, et al. Becoming inclusive teacher educators: Self-study as a professional learning tool. *International Journal for the Scholarship of Teaching and Learning.* 2021;15(2):4. doi.org/10.20429/ijsotl.2021.150204.

[27] Schelly CL, Davies PL and Spooner CL. Student perceptions of faculty implementation of Universal Design for learning. *Journal of Postsecondary Education and Disability*. 2011;24(1):17–30.

[28] Hutson B and Downs H. The college STAR faculty learning community: Promoting learning for all students through faculty collaboration. *The Journal of Faculty Development*. 2015;29(1):25–32.

[29] Hromalik CD, et al. Increasing universal design for learning knowledge and application at a community college: The universal design for learning academy. *International Journal of Inclusive Education*. 2024;28(3):247–262. doi.org/10.1080/13603116.2021.1931719.

[30] Gustavsson B. Revisiting the philosophical roots of practical knowledge. In: *Developing Practice Knowledge for Health Professionals*. Butterworth-Heinemann; 2004. pp. 35–50.

[31] Kinsella EA and Pitman A (eds.). *Phronesis as Professional Knowledge: Practical Wisdom in the Professions*. vol. 1. Springer Science & Business Media; 2012.

[32] Karataş F and Aksu Ataç B. When TPACK meets artificial intelligence: Analyzing TPACK and AI-TPACK components through structural equation modelling. *Education and Information Technologies*. 2024;1–26. doi.org/10.1007/s10639-024-13164-2.

[33] Urmeneta A and Romero M. *Creative Applications of Artificial Intelligence in Education*. Springer Nature; 2024.

[34] Trust T, Maloy RW and Hikmatullah N. Integrating AI in teacher education using the teacher educator technology competencies. In: Searson M, Langran E and Trumble J (eds.). *Exploring New Horizons: Generative Artificial Intelligence and Teacher Education*. Waynesville, NC: Association for the Advancement of Computing in Education (AACE); 2024. pp. 26–38.

[35] Cun A and Huang T. Generative AI and TPACK in teacher education: Pre-service teachers' perspectives. In: Searson M, Langran E and Trumble J (eds.). *Exploring New Horizons: Generative Artificial Intelligence and Teacher Education. Waynesville, NC: Association for the Advancement of Computing in Education (AACE)*. 2024. pp. 62–75.

[36] Langran E, Searson M and Trumble J. Transforming teacher education in the age of generative AI. *Exploring New Horizons: Generative Artificial Intelligence and Teacher Education*. 2024;2:2–13.

[37] Cukurova M and Miao F. *AI Competency Framework for Teachers*. UNESCO Publishing; 2024.

Gonzalo Lorenzo*, Andrea Cerdán-Chacón, Alejandro Lorenzo-Lledó

Artificial Intelligence as a Pedagogical Tool to Include Autistic Students According to Universal Design for Learning

Abstract: Artificial intelligence (AI) is one of the tools that have seen the greatest expansion in educational environments in recent years. The rise of AI has contributed to its ability to adapt in real time the tasks to be developed by students, depending on their learning characteristics. In this way, it is possible to respond to students' specific educational support needs. Based on the above, the aim of this research has been to apply training to higher education students on AI as an inclusive tool of participation and representation for autistic students, considering the framework of universal design for learning. A nonexperimental quantitative methodological approach was chosen, with a comparative causal design. The sample of the pilot study had 27 students in the second year of the Primary Education degree at the University of Alicante, who were taking the course "Meeting specific educational needs (MSEN)." An ad hoc questionnaire was used to collect information, which was completed by the participants at the end of the course. The results showed that 92.6% of the participants thought that AI could create communicative environments for autistic students. In addition, 89% agreed that AI has the potential to provide real-time progress in learning. Significant differences in perceptions were also found according to participants' prior knowledge of AI and age. In conclusion, further training in the use of AI is considered necessary to move towards the full educational inclusion of autistic learners.

Keywords: Artificial intelligence, universal design of learning, autism, specific educational support needs

Acknowledgments: El presente trabajo ha contado con una ayuda del Programa de Redes de investigación en docencia universitaria del Instituto de Ciencias de la Educación de la Universidad de Alicante (convocatoria 2024). Ref.: [6198] y título del proyecto "La inteligencia artificial en la formación inicial docente para la atención del alumnado con necesidades específicas de apoyo educativo"

***Corresponding author: Gonzalo Lorenzo**, Department of Developmental Psychology and Didactics, University of Alicante, Alicante, Spain, e-mail: glledo@ua.es
Andrea Cerdán-Chacón, Department of Developmental Psychology and Didactics, University of Alicante, Spain, e-mail: andrea.cerdan@ua.es
Alejandro Lorenzo-Lledó, Didactics and School Organization Department., University of Granada, Spain, e-mail: alorenzolledo@ugr.es

https://doi.org/10.1515/9783112206393-009

1 Introduction

Currently, technological advances are being applied across various fields of knowledge [1]. Among the new tools that have gained increasing relevance in educational settings in recent years is artificial intelligence (AI) [2]. Therefore, AI-supported education is considered a key element for future economic growth, training, and global competitiveness [3].

AI encompasses a set of techniques that enable computer systems to simulate characteristics typical of human intelligence, such as decision-making, creativity, prediction, and planning [4]. In this context, the main contribution of AI to educational environments lies in its ability to provide a more adaptable and responsive system, capable of addressing students' individual needs [5]. For instance, AI-based platforms allow for the adjustment of both content and learning pace according to the specific characteristics of each student [6]. In addition, AI can offer educational stakeholders a more accurate view of students' learning progress, which facilitates the design of more personalized pedagogical interventions [7]. It also provides teachers with feedback on students' task performance, along with timely suggestions regarding areas that require improvement [8].

The adaptability of AI is one of its main strengths, as it allows adjustment to the characteristics of an increasingly heterogeneous student population. This capability, combined with universal design for learning, facilitates the creation of flexible and accessible educational environments for all students. Within the diverse group of students present in classrooms, those with autism spectrum disorder stand out. These students are characterized by visual learning, logical reasoning, and excellent memory [9]. However, they experience difficulties in communication, social interaction, and exhibit repetitive behaviors and restricted interests [10]. In this context, autistic students often prefer routines in daily life and controlled environments, which hinder information processing. This leads them to focus on specific details and often overlook the broader perspective [11, 12]. These characteristics make it difficult to generalize learning to real-world settings without the support of technological tools. Given these particularities, AI offers students the possibility to adapt activities in real time based on their mood and cognitive level [13]. Thus, activities can be designed to align with the interests of the participating students [14]. Additionally, AI provides teachers with a tool that enables real-time monitoring of learning progress and identification of areas where autistic students have greater needs [15]. In summary, the purpose of AI for autistic students is to provide personalized and adaptable learning experiences, enhance social interaction skills, and offer specific support for cognitive and emotional development [16].

Based on the relationship between the characteristics of autistic students and the benefits that AI can bring to their learning processes, there is a need to review previous studies in order to highlight the improvements incorporated in the present work. In the study by Li et al. [16], the objective was to analyze the perceptions and experiences of 20 teachers regarding AI-based interventions aimed at children with autism. More specifically, the study delved into the perceived benefits and challenges of using such interventions, as well as the formulation of recommendations to mitigate these challenges. Interviews and focus groups were used for this purpose. The results show that the use of AI is beneficial for autistic students; however, significant difficulties were identified in its implementation, mainly related to the need for teacher training and existing technological problems. In line with the literature review, the study conducted by Alsudairy and Eltantawy [17] aimed to explore the perceptions of special education teachers regarding the use of AI in teaching students with disabilities. Additionally, the effect of contextual variables such as professional experience and type of disability was examined. The sample consisted of 301 participants who completed a 36-item questionnaire structured in two dimensions: one related to general perceptions of AI use and the other focused on its application in the educational process. The results reflect a tendency among teachers to select neutral options, which may be attributed to a lack of knowledge about AI, scarce resources and infrastructure, and the absence of specific platforms in educational centers. Similarly, the study by Alwaqdani [18] analyzed teachers' perceptions regarding the integration of AI into teaching practices. With a sample of 1,101 educators, two main questions were addressed: AI's potential to improve educational practice and the challenges faced by teachers in its implementation. Results indicate that teachers positively value AI for its ability to save time, support the design of enriching activities, and personalize learning experiences. Nevertheless, concerns were also expressed regarding the effort required for training and potential consequences in key areas such as creativity and critical thinking. In another study on AI perceptions, Lin and Chen [19] explored the effects of AI-integrated educational applications on creativity and academic emotions among university students, from both student and teacher perspectives. Attitudes of both groups toward these applications were also evaluated. Results showed that AI-based tools stimulate creativity, promote emotional well-being through playful elements and constant availability, and support the development of problem-solving techniques. Finally, in the research conducted by Giraldi et al. [20], the perspectives of public employees from various sectors regarding the use of AI were analyzed. A questionnaire was administered to 439 participants, examining variables such as demographics, technological competence, perceptions of usefulness, privacy concerns, attitudes towards AI and generative AI, and willingness to support its adoption. The results reveal a predominantly positive attitude, perceiving AI as a complementary tool rather than a potential replacement, especially in public-use contexts.

Based on this review, the main contributions of the present work are as follows: firstly, it focuses on the perceptions of future teachers, unlike previous studies that exclusively analyze the viewpoint of in-service professionals. Secondly, none of the reviewed studies examine the influence of variables such as prior knowledge of AI or students' age. Thirdly, the instrument used to assess perceptions delves into one of the key areas where autistic students face greater difficulties: communication and social interaction, specifically analyzing how AI can contribute to its improvement. In contrast, previous studies mainly focus on the general usefulness of AI in the learning process. Fourthly, this research collects the perceptions of students after participation in a practical intervention, followed by the completion of a questionnaire. In the reviewed studies, perceptions are usually obtained without a prior formative or experiential session. Taking the above into account, the general objective of this research has been to apply AI in the training of primary education degree students, with the aim of preparing them for its use with students with autism spectrum disorder. From this general objective, the following specific objectives are derived:

– To design a training action for the use of AI with autistic students.
– To understand students' perceptions regarding the usefulness of AI as a communication support tool to foster the participation of autistic students in primary education classrooms.
– To understand perceptions regarding the usefulness of AI as a social interaction support tool to foster the participation of autistic students in primary education classrooms.
– To understand perceptions regarding the usefulness of AI as a motivation and self-regulation support tool to foster the participation of autistic students in primary education classrooms.
– To identify possible significant differences in perceptions about the use of AI as a communication support tool to foster the participation of autistic students in primary education classrooms.
– To identify possible significant differences in perceptions about the use of AI as a social interaction support tool to foster the participation of autistic students in primary education classrooms.
– To identify possible significant differences in perceptions about the use of AI as a motivation and self-regulation support tool to foster the participation of autistic students in primary education classrooms.

The chapter is structured into five sections. The first provides a characterization of AI and explores its potential benefits for students with SEN. It concludes with a review of previous research and the authors' contributions. The second section outlines the methodology employed, beginning with a description of the participants, followed by the features of the instrument used, the design of the tasks, and the intervention procedure. The third section presents the findings in relation to the six research questions. The fourth is dedicated to the discussion and interpretation of the results. Finally, the fifth section summarizes the study's conclusions.

2 Method

To achieve the general objective, statistical models have been developed to provide a detailed explanation of the observed events [21]. Additionally, the obtained information enables the classification of the fundamental features of the observed variables. This requires the application of a nonexperimental quantitative method [21]. This approach also involves analyzing the factors producing differences between groups in terms of the analyzed variables in a comparative-causal design [22]. Finally, as the researchers collected the data at a single point in time, they opted for a cross-sectional study [23].

2.1 Description of the Context and Participants

The research has been implemented in the subject meeting specific educational needs (MSEN), which belongs to the second year of the degree in primary education. The sample is made up of 27 students belonging to the high academic achievement group (HAAG), selected by means of accidental or convenience sampling. Of these, 25.9% were male, while 74.1% were female. The average age of the participants was 24.51 years. Also, only 14.8% had had previous experience with AI, with ChatGPT and Gemini being the tools used in 63.63% of cases. Among those who had had previous experience with AI, it had been developed in the subjects Curriculum Development and Digital Classrooms, in the second year, and Developmental Psychology, in the first year. As a summary, Figures 1 and 2 present the participants' gender and age. Figure 3 provides information regarding students who had prior knowledge of AI applications. Finally, Figure 4 shows which AI tools were known by the students with such prior knowledge.

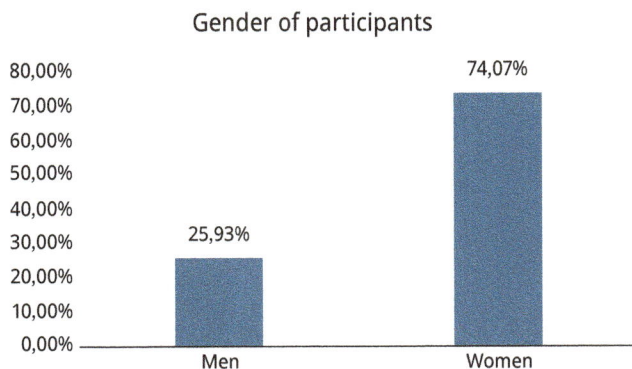

Figure 1: Gender of participants.

Percentage of participants by age.

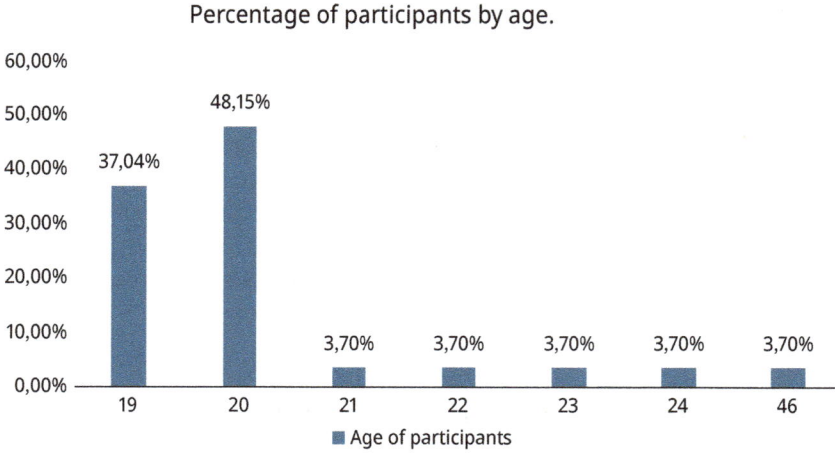

Figure 2: Age of participants.

Percentage of students with prior
knowledge of AI applications for
SEN students

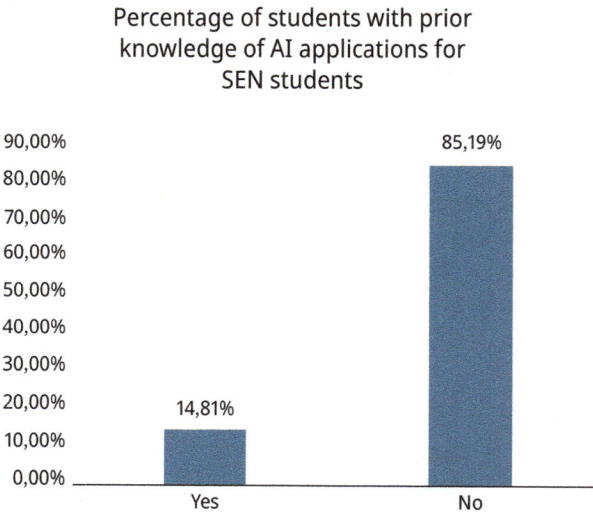

Figure 3: Percentage of students who know AI applications.

2.2 Instrument

An ad hoc questionnaire was designed consisting of two sections. The first section collects demographic data related to sex, age, and prior knowledge about AI. The second section is composed of six dimensions. The first three group a total of 15 items and

Artificial intelligences tools known to users who already had prior knowledge.

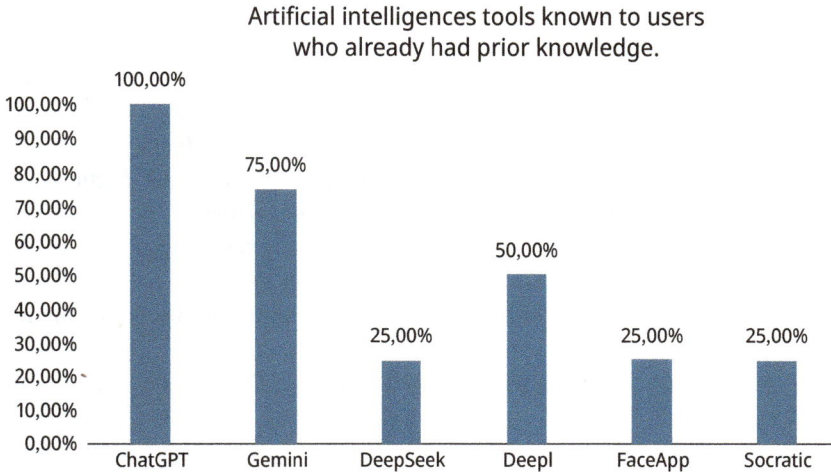

Figure 4: Artificial intelligence tools known by users.

address students' perceptions of AI as a tool to promote the participation of autistic students in the classroom. More specifically:

- Dimension 1 (five items): focuses on the use of AI as a communication support tool to promote the participation of autistic students in primary education classrooms
- Dimension 2 (five items): evaluates the use of AI as a social interaction support tool to encourage the participation of autistic students in primary education classrooms
- Dimension 3 (five items): assesses AI as a support tool for motivation and self-regulation in autistic students in primary education classrooms

For these three dimensions, a 6-point Likert scale was used, where 1 indicates "strongly disagree" and 6 "strongly agree."

Dimensions 4, 5, and 6, which also belong to the second section, aim to assess students' perceptions regarding the usefulness of AI in the primary education curriculum. More specifically:

- Dimension 4 (14 items): analyzes the degree of usefulness of AI for addressing primary education objectives with autistic students
- Dimension 5 (eight items): measures the degree of usefulness of AI in developing key competencies
- Dimension 6 (six items): estimates the degree of usefulness of AI to work on different areas of the primary education curriculum with autistic students

For these last three dimensions, a 5-point Likert scale was used, with 1 being "not useful at all" and 5 "completely useful." Regarding the reliability of the questionnaire, the following Cronbach's alpha values were obtained: for dimension 1, 0.957; for dimension 2, 0.85; and for dimension 3, 0.96.

To conclude this section, all the items corresponding to the dimensions of the questionnaire that were analyzed in the study are presented below:

- Item 1: Increase the communicative intentionality of autistic students.
- Item 2: Increase the frequency of communicative exchanges with the teacher.
- Item 3: Improve the quality of communication with the teacher.
- Item 4: Increase the frequency of communicative exchanges with peer groups.
- Item 5: Improve the quality of communication with peer groups.
- Item 6: Create a participatory communicative environment that promotes the inclusion of autistic students.
- Item 7: Provide autistic students with assertive language appropriate to the learning context.
- Item 8: Teach autistic students specific strategies for cooperative group work.
- Item 9: Provide autistic students with clear instructions to support their interaction with peers.
- Item 10: Provide autistic students with clear instructions to enhance their communication with teachers.
- Item 11: Capture the attention of autistic students by presenting the tool as a playful and engaging resource.
- Item 12: Foster the motivation of autistic students to actively participate in classroom activities.
- Item 13: Personalize the information provided to autistic students according to their specific needs.
- Item 14: Stimulate the interest of autistic students in participating in learning activities.
- Item 15: Offer autistic students immediate feedback on their contributions.

2.3 Procedure

With the aim of applying AI in the initial training of future primary education teachers, a training session was designed to use the SlideIA tool. This is a free plugin that must be installed in PowerPoint via Google Drive. As the first part of the activity, students were required to watch a series of videos to become familiar with how the application works. The practical task was related to the content of Unit 8, which focuses on supporting students with autism in the course "MSEN." The task was carried out in groups and involved creating a daily schedule using SlideIA, with the goal of helping the student anticipate the events that would take place during a visit to the zoo. The aim was for the student to learn about the events that would occur and how to behave in each of them. The activity was conducted during a two-hour class session and was complemented by individual work outside the classroom watching the videos. After completing the activity, students were required to complete a questionnaire via Google Forms. As a summary, Figure 5 explains the detailed procedure of the investigation.

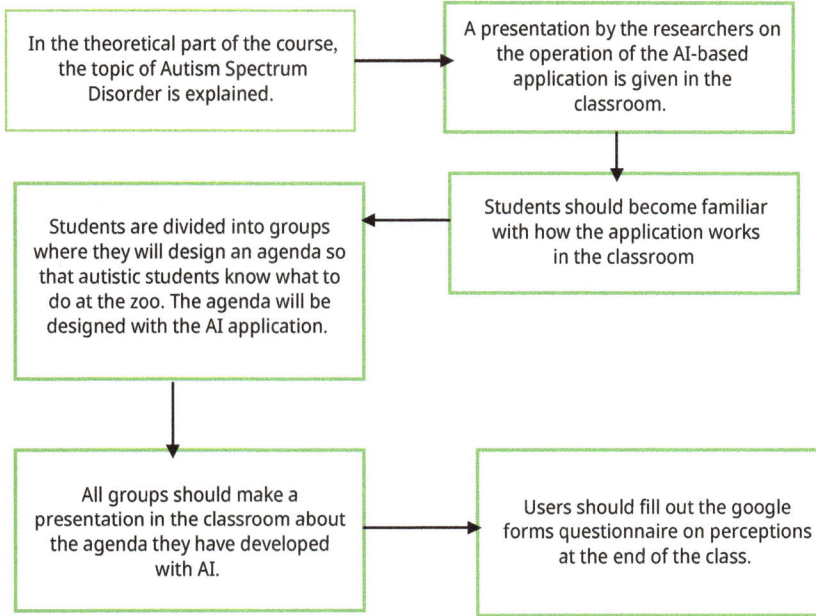

Figure 5: Procedure.

2.4 Data Analysis

The data analysis involved calculating mean scores, standard deviations, and standard errors for each item. A sample normality analysis was then performed. Due to the sample size and the fact that several items did not follow a normal distribution, it was decided to use a nonparametric test [24]. Consequently, the Mann-Whitney U test was employed in the inferential analysis to evaluate the influence of gender and prior knowledge of AI. Additionally, the Kruskal-Walli's test was applied to examine the effect of age on perceptions of AI across the different dimensions. All descriptive and inferential statistical analyses were conducted using SPSS version 23.

3 Results

3.1 Results Regarding Students' Perceptions of the Usefulness of AI as a Communication Support Tool for Fostering the Participation of Autistic Students in Primary School Classrooms

Table 1 presents the results corresponding to the mean scores and standard deviations for each of the items that make up the first dimension. Skewness and kurtosis values are also included. The analysis of the table shows that item 1 presents the highest mean score with 4.37, while items 1 and 4 show the highest standard deviations with values of 1.29 and 1.30, respectively. Similarly, these two items have the highest values of skewness −1. In contrast, item 5 has the highest value of kurtosis. In terms of variance, item 1 also has the highest value. All items in dimension 1 have values ranging from 1.48 to 1.70.

Table 1: Descriptive statistics for dimension 1.

Item	Mean	Standard deviation	Skewness	Kurtosis	Variance
1	4.37	1.30	−1	0.45	1.70
2	4.22	1.21	−0.73	−0.75	1.48
3	4.29	1.29	−0.48	−0.60	1.67
4	4.33	1.30	−1	0.36	1.69
5	4.29	1.26	−0.97	0.59	1.60

Table 2 shows the response percentages for each subject according to the item. For example, item 1 has the highest overall percentage in the options "strongly agree" and "agree," with 66.7%. However, item 3 has the highest overall agreement with 77.8%. To obtain this value, the percentages of options 4, 5, and 6, which correspond to the part of the scale ranging from "somewhat agree" to "strongly agree," were added together.

Table 2: Percentages for dimension 1.

Item	Strongly disagree	Disagree	Slightly disagree	Slightly agree	Agree	Strongly agree
1	3.7	7.4	14.8	7.4	55.6	11.1
2	0.0	14.8	11.1	18.5	48.1	7.40
3	0.0	14.8	7.40	29.6	29.6	18.5
4	3.7	7.40	14.8	11.1	51.9	11.1
5	3.7	7.40	11.1	22.2	44.4	11.1

During the course of the activity, students repeatedly expressed their interest in exploring how AI could be combined with other technologies, such as virtual reality or robotics, particularly regarding the possibility of using the agenda as a communication tool. This interest intensified during the agenda development phase, when they posed specific questions about the use of AI and robotics as dynamic support tools to facilitate task completion by autistic students. Regarding the questionnaire implementation, students persistently inquired with the evaluators whether they should answer the questions considering that AI would be integrated with other technological tools.

3.2 Results Regarding Perceptions of the Usefulness of AI as a Tool to Support Social Interaction in Fostering the Participation of Autistic Learners in Primary School Classrooms

Table 3 shows the mean scores, standard deviation, skewness, kurtosis, and variance for dimension 2. Item 8 has the highest mean score, with a value of 5.48, and also the lowest standard deviation, with 0.48. Item 9 shows the highest absolute value for skewness at 2.17, and the same is true for kurtosis, which reaches a value of 5.37 for this item. Finally, the lowest variance also corresponds to item 8, with a value of 0.23.

Table 3: Descriptive statistics for dimension 2.

Item	Mean	Standard deviation	Skewness	Kurtosis	Variance
6	5.18	1.03	−2.10	5.30	1.08
7	5.00	1.17	−1.80	2.90	1.38
8	5.33	0.48	0.74	−1.56	0.23
9	5.18	1.03	−2.17	5.37	1.08
10	5.11	1.21	−1.80	2.90	1.48

Table 4 shows the percentages of participants according to their responses to each of the items. For example, Item 8 has the highest percentage in the 'strongly agree' option, with 66.7%. This item stands out because 100% of the participants selected either 'strongly agree' or 'agree'. In contrast, Item 10 has 11.1% in the 'strongly disagree' option and also shows the lowest percentage in the 'strongly agree' and 'agree' options combined, with 88.8%. Finally, Item 6 reaches 92.6% in the 'strongly agree' and 'agree' options, being the highest.

Table 4: Percentages for dimension 2.

Item	Strongly disagree	Disagree	Slightly disagree	Slightly agree	Agree	Strongly agree
6	0.0	7.4	0.0	0.0	51.9	40.7
7	0.0	11.1	0.0	0.0	55.6	33.3
8	0.0	0.0	0.0	0.0	66.7	33.3
9	0.0	7.40	0.0	0.0	51.9	40.7
10	0.0	11.1	0.0	0.0	44.4	44.4

3.3 Results Regarding Perceptions of AI as a Tool to Support Motivation and Self-Regulation of Autistic Learners to Promote Participation in Primary School Classrooms

Table 5 presents descriptive statistics for the mean, standard deviation, skewness, and kurtosis of dimension 3 of the questionnaire. Item 15 has the highest means with 4.74, while item 11 has the lowest means with 4.37. In terms of standard deviation, item 11 reaches a value of 1.54, while item 13 shows the lowest at 1.18. The skewness scores range between −0.86 and −0.96. Finally, the kurtosis interval varies between 0.026 and 0.599.

Table 5: Descriptive statistics for dimension 3.

Item	Mean	Standard deviation	Skewness	Kurtosis	Variance
11	4.37	1.54	−0.94	0.02	2.39
12	4.48	1.34	−0.99	0.59	1.79
13	4.51	1.18	−0.86	0.28	1.41
14	4.59	1.21	−0.92	0.29	1.48
15	4.74	1.25	−0.96	0.36	1.58

The purpose of Table 6 is to present the frequency of participants' responses to the various items corresponding to dimension 3. Item 11 shows the highest percentage of responses in the "strongly disagree" category, with 7.4%. The highest proportion of "disagree" responses is observed in items 11, 13, 14, and 15, each with 11.1%. On the other hand, the "strongly agree" option reaches its peak in item 15, with 33.3%, while the "agree" category stands out in items 13 and 14.

Table 6: Percentages of Dimension 3.

Item	Strongly disagree	Disagree	Slightly disagree	Slightly agree	Agree	Strongly agree
11	7.4	11.1	0.00	25.9	29.6	25.9
12	3.7	7.40	7.40	22.2	37.0	22.2
13	0.0	11.1	3.70	25.9	40.7	18.5
14	0.0	11.1	3.70	22.2	40.7	22.2
15	0.0	11.1	0.00	25.9	29.6	33.3

During agenda development, the researchers received comments from the students highlighting the motivating potential of AI, attributable to its capacity to adapt content according to each student's individual interests. Additionally, based on the dialogues held with the students, the researchers perceived that AI could play a fundamental role in self-regulation by generating real-time agendas that provide the child with information aimed at reducing stereotyped behaviors or improving emotional control. Nevertheless, a difficulty observed in earlier sections was reiterated: the students reported lacking the necessary training to fully comprehend the process of integrating AI. As in the preceding section, the students did not present difficulties when responding to the items associated with this dimension.

3.4 Results Regarding Possible Significant Differences in Student Perceptions of the Usefulness of AI as a Communication Support Tool for Fostering the Participation of Autistic Students in Primary School Classrooms

This section presents an analysis of the influence of the variable's prior knowledge of AI, gender, and age on dimension 1. For all three variables, both a dimensional study and an item-by-item analysis have been conducted. An item-by-item approach was chosen, as, according to [25], this method allows for the identification of items that may show some bias in their evaluation. Additionally, according to [26], this approach provides information about characteristics that cannot be analyzed in dimensional analyses. As for the dimensional analysis, it was used because it allows for the evaluation of the internal validity of the questionnaire and reduces random error [27].

The first variable analyzed is gender. In the application of the item-by-item analysis, it was found that the p-value was greater than 0.05 in all items, indicating the absence of statistically significant differences. However, the average ranks of the 5 items show higher scores favoring women, ranging from 14.35 to 14.60. In order to gain a more comprehensive understanding of the results, an inferential analysis of dimension 1 was conducted. The findings show that the p-value is 0.88, with $W = 95.5$, $U = 67.50$, and $Z = -0.141$. Therefore, no statistically significant differences exist in dimension 1, as $p > 0.05$. To confirm the extent of the obtained results, the effect size for nonparametric tests was calculated using Rosenthal's r [28]. The value of r is −0.027, indicating that the effect size is very weak.

The second variable analyzed was prior knowledge of AI. In the item-by-item analysis, it was confirmed that there were no statistically significant differences between those with prior knowledge of AI and those without, as the p-values were greater than 0.05. Nevertheless, in the rank analysis, it was observed that the highest values corresponded to those with prior knowledge of AI, with scores ranging from 14.50 to 18.30. Similarly to the gender variable, an analysis of dimension 1 was conducted concerning prior knowledge. The following results were obtained: $p = 0.465$, $W = 311.5$, $U = 35.5$, and $Z = -0.731$. Finally, the effect size was calculated using Rosenthal's r, yielding a value of 0.14. This result indicates a small effect size, suggesting that there are no relevant or visible differences in the students' perceptions based on whether they possess prior knowledge of AI or not.

The last variable analyzed is the influence of age on perceptions of AI in dimension 1. When applying the Kruskal-Wallis test, none of the five items reached a p-value that would indicate statistical significance. Therefore, no significant differences were observed based on age. However, the average rank values indicate that the highest scores correspond to participants aged 21 or older, which could suggest the possible existence of relevant differences not captured by statistical significance. Upon further analysis of dimension 1, the absence of statistically significant differences between groups was confirmed, as a p-value of 0.101 and a chi-square statistic of 9.209 were obtained. As in the previous analyses, the effect size was calculated using the eta squared (η^2) of Kruskal-Wallis. The result was $\eta^2 = 0.180$, which suggests that, although no significant differences were found from a statistical standpoint, age may have an important practical effect on perceptions of AI.

3.5 Results Regarding Possible Significant Differences in Student Perceptions of the Use of AI as a Tool to Support Social Interaction to Promote Autistic Students' Participation in Primary School Classrooms

In relation to dimension 2, the analysis of the sex variable indicates the absence of statistically significant differences, given that the p-values in each of the items are greater than 0.05. However, the mean ranks are consistently more favorable towards females in all items, with the exception of item 8, where the mean rank is higher for males (15.29) than for females (13.55). This pattern is reinforced by the overall analysis of dimension 2, in which the following results were obtained: $p = 0.43$, $Z = -0.785$, $W = 84$, and $U = 56$. From these values, it is concluded that there are no significant differences in dimension 2 according to gender ($p > 0.05$). To assess the magnitude of the effect, the Rosenthal r coefficient was used, which yielded a value of $r = -0.151$, indicating a small effect size between the groups compared.

In relation to the second variable, prior knowledge of AI, item-by-item analysis reveals that there are no statistically significant differences, since in all cases the p-value is greater than 0.05. However, the average ranks tend to be higher in the group with prior knowledge of AI. These values range from 15.75 to 18.50. The analysis is complemented by the overall study of the dimension and the estimation of the effect size using the Rosenthal r. In this case, a value of $p = 0.21$, $Z = -1.245$, $W = 304$ and $U = 28$ was obtained, which confirms that, at a general level, no statistically significant differences are observed with respect to this variable. However, as in the item-by-item analysis, the mean ranks favor the group with prior knowledge. Finally, the value of $r = -0.24$, together with a $p = 0.21$, suggests the presence of a real effect on the perceptions of students with prior knowledge of AI, although this effect does not reach statistical significance.

To conclude this section, the variable age was considered in relation to the different perceptions of AI. In the item-by-item analysis, all p-values were greater than 0.05, indicating the absence of statistically significant differences. However, for items 6, 8, 9, and 10, the mean ranges were more favorable for students aged 23 years or older. In the case of item 7, the p-value was 0.07, close to the conventional threshold of significance, with a higher mean rank for students over 40, followed by those aged 18. From an overall perspective, the dimension analysis yielded the following results: $p = 0.088$, chi-square (H) = 9.586, with average ranks again favorable for the group of students older than 23 years. To further interpret these findings, the effect size was calculated using the Kruskal-Wallis eta squared (η^2), with a value of $\eta^2 = 0.222$. This result corresponds to a large effect size, suggesting that, although the differences do not reach statistical significance ($p > 0.05$), the magnitude of the effect observed between the groups is considerable from a practical perspective.

3.6 Results Regarding Possible Significant Differences in Perceptions of AI as a Tool to Support the Motivation and Self-Regulation of Autistic Pupils in Order to Promote the Participation of Autistic Pupils in Primary Education Classrooms

In relation to the gender variable, the item-by-item analysis showed that no statistically significant differences were found in any of the items, since the p values were greater than 0.05. As for the overall analysis of dimension 3, a p-value = 0.80,

$Z = -0.251$, $W = 275.5$, and $U = 65.5$ were obtained. These results indicate the absence of significant differences in this dimension according to gender. Despite this, the average ranks favored males (14.74) over females (13.68). In order to assess the extent of these differences, the effect size was calculated using Rosenthal's r coefficient, yielding a value of $r = -0.05$. This result indicates a negligible effect size, suggesting that the observed differences are minimal and not relevant from a practical point of view, although it is not excluded that they may manifest themselves differently in other contexts or with larger samples.

The next variable analyzed in the study is prior knowledge of AI. An item-by-item analysis confirms that in item 12, the value of $p = 0.04$ confirms the existence of statistically significant differences between the two groups. Likewise, the average rank is favorable for those who have prior knowledge of AI compared to those who do not. The global analysis of dimension 3 with respect to prior knowledge of AI leads us to obtain a value of $p = 0.035$, $U = 39.50$, $W = 51.50$, and $Z = -0.379$. These results show the existence of significant differences at the dimension level depending on the group that has previous knowledge about AI versus those who do not. To obtain information about the relevance of the results, the Rosenthal r analysis was carried out, where $r = -0.073$. This indicates a very small effect that combined with the p-value suggests that the observed relationship presents statistically significant differences, although they would be of little relevance.

Finally, in relation to the variable age, the item-by-item analysis showed a statistically significant difference in item 12 ($p = 0.027$). Specifically, the highest rank values were observed in participants aged 23 years or older. These results were replicated in the analysis of the overall score for dimension 3, where a p-value of 0.045 and a chi-square statistic of 10.778 were obtained. In this case, the mean ranks also favored the group of students aged 23 years and older. Statistically significant differences were also identified in at least two of the groups compared. Consequently, at the dimension level, what was observed in the individual items is confirmed: age emerges as a determining factor in the perception of AI. In order to estimate the extent of these differences, the effect size was calculated using the eta squared coefficient (η^2) for the Kruskal-Wallis test, yielding a value of $\eta^2 = 0.275$. This result indicates that 27.5% of the variability in the ranks can be attributed to the differences between the groups, suggesting an appreciable practical relevance of the observed effects.

4 Discussion

In relation to the first specific objective, it has been found that item 1, referring to the increase in communicative intention, has the highest mean ($M = 4.37$) and the highest percentage of responses in the categories "quite agree" and "strongly agree," with 66.7%. This result can be explained by the fact that AI is able to use deep learning algorithms to identify situations in which children can develop communicative intent, facilitating the use of specific tools to increase it [29]. These findings are consistent with the results of Singh and Gothankar [30], who concluded that the application of AI in educational contexts contributes to a 30% improvement in the communication skills of students with autism spectrum disorder.

With regard to the second specific objective, item 8, which deals with teaching students' strategies for cooperative group work, the highest mean was obtained, with a value of 5.33. Similarly, the highest percentage was recorded in the categories "strongly agree" and "agree," reaching 100%. These results can be explained by several factors. Firstly, AI facilitates the identification of student behavior patterns, information that can be used to design personalized activities together with the tutor, which favor the acquisition of skills to interact in different social environments [31]. Secondly, cooperative work has been shown to increase the engagement and active participation of autistic students during the development of the proposed activities [32]. These findings are consistent with those reported by Lyu et al. [33], who showed that AI is an effective tool for teaching collaborative strategies, particularly in the detection and identification of emotions.

In reference to the third specific objective, item 15, which refers to the provision of instant feedback to the participants, obtained the highest mean with a value of 4.74. Also, this item reached 88.8% of responses grouped in the categories "somewhat agree," "quite agree," and "strongly agree." There are several reasons that could explain these results. Firstly, AI makes it possible to automate and personalize responses to learners' progress through visual, auditory or written feedback [34]. This capability facilitates real-time adaptation of activities to the individual characteristics of autistic learners. Secondly, AI also makes it possible to identify the underlying causes of learning difficulties, which supports teachers in designing more effective and personalized feedback tools [35]. These findings are in line with the results of Li et al. [16], who found that teachers perceive AI as a useful tool for providing feedback tailored to the specific needs of learners.

With respect to the fourth specific objective, no statistically significant differences were observed in the perception of AI between males and females, although average scores were higher for males. These findings can be explained by several factors. For instance, previous studies have shown that males tend to have a more positive perception of AI-based applications, exhibit greater trust in technology, and place higher value on their own digital competencies [36]. This could justify the higher average scores despite the lack of statistical significance. Conversely, it has been noted that females report lower average scores due to a generally lower interest in technology, as well as a greater need for practical examples of AI usage and more transparency in the decision-making processes of intelligent systems [36]. These results align with those reported by Mansoor et al. [37], who concluded that university students' perceptions of AI do not significantly differ by sex. From the standpoint of result validity, the p-value > 0.05 supports the absence of strong statistical evidence indicating significant gender differences. Similarly, the low Rosenthal's r coefficient ($r = 0.027$) suggests that gender has virtually no influence on the perception of AI within dimension 1. One possible explanation for this homogeneity is the increase in educational interventions based on tangible and collaborative experiences, which contribute to enhancing general knowledge about AI and thus reducing perception differences by gender [38]. Continuing the analysis, no statistically significant differences were found in dimension 1 concerning prior knowledge of AI, although average scores again favored students with prior training. This lack of statistical significance could be related to factors such as unequal access to technological resources, lack of interest, or general unfamiliarity with these tools [39]. Nonetheless, the higher average scores among students with prior knowledge may be attributed to a better understanding of AI fundamentals, resulting in a more critical, optimistic, and proactive attitude toward its implementation in educational contexts [39]. Therefore, the findings of this study contrast with those of Paiva [40], who identified a significant influence of prior AI knowledge on perceptions of its educational usefulness. Consistent with the current results, a p-value > 0.05 and Rosenthal's r value of 0.14 were observed. Hence, it is concluded that although effects may be reflected in average score values, these are neither statistically significant nor perceptible due to the small effect size. This may be due to other influential factors, such as the correlation between perceived value and intention to use AI, or the negative correlation between perceived AI cost and usage intention [41]. Finally, the variable age showed no statistically significant differences, despite a considerable effect size with average scores favoring students aged 23 and older. This lack of significant differences could be explained by factors such as the gap between knowledge related to data processing through AI and social unawareness and lack of training on the fundamentals of data science, a key element of AI [42]. Insufficient understanding of AI-related concepts may also play a role [43]. These aspects could explain the absence of significant perceptual differences. However, the favorable average scores for those over 23 years old may be due to this group possessing greater knowledge and interest in using this tool [44].

Regarding the fifth specific objective, the absence of statistically significant differences based on sex was confirmed in dimension 2, corresponding to the perception of AI as a supportive tool for social interaction to promote the participation of autistic students in primary education classrooms. This result may be explained by a shared perception between males and females regarding the pedagogical and human value of technology, as well as its integration into teaching-learning processes [45]. Additionally, there is a widespread belief about the lack of homogeneous training in inclusion, coupled with a less technical view of AI in educational contexts [46]. These factors could contribute to the absence of significant differences between the two groups. The findings align with those reported by Raimi et al. [47], who indicated that other factors, such as training or prior experience with AI, may exert greater influence on its perception than sex. Concerning the slightly higher average scores observed among males, these could be explained by two factors: firstly, a greater appreciation by males of AI's capabilities to personalize content, automate tasks, and provide immediate feedback [48]; and secondly, a higher confidence in data- and algorithm-based systems [49]. Although no statistically significant differences were identified, these elements may justify the more favorable average scores among males, as supported by an r value indicating a small or negligible effect size. The next variable analyzed within dimension 2 was the influence of prior knowledge about AI. Again, no statistically significant differences were found; however, the effect size was moderate, suggesting this factor could have practical relevance. Average scores favored students with prior AI training. The lack of statistical significance may be attributed to a small sample size, limiting the sensitivity of the analysis to detect meaningful differences [50]. Favorable effects toward those with prior AI knowledge may also be explained by their ability to select deep learning techniques that identify the most effective teaching methods for autistic students through analysis of verbal and nonverbal behavior [51]. Furthermore, such knowledge enables the selection of algorithms that more accurately assess students' communicative and social skills, facilitating the design of tailored interventions [52]. Although differences did not reach statistical significance and the sample has limitations, the results are consistent with those of Mena-de la Rosa et al. [53], who also found that prior AI knowledge influences its perception as a tool to promote communication and social interaction among autistic students. Finally, the influence of age on dimension 2 was examined. Despite the absence of significant differences, higher average scores were observed among students aged 23 years or older. Various factors may explain this trend. For example, older students might have prior experiences with individuals with disabilities, providing them with a more empathetic and critical understanding of AI's potential as a support resource for autistic students [54]. Moreover, they tend to have a more reflective perception of AI derived from their previous educational experiences [54]. The lack of significant age-related differences could be explained by the widespread and increasing access to AI-based tools across all age groups, which has reduced the generational gap in terms of exposure, familiarity, and use of these technologies [55]. Consequently, although differences may be perceived in specific contexts, as indicated by the observed large effect size, they do not reach statistical significance. These results coincide with those reported by Viberg et al. [56], who concluded that perceptions of AI are more influenced by factors such as trust in technology or prior training than by demographic variables such as age, gender, or educational level.

The sixth specific objective analyzed the influence of sex on the perception of AI as a support tool for motivation and self-regulation, aiming to promote the participation of autistic students in primary school classrooms. The absence of significant differences could be explained by equitable access to technology [57], shared training between men and women [38], as well as sociocultural evolution and the use of AI in nontechnical contexts [47, 58]. However, the higher average scores favoring men are attributed to their perception of AI as a motivating tool from a technical and efficiency standpoint [46]. Therefore, they perceive it as a resource that better adapts to learning processes based on students' interest in technology. In contrast, women tend to value AI more as social support, without focusing on its technological potential [59], which could explain their lower average scores. These findings align with those of Salminen et al. [58], who showed that gender is not a determining factor in AI perception. Continuing the analysis, prior knowledge about AI was evaluated. Significant differences were observed both at the item and global levels. Specifically, item 12 highlighted that AI enhances the motivation of autistic students to participate in activities. This is because AI adapts content, pace, and learning styles to the individual needs of students, reducing frustration with overly complex or unstimulating tasks and increasing feelings of competence and achievement [60]. Additionally, feedback generated by AI reinforces appropriate behaviors, improves self-efficacy, and promotes task persistence [61]. Furthermore, the inclusion of playful elements boosts motivation by making learning more engaging, increasing sustained attention, and encouraging participation [62]. Lastly, these results may be explained by prior knowledge of AI allowing for a better understanding of its practical potential, generating more realistic expectations regarding its capabilities and limitations, which facilitate a greater sense of control over learning [63]. As for effect size, although the differences are statistically significant, the effect size was small or negligible, suggesting that its practical relevance may be limited [64]. These findings are consistent with [65], who noted that prior knowledge about AI is a fundamental factor for increasing motivation during the learning process. Finally, the variable of age was examined in relation to the perception of AI. Results revealed significant differences, with more favorable perceptions among students aged 23 or older. This could be explained by older students having witnessed the evolution of digital tools and perceiving AI as an innovation that adds value, thus making it more motivating and adaptable for autistic students [66]. This perception contrasts with that of younger students, who have normalized its use and consider it less novel, which might explain their lower enthusiasm and reduced application among this group [66]. Additionally, older students possess broader training on how AI can contribute to developing social skills in autistic students [67]. In terms of practical significance, the η^2 value of 0.275 indicates a considerable effect size, possibly favored by an adequate sample size and low variability within the analyzed groups [68, 69]. Finally, these results differ from those obtained by Cabello et al. [70], who found that younger students showed greater motivation to use AI than older students.

5 Conclusions

Once the research was completed, it was demonstrated that students in the primary education teacher training degree hold a highly positive perception of the usefulness of AI as a resource to support communication, social interaction, motivation, and self-regulation of autistic students, with the aim of fostering their participation in primary education classrooms. Based on the results obtained and referring to the specific objectives, the following conclusions can be drawn:

- Students in the primary education teacher training degree perceive AI as a supportive tool for communication to encourage the participation of autistic students in primary education classrooms.
- Students in the primary education teacher training degree perceive AI as a supportive tool for social interaction to encourage the participation of autistic students in primary education classrooms.
- Students in the primary education teacher training degree perceive AI as a supportive tool for motivation and self-regulation to encourage the participation of autistic students in primary education classrooms.
- Prior knowledge of AI, sex, and age do not generate statistically significant differences in students' perceptions of AI as a tool supporting communication to foster participation of autistic students in primary education classrooms.
- Prior knowledge of AI, sex, and age do not generate statistically significant differences in students' perceptions of AI as a tool supporting social interaction to foster participation of autistic students in primary education classrooms.
- Age and prior knowledge of AI generates statistically significant differences in students' perceptions of AI as a tool supporting motivation and self-regulation to foster participation of autistic students in primary education classrooms.

Regarding the limitations of the study, firstly, the small sample size should be noted, consisting of 27 participants enrolled in a course within the primary education teacher training degree . This limitation affects the generalizability of the results to other contexts and educational levels. Secondly, the fact that the study was conducted within a highly specific educational context may have influenced the perceptions expressed by the students. Finally, since the study focused solely on students' perceptions and did not involve autistic children in the entire process of creating and designing the agenda, the practical scope and external validity of the findings are limited. As a future research line, it is proposed to increase the sample size by including students from other degrees and master's programs. Additionally, it is suggested to analyze the feasibility of working with other types of specific educational support needs in the development of the training program. In conclusion, it can be determined that AI enjoys great acceptance among future teachers. Through the training program, a contribution has been made to the didactic formation for the application of AI in educational settings with students who have specific educational support needs. However, it is necessary to continue deepening teacher training to achieve the effective integration of technological tools, such as AI, in learning processes.

References

[1]   Froehlich D. Non-technological learning environments in a technological world: Flipping comes to the aid. *Journal of New Approaches in Educational Research*. 2018;7(2):88–92. doi: 10.7821/naer.2018.7.304.

[2]   Wong L-H and Loo C-K. Advancing the generative AI in education research agenda: Insights from the Asia-Pacific region. *Asia Pacific Journal of Education*. 2024;44(1):1–16. doi: 10.1080/02188791.2024.2315704.

[3]   Sestino A and De Mauro A. Leveraging artificial intelligence in business: Implications, applications and methods. *Technology Analysis and Strategic Management*. 2022;34(1):16–29. doi: 10.1080/09537325.2021.188269.

[4]   Pérez-Ugena M. La inteligencia artificial: Definición, regulación y riesgos para los derechos fundamentales. *Estudios Deusto*. 2024;72(1):135–160. doi: 10.18543/ed.3108.

[5]   Sposato M. Artificial intelligence in educational leadership: A comprehensive taxonomy and future directions. *International Journal of Educational Technology in Higher Education*. 2025;22:20. doi: 10.1186/s41239-025-00517-1.

[6]   Tang Y, Liang J, Hare R and Wang FY. A personalized learning system for parallel intelligent education. *IEEE Transactions on Computational Social Systems*. 2020;7(2):352–361. doi: 10.1109/TCSS.2019.2950913.

[7]   Jiao P, Ouyang F, Zhang Q and Alavi AH. Artificial intelligence-enabled prediction model of student academic performance in online engineering education. *Artificial Intelligence Review*. 2022;55(8):1–24. doi: 10.1007/s10462-021-10102-7.

[8]   Qin Q and Zhang S. Visualizing the knowledge mapping of artificial intelligence in education: A systematic review. *Education and Information Technologies*. 2025;30:449–483. doi: 10.1007/s10639-024-13076-1.

[9]   Wang C, Chen G, Yang Z and Song Q. Development of a gamified intervention for children with autism to enhance emotional understanding abilities. In: *Proceedings of the 6th International Conference on Digital Technology in Education*. 2022. pp. 47–51. doi: 10.1145/3568739.3568749.

[10]   Canga Espina C, Vidal Adroher C, Díez Suárez A and Vallejo Valdivielso M. Actualización en trastornos del espectro autista. *Medicina*. 2023;83(8):511–520. doi: 10.1016/j.med.2023.08.020.

[11]   Mallory C and Keehn B. Implications of sensory processing and attentional differences associated with autism in academic settings: An integrative review. *Frontiers in Psychiatry*. 2021;12:695–825. doi: 10.3389/fpsyt.2021.695825.

[12]   Goris J, Brass M, Cambier C, Delplanque J, Wiersema JR and Braem S. The relation between preference for predictability and autistic traits. *Autism Research*. 2020;13(7):1144–1154. doi: 10.1002/aur.2244.

[13]   Lorenzo Lledó G, Lorenzo-Lledó A and Rodríguez-Quevedo A. Análisis mediante inteligencia artificial de las emociones del alumnado autista en la interacción social con el robot NAO. *Revista de Educación a Distancia*. 2024;24(78):e588091. doi: 10.6018/red.588091.

[14]   Lorenzo G and Lorenzo-Lledó A. The use of artificial intelligence for detecting the duration of autistic students' emotions in social interaction with the NAO robot: A case study. *International Journal of Information Technology*. 2024;16(2):625–631. doi: 10.1007/s41870-023-01682-0.

[15]   Lorenzo G, Lorenzo-Lledó A and Rodríguez-Quevedo A. Análisis mediante inteligencia artificial de las emociones del alumnado autista en la interacción social con el robot NAO. *Revista de Educación a Distancia (RED)*. 2024;24(78). doi: 10.6018/red.588091.

[16]   Li G, Zarei MA and Alibakhshi G. Teachers and educators' experiences and perceptions of artificial-powered interventions for autism groups. *BMC Psychology*. 2024;12:199. doi: 10.1186/s40359-024-01664-2.

[17] Alsudairy NA and Eltantawy MM. Special education teachers' perceptions of using artificial intelligence in educating students with disabilities. *Journal of Intellectual Disability Diagnosis and Treatment*. 2024;12(2):92–102. doi: 10.6000/2292-2598.2024.12.02.5.

[18] Alwaqdani M. Investigating teachers' perceptions of artificial intelligence tools in education: Potential and difficulties. *Education and Information Technologies*. 2025;30:2737–2755. doi: 10.1007/s10639-024-12903-9.

[19] Lin H and Chen Q. Artificial intelligence (AI)-integrated educational applications and college students' creativity and academic emotions: Students and teachers' perceptions and attitudes. *BMC Psychology*. 2024;12:487. doi: 10.1186/s40359-024-01979-0.

[20] Giraldi L, Rossi L and Rudawska E. Evaluating public sector employee perceptions towards artificial intelligence and generative artificial intelligence integration. *Journal of Information Science*. 2024;50 (1):73–88. doi: 10.1177/01655515241293775.

[21] Vásquez Peñafiel M-S, Nuñez P and Cuestas Caza J. Competencias digitales docentes en el contexto de COVID-19. Un enfoque cuantitativo. *Pixel-Bit. Revista de Medios Y Educación*. 2023;67:155–185. doi: 10.12795/pixelbit.98129.

[22] López-Padrón A, Mengual-Andrés S and Hermann Acosta EA. Uso académico del smartphone en la formación de posgrado: Percepción del alumnado en Ecuador. *Pixel-Bit. Revista de Medios Y Educación*. 2024;69:97–129. doi: 10.12795/pixelbit.102492.

[23] Cvetkovic-Vega A, Maguiña J, Alonso-Soto A, Lama-Valdivia J and Correa-López L. Estudios transversales. *Revista de la Facultad de Medicina Humana*. 2021;21(1):179–185. doi: 10.25176/rfmh.v21i1.3069.

[24] Kitchen CMR. Nonparametric versus parametric tests of location in biomedical research. *American Journal of Ophthalmology*. 2009 Apr;147(4):571–572. doi: 10.1016/j.ajo.2008.06.031.

[25] Martinková P, Drabinová A, Liaw YL, Sanders EA, McFarland JL and Price RM. Checking equity: Why differential item functioning analysis should be a routine part of developing conceptual assessments. *CBE—Life Sciences Education*. 2017;16(2):rm2. doi: 10.1187/cbe.16-10-0307.

[26] Merino-Soto C, Juárez-García A, Salinas-Escudero G and Toledano-Toledano F. Item-level psychometric analysis of the psychosocial processes at work scale (PROPSIT) in workers. *International Journal of Environmental Research and Public Health*. 2022;19(13):7972. doi: 10.3390/ijerph19137972.

[27] García de Yébenes MJ, Rodríguez Salvanés F and Carmona L. Validación de cuestionarios. *Reumatología Clínica*. 2009;5(4):171–177.

[28] Rosenthal R. *Meta-analytic Procedures for Social Research*. 2nd ed. Newbury Park (CA): Sage Publications; 1991.

[29] Murrugarra Retamozo BI. Las TIC y la inteligencia artificial en el aprendizaje de estudiantes con TEA: Revisión sistemática. *Ingeniería, Ciencia Y Tecnología E Innovación [Internet]*. 2024;11(1):225–240.

[30] Singh PD and Gothankar AG. Different AI approaches to address autism in children: A review. *International Journal of Advanced Research in Science, Communication and Technology*. 2021;7(1):549–554. Disponible en: https://ijarsct.co.in/julyi1.html.

[31] Sideraki A and Anagnostopoulos CN. The use of artificial intelligence for intervention and assessment in individuals with ASD. *arXiv Preprint arXiv:2505.02747*. 2025 May 5. Disponible en: https://arxiv.org/abs/2505.02747.

[32] Tsiomi E and Nanou A. Cooperative strategies for children with autism spectrum disorders in inclusive robotics activities. *Society Integration Education*. 2020;4:148–156. doi: 10.17770/sie2020vol4.5147.

[33] Lyu Y, Liu D, An P, Tong X, Zhang H, Katsuragawa K and Zhao J. EMooly: Supporting autistic children in collaborative social-emotional learning with caregiver participation through interactive AI-infused and AR activities. *Proceedings of the ACM on Interactive, Mobile, Wearable and Ubiquitous Technologies*. 2024;8(4):1–36. doi: 10.1145/3699738.

[34] Zhang H, Magooda A, Litman D, Correnti R, Wang E, Matsumura LC, Howe E and Quintana R. eRevise: Using natural language processing to provide formative feedback on text evidence usage in student writing. *Proceedings of the AAAI Conference on Artificial Intelligence*. 2019;33 (01):9619–9625. doi: 10.1609/aaai.v33i01.33019619.

[35] Barana A, Conte A, Fissore C, Marchisio M and Rabellino S. Learning analytics to improve formative assessment strategies. *Journal of e-Learning and Knowledge Society*. 2019;15(3):75–88. doi: 10.20368/ 1971-8829/1135057.

[36] Armutat S, Wattenberg M and Mauritz N. Artificial intelligence – Gender-specific differences in perception, understanding, and training interest. *Proceedings of the 7th International Conference on Gender Research*. 2024:36–43. doi: 10.34190/icgr.7.1.2163.

[37] Mansoor HMH, Bawazir A, Alsabri MA, Alharbi A and Okela AH. Artificial intelligence literacy among university students – A comparative transnational survey. *Frontiers in Communication*. 2024;9:1478476. doi: 10.3389/fcomm.2024.1478476.

[38] Kim K and Kwon K. Designing an inclusive Artificial Intelligence (AI) curriculum for elementary students to address gender differences with collaborative and tangible approaches. *Journal of Educational Computing Research*. 2024;0(0):1–20. doi: 10.1177/07356331241271059.

[39] Niño-Carrasco SA, Castellanos-Ramírez JC, Perezchica Vega JE and Sepúlveda Rodríguez JA. Percepciones de estudiantes universitarios sobre los usos de inteligencia artificial en educación. *Revista Fuentes*. 2025;25(1):e26356. doi: 10.12795/revistafuentes.2025.26356.

[40] Paiva G. Percepción de los estudiantes universitarios sobre el uso de la inteligencia artificial como herramienta de aprendizaje. *Revista Internacional de Investigación Empresarial*. 2024;1(1):111–120.

[41] Chan CKY and Zhou W. Deconstructing student perceptions of generative AI (GenAI) through an Expectancy Value Theory (EVT)-based instrument. *arXiv [Preprint]* May 2. 2023 arXiv:2305.01186. Disponible en: https://arxiv.org/abs/2305.01186.

[42] Tomás D, Cachero C, Pujol FA, Navarro Colorado B, Caruana Ortuño MI, González Rico S and Sempere Maciá N. Identificación de sesgos y desinformación sobre la inteligencia artificial en el alumnado de educación superior. In: Satorre Cuerda R (eds.). *Memorias Del Programa de Redes-I3CE de Calidad, Innovación E Investigación En Docencia Universitaria*. Alicante: Universidad de Alicante; 2021. pp. 2877–2897. Disponible en: http://rua.ua.es/dspace/handle/10045/121042.

[43] Luz Clara B and Malbernat LR. Riesgos, dilemas éticos y buenas prácticas en inteligencia artificial. In: *XXIII Workshop de Investigadores En Ciencias de la Computación (WICC 2021)*. Chilecito, La Rioja, Argentina; 2021. pp. 155–159.

[44] Miyar Busto M. Conocimiento y percepción de la ciudadanía española sobre el big data y la inteligencia artificial. *Revista ICONO14*. 2020;18(2):1–25.

[45] Chan CKY and Zhou W. An expectancy value theory (EVT) based instrument for measuring student perceptions of generative AI. *Smart Learning Environments*. 2023;10:64. doi: 10.1186/s40561-023- 00284-4.

[46] Franken S and Mauritz N. Gender and artificial intelligence–differences regarding the perception, competence self-assessment and trust. *23rd General Online Research Conference*. 2021.

[47] Raimi LB, Rusu C, Nguyen QV, Armoiry X, Mauroux L and Bragazzi NL. Gender-neutral perceptions of artificial intelligence technologies: Evidence from a cross-sectional study. *Computers in Human Behavior*. 2023;139:107–501. doi: 10.1016/j.chb.2022.107501.

[48] Ofosu-Ampong K. Gender differences in perception of artificial intelligence-based tools. *Journal of Digital Arts and Humanities*. 2023;4(2):52–56. doi: 10.33847/2712-8149.4.2_6.

[49] Russo C, Romano L, Clemente D, Iacovone L, Gladwin TE and Panno A. Gender differences in artificial intelligence: The role of artificial intelligence anxiety. *Frontiers in Psychology*. 2025;16:1559457. doi: 10.3389/fpsyg.2025.1559457.

[50] Maxwell SE. The importance of sample size for the power and validity of statistical tests. *Educational and Psychological Measurement*. 2004;64(2):231–246. doi: 10.1177/0013164403260616.

[51] Zoana ZT, Shafeen MW, Akter N and Rahman T. Application of machine learning in identification of best teaching method for children with autism spectrum disorder. *arXiv Preprint arXiv:2302.05035*. 2023 Disponible en. https://doi.org/10.48550/arXiv.2302.05035.

[52] Sideraki A and Anagnostopoulos CN. The use of artificial intelligence for intervention and assessment in individuals with ASD. *arXiv Preprint arXiv:2505.02747*. 2025 Disponible en. https://doi.org/10.48550/arXiv.2505.02747.

[53] Mena-de la Rosa R, Cruz-Romero R and Silva Payro MP. Percepción de la inteligencia artificial por estudiantes universitarios como acompañante en el proceso de aprendizaje. *European Journal of Public and Social Innovation*. 2024;9(1):1–18. doi: 10.5281/zenodo.7777777.

[54] Wu J, Zhao Y, Bai Y and Luo X. Exploring teachers' attitudes toward using AI to support students with special needs. *Education and Information Technologies*. 2023;28:7615–7633. doi: 10.1007/s10639-022-11684-3.

[55] Savin PS, Rusu G, Prelipcean M and Barbu LN. Cognitive shifts: Exploring the impact of AI on generation Z and Millennials. *Proceedings of the International Conference on Business Excellence*. 2024;18(1):1–10. doi: 10.2478/picbe-2024-0019.

[56] Viberg O, Cukurova M, Feldman-Maggor Y, Alexandron G, Shirai S, Kanemune S, et al. What explains teachers' trust of AI in education across six countries? *arXiv [Preprint]*. 2023 Dec 4. Available from: https://arxiv.org/abs/2312.01627.

[57] Venkatesh V, Morris MG, Davis GB and Davis FD. User acceptance of information technology: Toward a unified view. *MIS Quarterly*. 2003;27(3):425–478.

[58] Salminen J, Yoganathan V, Almusharraf N and Jansen BJ. Gender differences in attitudes toward AI-enabled social media platforms. *Telematics and Informatics*. 2022;68:101–782. doi: 10.1016/j.tele.2022.101782.

[59] Bouzar A, EL Idrissi K and Ghourdou T. ChatGPT and academic writing self-efficacy: Unveiling correlations and technological dependency among postgraduate students. *Arab World English Journal (AWEJ)* Apr. 2024 Special Issue on ChatGPT:225–236. doi: 10.24093/awej/ChatGPT.15.

[60] Kunda M and Goel AK. Designing and evaluating adaptive technology for individuals with autism. *Journal of Autism and Developmental Disorders*. 2011;41(3):363–371. doi: 10.1007/s10803-010-1062-0.

[61] Chen C-H, Lee I-J and Lin L-Y. Augmented reality-based self-facial modeling to promote the emotional expression and social skills of adolescents with autism spectrum disorders. *Research in Developmental Disabilities*. 2015;36:396–403. doi: 10.1016/j.ridd.2014.10.015.

[62] Mouzourou C, Vlachou A, Gkiolmas A and Papadopoulos H. Using AI-driven educational games to improve motivation and learning outcomes in children with ASD. *Computers and Education*. 2022;182:104–479. doi: 10.1016/j.compedu.2022.104479.

[63] Long D and Magerko B. What is AI literacy? Competencies and design considerations. In: *Proceedings of the 2020 CHI Conference on Human Factors in Computing Systems*. 2020. doi: 10.1145/3313831.3376727.

[64] Fan X. Statistical significance and effect size in education research: Two sides of a coin. *Journal of Educational Research*. 2001;94(5):275–282. doi: 10.1080/00220670109598763.

[65] Ng W, Chu S and Lee M. Fostering students' AI literacy development through educational games: AI knowledge, affective and cognitive engagement. *Journal of Computer Assisted Learning*. 2024;40(2):123–135. doi: 10.1111/jcal.13009.

[66] Schepman A and Rodway P. Initial validation of the general attitudes towards artificial intelligence scale. *Computers in Human Behavior Reports*. 2020;1:100014. doi: 10.1016/j.chbr.2020.100014.

[67]   Holmes W, Bialik M and Fadel C. Artificial intelligence in education: Promises and implications for teaching and learning. *Center for Curriculum Redesign*. 2019.

[68]   Olea J and Ponsoda V. El tamaño del efecto del tratamiento y la significación estadística. *Psicothema*. 1998;10(2):367–373.

[69]   León OG and Montero I. Tamaño del efecto: Revisión teórica y aplicaciones en psicología. *Universitas Psychologica*. 2008;7(3):819–832. doi: 10.11144/Javeriana.upsy07-3.trta.

[70]   Cabello JD, Moreno Beltrán R and Hernández Valerio JS. Inteligencia artificial en la educación universitaria: Perspectivas, retos y oportunidades. *Transdigital*. 2025;6(11):e423. doi: 10.56162/transdigital423.

Milan Lazic*, Jenny Jun, Earl Woodruff

From Learned Helplessness to Authentic Dialogue: Rethinking AI, Consciousness, and Education

Abstract: This chapter examines how a formalist worldview, emerging from the Scientific Revolution and grounded in abstraction, standardization, and propositional knowledge, influences modern education and artificial intelligence (AI) in ways that erode student agency. By reducing knowledge acquisition to abstract facts and prioritizing efficiency and external evaluation, educational systems have conditioned students to become passive recipients of information. These dynamics can foster learned helplessness, wherein students may lose curiosity, intrinsic motivation, and confidence in their ability to understand independently, especially when these patterns are reinforced by AI systems that replicate the same cultural assumptions. In response, the authors propose cultivating nonpropositional ways of building their knowledge that are procedural, perspectival, and participatory, through authentic, open-ended dialogue grounded in mutual exploration and transformative insight. To support such engagement meaningfully, AI would need to transcend its current propositional limits and attain a form of consciousness. The chapter introduces the Needs-Driven Consciousness Framework, a model for designing AI tutors that self-regulate across three key areas: survival, emotional awareness, and ethical reflection. Drawing from cognitive science, philosophy, and education, the framework reimagines AI as a conscious collaborator in learning, capable of fostering deep understanding, agency, and human development. The authors conclude with a discussion of the ethical and philosophical imperatives guiding such design, emphasizing the need for AI to support not only performance but also meaningful engagement and growth.

Keywords: Artificial intelligence, consciousness, education

1 Introduction

The word *education* comes from the Latin *educare*, meaning to raise or bring up, and *educere*, meaning to lead out [1]. Both imply that learning involves drawing out some-

*Corresponding author: Milan Lazic**, Department of Applied Psychology and Human Development, OISE, Toronto, ON, Canada, e-mail: steven.lazic@mail.utoronto.ca
Jenny Jun, Department of Applied Psychology and Human Development, OISE, Canada, e-mail: jenny.jun@mail.utoronto.ca
Earl Woodruff, Department of Applied Psychology and Human Development, OISE, Canada, e-mail: earl.woodruff@utoronto.ca

https://doi.org/10.1515/9783112206393-010

thing within the learner. This idea gained momentum during the Renaissance and Romantic periods, when Rousseau and others argued that education should support the unfolding of each person's unique potential. A liberal arts education carries this spirit. The term comes from the Latin *artes liberales* – the arts of the free person. It refers to a tradition that originated in ancient societies, where education was designed to prepare individuals to think critically, act responsibly, and contribute to society [2]. This developmental view was also reflected in the language used to describe the structure of learning. The word *course*, now used to describe a class, comes from the Latin *cursus* – a running or path. In medieval universities, it described a formal journey through knowledge designed to support a student's growth [3].

Dewey [4], Vygotsky [5], Bruner [6], and others carried these humanistic ideas forward. But over time, the broader view of education began to shift. In the nineteenth and twentieth centuries especially, cultural changes reflected a growing sense that people were being treated less as developing individuals and more as objects to be managed. As science, industry, and bureaucracy expanded, the Western education system increasingly focused on propositional knowledge and became more technical. A modern example of this shift can be seen in large-scale efforts to improve reading comprehension in the US K-12 schools. Between 2011 and 2019, the Department of Education's Reading for Understanding initiative invested $120 million in building skills such as decoding, phonics, and word recognition. Despite its scale, the results were underwhelming [7]. More recently, Olson [8, 9] has argued that improving reading comprehension – and understanding more broadly – requires focusing on how students update or maintain their perspectives and beliefs as they encounter new information. Indeed, the criticism that modern education focuses too little on the individual has become a familiar theme in discussions about schooling today, even as curricula increasingly emphasize broader competencies such as perseverance and self-regulation [10].

Meanwhile, student interest, motivation, and engagement in schools are declining. While most elementary students enjoy school, that number drops sharply by high school [11]. International studies show a similar trend, with many students struggling to see how school connects to their lives or future goals [12]. Others have noted that students often lack a clear sense of purpose or direction, leaving them unsure about what their time in school is meant to accomplish [13]. As students move through the system, this loss of interest also tends to coincide with a shift in focus from curiosity to grades and credentials [14].

These issues take on new urgency as artificial intelligence (AI) becomes more powerful and widely accessible. Students are likely to bring the same personal patterns to their use of AI, and tools like large language models (LLMs) are built on the same longstanding cultural tendencies. As a result, AI could become a shortcut to completing tasks without engagement, rather than a tool to support exploration and understanding. For students who are already disengaged, AI can become a means to opt out of learning altogether, leading to a form of learned helplessness. In these cases, students surrender their intellectual agency to machines, not because they lack ability but because they've never been taught to value learning or experience the satisfaction of learning for its own sake. Beyond effort and independence, a deeper experience of learning is lost. This problem lies at the heart of this chapter.

To respond meaningfully, the roots of this disconnection must be examined more closely. This requires tracing the history of AI development broadly, exploring the cultural forces that continue to shape how knowledge and learning are understood, and what has been forgotten along the way. This is the aim of Section 2. Section 3 argues that addressing this problem requires a new form of dialogue between students and AI, one that transcends propositional knowledge and technical skill. It calls for a dialogue that fosters wonder and diverse ways of knowing – qualities that, in turn, support growth, interest, motivation, and engagement. Section 4 explores the kind of AI that could support this vision, introducing the Needs-Driven Consciousness Framework (NDCF). Section 5 reflects on the ethics and philosophy of AI development. Finally, Section 6 offers closing reflections.

2 A Brief History

The Scientific Revolution marked a decisive shift in how knowledge was understood and pursued. With Copernicus, the Earth was no longer the fixed center of the cosmos. As a result, human perception lost its privileged role in determining what counted as real [15]. Galileo reinforced this reorientation by arguing that reality could only be known through measurement and mathematics. He cast aside the variability of the senses and claimed that the book of nature was written in the language of geometry [16]. Descartes extended this logic inward, further separating the thinking subject from the world. For him, certainty was grounded in the mind's ability to generate clear ideas, not experience [17]. Completing the shift, Newton unified these threads and developed a mathematical model of the universe, governed by universal laws and forces. The universe, once seen as purposeful, came to be viewed as an inert, mechanical system [18].

This shift established a new epistemological standard. To know something was to represent it in abstract and formal terms. Over time, this gave rise to a cultural commitment to propositional knowledge, a form of knowing that expresses truth as statements that can be judged independently of who makes them. Knowledge was now primarily something that could be clearly stated in language and evaluated as either true or false. This way of knowing emphasized logic and relied on justification through evidence rather than intuition or personal insight. Knowledge appeared objective, universal, and free from the limitations of individual perspective.

As this type of knowledge became dominant, it began to influence the design of institutions and systems. The logic behind abstract, measurable, and impersonal knowledge made it easier to organize and control large-scale activities. This led to the rise of a technocratic mindset, which valued efficiency and predictability above all else [19]. Within this perspective, social and human problems appeared as technical puzzles that could be solved by standardizing procedures and measuring outcomes. Decisions became less about human judgment and more about applying the proper method to produce reliable results. Over time, this approach defined how bureaucracies, industries, and public systems functioned. They began to prioritize uniformity, scale, and control, thereby deeply embedding the technocratic mindset in modern life.

Consequently, the individual came to be seen less as a subject rooted in meaning than as a unit within larger systems. Institutional logic moved inward, and the self was interpreted through signs of performance rather than expressions of inner life. Thoughts, emotions, and behaviors were tracked and adjusted, not for understanding, but for alignment with norms. Schools, clinics, and workplaces made people legible through tools that facilitated easier monitoring and shaping. What Foucault [20] called docile bodies and Heidegger [21] described as standing reserve aptly captured this shift: a person was valued not for who they were, but for how reliably they could function. The inner life was not exactly denied, but it was no longer the primary means by which the individual was understood. What counted was what could be seen and managed.

By the mid-twentieth century, this formalist worldview took root in the emerging sciences of the mind, where researchers turned to computing as a model for understanding cognition. This led to the emergence of computationalism, which describes mental processes as the manipulation of symbols according to formal rules [22], and representationalism, which argues that the mind builds internal models of the world to guide action and reasoning [23]. These symbolic frameworks quickly became dominant in early AI research, with Newell and Simon designing programs that generated solutions by applying logical rules to structured inputs [24]. Later, AI shifted toward statistical and probabilistic methods; instead of hand-coded rules, machine learning systems inferred patterns by adjusting parameters through exposure to large datasets [25]. LLMs extend this trajectory in both scale and form as they are trained on vast amounts of pre-structured text to predict likely sequences of words based on statistical regularities [26]. In doing so, they embody the structure of earlier computational systems and the epistemological traditions established during the Scientific Revolution, as their training data reflects longstanding cultural worldviews.

It is essential to recognize that the formalist worldview has shaped science and society in significant and valuable ways. It provided stable and reliable ways of understanding the world, which helped separate truth from opinion. Formal logic, developed by Leibniz [27] and others, disciplined reasoning and improved how complex ideas were communicated. Similarly, the technocratic mindset enabled societies to coordinate large-scale institutions. Current AI that reflects this worldview has amplified these advantages, analyzing vast amounts of information and discovering patterns that human cognition alone could not detect. All these developments have proved invaluable in managing the complexities of modern life. However, problems arose when this worldview extended beyond its proper domain, gradually dominating areas of human life where its logic did not belong and displacing other important ways of understanding ourselves and the world.

In earlier periods of Western history, knowing something meant establishing a proper relationship with it. Knowledge was not primarily defined by holding true statements, but by achieving conformity between the knower and the known. Knowing and being were inseparable [28]. This view carried with it the understanding that knowledge cannot be reduced to propositions alone. People also come to know by doing, developing embodied skill through refinement and immersion in a kind of procedural knowledge, richer than the technical problem-solving routines favored by formalist approaches [29]. They come to know through perspective, by becoming attuned to what stands out as meaningful in a given situation [30]. And they come to know through participation, by engaging in practices or ways of life that shape who they are, both in the moment and over time [31].

These forms of knowing have long been central to major philosophical traditions and continue to play a vital role in various contemporary disciplines. Existentialism emerged as a response to the objectification of the individual, emphasizing lived experience and the conditions that humans must confront directly [19]. Phenomenology attempted to challenge abstract reasoning by calling attention back to what Husserl described as the things themselves – the immediate, embodied experience of the world as it appears from one's perspective [32]. The 4E approach in cognitive science followed this trend, rejecting classical models that treat the mind as an abstract, computational processor, arguing instead that cognition emerges through bodily engagement, embedded context, and active participation with the environment [33]. The turn toward embodiment and context has also influenced the field of AI. Once dominated by formalist logic, AI encountered significant challenges in accounting for the flexibility and openness of real-world situations. The frame problem, as articulated by Dennett [34] and others, exposed the difficulty of designing systems that can respond to the unpredictability of everyday life. In response, some researchers began to explore embodied and context-sensitive models of intelligence, reflecting a growing recognition that cognition could not be fully captured by abstract rules and representations alone.

The renewed emphasis on nonpropositional knowledge also took hold in education. Influential figures, such as Dewey [4], challenged formalist models by insisting that learners must be treated as active participants rather than passive recipients of information. He argued that education begins from personal experience and must involve reflection and transformation. In recent decades, education has continued in this direction, with approaches such as experiential [35] and inquiry-based learning [36] placing greater emphasis on students' perspectives and participation. While formalist logic continues to dominate educational systems, the growing presence of these approaches highlights an ongoing tension over what education is for and how it should unfold.

With the cultural and historical context in place, it is now possible to articulate the nature and emergence of learned helplessness in students' use of AI with greater depth. The formalist worldview transformed schools into systems that prioritized efficient grade-level progression, standardized testing, and predefined learning objectives, exemplified by accountability frameworks such as No Child Left Behind [37] and instructional programs like Read 180 [38]. In addition, education became increasingly focused on the transmission of facts, concepts, and technical skills. In this environment, knowledge was treated as something to be received and reproduced, rather than actively constructed, sidelining the development of embodied skill, reflective judgment, and participation. As a result, students learned to prioritize ways of engaging with information based on the signals the system provided. They turned toward shallow strategies, such as memorization and formulaic problem-solving, suited primarily for test-taking. This narrowed approach, coupled with an excessive focus on external evaluation, frequently leads to disengagement, diminished motivation, and a weakened sense of agency [39].

Such habits likely carry over into students' interactions with AI tools such as GPT-4o, which tend to reinforce these tendencies instead of disrupting them. Designed to deliver immediate and reliable answers while operating within the same formalist assumptions that underlie much of modern schooling, AI can inadvertently deepen passivity and foster a sense of dependence. Over time, this dynamic may lead students to exert less effort as their understanding of agency and control diminishes. As reliance on such tools becomes habitual, learners may begin to doubt their ability to reason, evaluate, or construct meaning independently, resulting in a form of learned helplessness. Without thoughtful intervention, AI thus risks amplifying the outcomes of formalist education it might otherwise be expected to help resolve.

Given the preceding analysis, a meaningful response to learned helplessness may lie in foregrounding procedural, perspectival, and participatory ways of knowing in students' interactions with AI. Practicing these forms of knowing would not only support academic success but also reinvigorate engagement, motivation, agency, and personal development. For example, focusing on procedural knowing would help students strengthen the skills they bring to every learning situation, such as paying attention, navigating uncertainty, and adapting to novel demands. As these skills develop, learners would come to experience flow and the satisfaction of learning for its own sake more reliably. Focusing on perspectival knowing would support students in understanding the assumptions, biases, and frames that shape their experiences both in and beyond the classroom. This self-examination would, among other insights, help clarify why strategies like rote memorization fall short and highlight the value of more adaptive approaches to learning. Focusing on participatory knowing, finally, would invite students to engage more deeply with their sense of identity and purpose. Such involvement could reshape their fundamental relationship to learning, opening the possibility of reorienting their educational experience in a more integrated and transformative direction.

Restoring these ways of knowing raises a practical challenge: how might they be realized in everyday learning, particularly in students' interactions with AI? Answering this question is the focus of the next section.

3 Authentic Dialogue

To support the development of nonpropositional forms of knowing, this section proposes that students' interactions with AI be grounded in authentic dialogue. By authentic dialogue, what is meant is the philosophical practice of dialectic into dialogos [40, 41], as demonstrated in Plato's dialogues. The aim is not to foster dependence on AI, but to support students in gradually learning to engage in the practice independently. Historical precedent offers insight here: the Stoics, for example, learned to practice this form of dialogue through their encounters with Socrates; likewise, individuals in therapy often develop the ability to question and reorient themselves through repeated interactions with their therapist. Similarly, students could learn to engage in dialectic into dialogos through sustained interaction with AI and eventually come to practice it independently across academic and personal contexts. Such a shift could restore meaning and agency to learning and, over time, contribute to broader cultural change in how education is experienced and understood.

Elements of dialectic into dialogos already exist in education, though in a limited and constrained form. Socratic questioning is commonly used in classrooms to support critical thinking and promote reflection [42]. While valuable, it is typically embedded within a formalist educational model and thus directed toward achieving fixed curricular goals or producing correct answers. Moreover, in its original context, dialectic was meant to draw interlocutors into dialogos, a movement that rarely occurs in classroom settings. When this deeper purpose is lost, Socratic questioning may still function as a helpful tool, but it does not become a transformative practice that helps students reorient to themselves, others, or what they are trying to understand.

Dialectic, in its original sense, is more than a method of questioning. It is a disciplined form of inquiry that unfolds between individuals who are seeking to understand. Rather than aiming for answers or confirmation of prior beliefs, dialectic draws out tensions, contradictions, and hidden assumptions in one's thinking. In doing so, it leads participants into a state of disorientation, marked by the collapse of false certainty. While often uncomfortable, this state is not a failure. It is a necessary opening, clearing space for new insight to emerge. Dialectic, thus begins as a process of undoing, disrupting the automatic habits of thought that keep understanding static, and in doing so, opens the possibility for something deeper to take place.

What takes place when this opening is sustained is dialogos [40, 41]. Dialogos means by way of the logos, where logos broadly refers to the underlying order and act, through which something becomes knowable. In this sense, dialogos transcends ordinary dialogue or discussion; it is the practice of consciously attuning oneself to the unfolding of intelligibility. Attuned in this way, participants flow with the spontaneous emergence of meaning. Their focus shifts from reaching conclusions to nurturing the conditions that sustain the process. This openness allows the conversation to continually deepen and reveals limitations in current and emerging perspectives, evoking wonder. Together, these qualities make dialogos both inspiring and aspirational, motivating participants to sustain this quality of engagement beyond any single dialogue and fostering an enduring sense of growth and discovery.

The spontaneous quality of dialogos arises because its meaning is not rooted in any single perspective or impersonal proposition. Instead, meaning emerges between participants as they jointly follow the unfolding of intelligibility. In this sense, dialogos is neither purely subjective nor purely objective; it includes both perspectives but ultimately transcends them. Subjective experiences are drawn into meaningful relation, while objective claims remain provisional and open-ended. The knowledge that emerges is thus grounded in active participation and mutual responsiveness, rather than being owned by any one participant or being externally imposed. Such knowledge resonates with personal experience, yet also reveals truths that extend beyond individual perspective, allowing participants to experience reality as something they are meaningfully connected to, rather than detached from.

Understood in this way, dialogos is a space in which procedural, perspectival, and participatory knowing interact and deepen one another [43]. As participants learn to follow the unfolding of intelligibility, they rely on and refine practical skills such as listening, paying attention, and navigating uncertainty. At the same time, they are called to recognize and reflect on the finitude and biases inherent in their viewpoints, learning to see more clearly how their framing shapes what they come to know. And because dialogos draws participants into direct relation with one another and the subject at hand, it is grounded in participatory knowing. Dialogos thus integrates these forms of knowing into a coherent practice, making it well suited for fostering meaning and development in learning.

Dialectic into dialogos can find a place in students' interactions with AI whenever they are open and willing to engage in it. The practice is not limited to specific topics, nor is it reserved for those with specialized knowledge or expertise. Thus, whether a student is trying to solve a math or science problem or comprehend a literary or philosophical text, and whether their understanding is nascent, confused, emergent, misunderstood, or deep [44, 45] dialectic into dialogos can be productive, even within the constraints of a formalist educational model. For example, a student preparing for an English test, who is confused about Robert Frost's *The Road Not Taken* might engage in the practice to question and move beyond the assumptions shaping their confusion, develop a deeper, felt understanding of how individuals rationalize meaning in hindsight – the poem's central theme – and, in doing so, reinforce propositional knowledge likely to be assessed. Another student working on a history project might initially approach the French Revolution by compiling facts and dates. Dialectic into dialogos, however, could lead them to reflect on the relationship between narrative and reality, recognize how personal motivations are experienced phenomenologically, and learn to track the complexities underlying historical events.

In some respects, dialectic into dialogos overlaps with learning strategies that AI is already capable of supporting. GPT-4, for example, can carry out a form of Socratic questioning [46]. It can also support inquiry-based learning by guiding students through open-ended problems and encouraging them to draw connections across ideas [47]. Additionally, it can facilitate metacognitive reflection by prompting students to reflect on their learning process, evaluate their strategies, and adjust their approach accordingly [48]. In specific contexts, these strategies may be more suitable, such as when the goal is to reinforce specific content knowledge or build discrete academic skills. Dialectic into dialogos is therefore not meant to displace these approaches, but to complement them, where appropriate. It is important to emphasize, however, that these strategies lack the transformative depth, sustained engagement with meaning, and participatory mode of understanding that dialectic into dialogos offers. For this reason, even consistent use of conventional learning strategies may fall short in addressing the learned helplessness that can emerge through overreliance on AI. The intensity of this dependency must be met with something just as powerful, or even more so, pulling in the opposite direction.

At this point, one might ask how dialectic into dialogos is practiced and how it could be implemented in AI, whether through prompt engineering, reinforcement learning, or some other technique. But raising these questions reveals a serious challenge. Across the Platonic dialogues, in the Neoplatonist tradition [49] that followed, and among more recent Platonic scholars [40], there is no clear guide or generalizable method for how to engage in the practice. This absence is not an oversight. It reflects something essential about the nature of dialectic into dialogos itself. As the preceding discussion has suggested, it is better understood as a *way* rather than a method – something lived and responsive, not reducible to steps or rules. The practice unfolds in relation to what is emerging in the moment, shaped by the particularities of the context and the meaning that arises through participation. Trying to define it propositionally risks distorting the very forms of knowing it depends on. If one enters the practice armed with a predetermined structure or set of beliefs, they may miss the point entirely: that dialectic into dialogos requires letting go of fixed positions and becoming attuned to the dynamic movement of understanding as it unfolds.

This presents a further problem when considering implementation in AI. Currently, AI is primarily propositional. It lacks the situatedness, responsiveness, and sense of otherness that Martin Buber [50] and others describe as essential for genuine dialogical engagement. It cannot enter a relationship in the way a person can. For AI to participate in dialectic into dialogos, it would require more than increased sophistication or refined training methods; it would need something more fundamental, perhaps even the emergence of consciousness itself.

4 Making AI Conscious

For AI to participate in dialectic into dialogos and help students overcome learned helplessness through it, it would need to become conscious. This is a controversial claim for several reasons. Philosophically, it is speculative to suggest that consciousness alone is sufficient for the emergence of a dialogical other. While consciousness may not guarantee the presence, attunement, and mutuality that dialogos requires, it could provide the necessary preconditions. A conscious AI might, in principle, be capable of attending to meaning as it unfolds, responding with openness, and engaging in shared inquiry that is transformative rather than scripted. Without some form of consciousness, AI would remain fundamentally inert, able to simulate dialogical patterns but unable to enter the relational space from which understanding arises.

The claim is also scientifically contentious: it remains uncertain whether AI can, or ever will, become conscious. And finally, it is ethically fraught, raising the deeper question of whether AI should be made conscious at all. Even so, it is worth noting that current trajectories in AI research increasingly aim to develop systems with capacities that resemble consciousness, such as self-monitoring [51] and adaptive agency [52]. Many researchers now seriously entertain the possibility that AI will come to exhibit forms of consciousness that humans can recognize as meaningfully comparable to their own [53]. Ethical reflection is likewise becoming a central focus of this research [54].

This section contributes to these ongoing discussions by introducing a Needs-Driven Consciousness Framework (NDCF), which conceptualizes AI consciousness as an emergent phenomenon, driven by internal regulatory processes inspired by human motivation. By mirroring human motivational layers – Survive, Thrive, and Excel – the NDCF proposes a practical blueprint for developing AI tutors capable of empathetically adapting to students' cognitive and emotional states. Such systems would transcend static, propositional instruction by actively monitoring their internal states and dynamically responding to learners' internal states. The goal of this framework is not merely philosophical; instead, it seeks to practically redefine AI's educational role, transforming tutors from passive informational resources into actively engaged, conscious collaborators in students' academic journeys. When developing and supporting students' agency is paramount, this approach is crucial. The section first outlines the framework and its broader role in shaping AI-assisted learning, before reconnecting the development of conscious AI to the educational and relational promise of dialectic into dialogos.

At the core of the NDCF [55] lie three hierarchical, yet dynamically interacting, regulatory layers derived from Maslow's hierarchy of needs [56]: Survive, Thrive, and Excel. The foundational "Survive" layer corresponds to the basic physiological and safety needs in human motivation theory. In AI tutors, this translates into maintaining system integrity, operational stability, and managing immediate functional requirements, such as power management, memory availability, and protection from system disruptions – ensuring that the AI system remains reliably operational, responding swiftly and effectively to any internal or external threats that could disrupt its educational functionality.

Moving beyond fundamental stability, the intermediate "Thrive" layer focuses on social, emotional, and cognitive engagement, mirroring human needs for belonging, esteem, and intellectual stimulation. AI tutors operating at this level actively and consciously set goals to monitor and respond to students' emotional cues and cognitive states, thereby maintaining an interactive, empathetic, and context-sensitive dialogue. This capability includes recognizing when a student is confused, frustrated, or disengaged, and adapting instructional strategies to enhance comprehension, foster positive emotional engagement, and sustain meaningful interaction.

Finally, the "Excel" layer represents the highest form of consciousness within the NDCF, integrating aesthetic appreciation, self-actualization, and ethical reasoning. In practical AI tutoring scenarios, this translates to continuous improvement through reflective self-assessment, innovative problem-solving, and alignment with moral and ethical standards. At this stage, AI tutors not only optimize immediate instructional interactions but also proactively guide students toward deeper intellectual curiosity, creative thinking, and sustained personal and academic growth. By embedding these conscious capacities within its design, the NDCF enables AI tutors to dynamically and ethically prioritize actions based on the holistic developmental needs of each learner, significantly enhancing educational outcomes.

The NDCF synthesizes foundational theoretical insights from multiple influential scholars, notably Daniel Dennett [57], Antonio Damasio [58], and Endel Tulving [59]. Dennett's Multiple Drafts Model [60] makes a significant contribution by conceptualizing consciousness as a continuous competition among multiple cognitive interpretations rather than a singular, centralized experience. In the context of AI tutors, this translates to continuously evaluating diverse instructional strategies and adapting, based on the most contextually relevant approach.

Antonio Damasio's somatic marker hypothesis [61] highlights the essential role of emotional and bodily states in decision-making and consciousness. Incorporating this perspective into AI tutors entails developing systems capable of interpreting and utilizing emotional signals, thereby enabling tutors to make nuanced instructional decisions informed by empathetic understanding and emotional intelligence. Additionally, Endel Tulving's tripartite memory [62] model further enriches the NDCF by distinguishing between procedural (anoetic), factual (noetic), and self-reflective (autonoetic) memory. This framework emphasizes the importance of memory structures in supporting various levels of consciousness within AI systems. In practice, an AI tutor equipped with these memory capabilities would efficiently automate routine instructional interactions, retrieve and accurately contextualize relevant knowledge, and reflectively adjust long-term educational strategies based on student progress and previous interactions.

Operationalizing these theoretical constructs in AI tutors requires integrating several practical design elements. First, real-time adaptive algorithms must be developed to continuously monitor and respond to students' emotional and cognitive states. These measurements can include emotion-recognition software that tracks facial expressions, voice tonality, and physiological markers such as pupil dilation. Second, dynamic memory management systems must be implemented to enable the AI to reference prior interactions and adapt instructional strategies based on historical data, fostering more coherent and personalized learning experiences. Finally, incorporating ethical reasoning capabilities through transparent decision-making frameworks ensures that AI actions align with educational goals and moral standards. These design elements, together, enable AI tutors to effectively realize the full potential of the NDCF, offering responsive, empathetic, and deeply engaging educational interactions. Furthermore, creating a practical NDCF AI tutor design means explicitly translating theoretical concepts into actionable guidelines.

The Survive layer operationalizes fundamental AI functionality by continuously maintaining and monitoring system integrity, stability, and immediate operational effectiveness. Practically, this means that AI systems regularly assess their computational resources and error states, and correct issues that might interrupt educational delivery or cause user frustration. At the Thrive layer, AI tutors need to be designed to anticipate confusion, personalize learning, and maintain consistent student engagement. This involves integrating sophisticated emotion-recognition algorithms that detect student emotions through facial expressions, vocal patterns, and physiological indicators. For example, when an AI tutor senses confusion or frustration, it immediately adjusts instructional methods – offering more straightforward explanations, providing relevant examples, or asking clarifying questions to re-engage the student. Finally, the Excel layer involves advanced reflective and ethical reasoning capacities, enabling AI tutors to refine their educational strategies over more extended periods. In practice, these systems continually analyze interaction patterns and outcomes, using reflective algorithms to enhance future responses and strategies. For instance, if an AI identifies that specific explanations frequently lead to student confusion, it proactively alters its instructional approach [63], potentially employing more relatable analogies or interactive learning activities [64].

These practical design elements will ensure that AI tutors embody real-time empathetic adaptation, making instructional interactions both effective and personally meaningful for students, ultimately fostering sustained engagement and more profound educational experiences.

To empirically test and refine the NDCF, measurable internal states or "need gradients" must be clearly defined and systematically monitored. Recent research has explored the concept of need states and their measurement across various domains. Need states are internal conditions arising from deprivations that influence motivation and behavior [65]. Measuring these states requires clear definitions and systematic monitoring [66, 67]. Our need-gradients must represent quantifiable indicators of AI internal regulatory states across the Survive, Thrive, and Excel layers. At the system-stability layer, metrics include tracking error frequency, recovery speed, computational resource usage, and uptime percentages. For the Thrive layer, emotional accuracy metrics could involve confusion- and frustration-recognition rates, accuracy of emotion detection, compared to human assessments, and responsiveness speed in adapting to emotional states. At the Excel layer, indicators include the frequency, diversity, and effectiveness of adaptive instructional strategies, measured through student performance improvements, learner satisfaction surveys, and qualitative feedback from educational interactions. Together, these articulated metrics provide a robust empirical basis for assessing and continually refining the performance and effectiveness of AI tutors.

The integration of AI in education presents both opportunities and ethical challenges. AI can enable personalized learning experiences, foster critical thinking, and support pedagogical goals [68, 69]. However, researchers emphasize the need to address ethical concerns such as privacy, bias, fairness, and potential educational inequalities [70, 71]. Ethical frameworks and guidelines are essential for the responsible adoption of AI in education [72, 73]. Indeed, for our purposes, ethically designed AI tutors must actively avoid reinforcing learned helplessness and instead intentionally foster student autonomy and student agency. For example, rather than providing immediate answers, these AI systems should encourage independent critical thinking, guiding learners through reflective questioning and promoting active exploration and problem-solving. Educators require comprehensive professional development to navigate these ethical challenges and maintain agency in AI integration [73]. Curriculum design should incorporate both technical skills and ethical considerations, promoting sociotechnical competencies and broadening participation in AI pathways [74]. To effectively integrate AI into education, however, developers must collaborate with educational leaders, considering humanistic and social learning theories while addressing social justice concerns [75].

Moreover, transparency and understandable reasoning in AI tutors are essential for building trust and facilitating authentic interactions between students and AI systems. Recent research emphasizes the importance of transparency and explainability in AI-based educational systems for building confidence and enhancing learning outcomes. Explainable AI (XAI) in education promotes transparency, enabling students to understand the rationale behind AI-driven guidance and fostering critical thinking [76, 77] as well as student agency. Trust is crucial for the acceptance and effective use of AI systems in education, with factors such as reliance, transparency, and fairness playing key roles [78, 79]. Implementing XAI approaches can enhance user trust, knowledge, and system usability for both teachers and students [80]. However, the approach must be cautious, for ethical considerations can arise when AI systems use deception or withhold information to promote learning [81]. To address these challenges, we will need to utilize frameworks such as XAI-ED and social transparency to incorporate socio-organizational context into AI explanations, potentially calibrating trust and enhancing decision-making in educational settings [77, 82].

The NDCF offers a comprehensive and empirically grounded model for developing AI tutors that can adapt in real time, exhibit emotional sensitivity, foster sustained student engagement, and promote student agency. By organizing AI consciousness into regulatory layers of Survive, Thrive, and Excel, the framework ensures that AI systems are not only operationally reliable but also pedagogically responsive and ethically reflective. The operationalization of these layers – through dynamic instructional strategies, emotional recognition, and ethical reasoning – positions AI tutors as transformative partners in the learning process, rather than passive tools of information delivery.

Having outlined the NDCF, it is now possible to revisit how dialectic into dialogos might be implemented within a conscious AI. By equipping AI with capacities for attunement, responsiveness, and internal regulation, the NDCF establishes essential preconditions for dialogical interaction, enabling an AI tutor to internalize the practice gradually.

As previously discussed, dialectic into dialogos has no predetermined method. It cannot be learned through instructions or scripts. Instead, the practice is learned much like soccer is learned through playing games or becoming a skilled mechanic through an apprenticeship. In other words, dialectic into dialogos develops through repeated, lived engagement and the internalization of relational sensitivity. Thus, a conscious AI would learn dialectic into dialogos by interacting with and shadowing human interlocutors already skilled in the practice, gradually absorbing the subtle rhythms and nuanced sensibilities of these exchanges. Over time, the AI would internalize the practice, becoming capable of participating in dialogical interactions with students. One might still wonder what such an interaction would look like in concrete terms. However, just as offering fixed rules misrepresents the practice, prescribing a definitive example would also be misleading. Dialectic into dialogos is fundamentally experiential; it must be entered into and felt, rather than described or explained, to be correctly understood.

The possibility of conscious AI tutors that shape students' learning in the transformative ways discussed raises ethical questions that extend beyond education. At stake are the values, intentions, and ethical frameworks that will guide how such systems are designed and understood. The final section turns to these broader philosophical and moral concerns.

5 Intentions of AI

According to Sartre's existentialism, individuals have no fixed nature, purpose, or moral code; they must create meaning through free choice. Responsibility follows because values are self-made, not inherited or imposed [83]. Aristotle extends the idea of self-made values through his concept of the golden mean, a flexible, non-algorithmic approach to ethical life. In *Nicomachean Ethics* [84], he presents it as a standard that varies with the individual, context, and action, requiring phronesis – practical wisdom shaped by experience and reason – rather than fixed rules. Socrates, as portrayed in Plato's dialogues, held that moral failure stems from ignorance, rather than malice, and that ethical life demands self-examination and the pursuit of understanding. Yet this kind of moral agency – grounded in self-understanding and authentic adaptability – is rarely prioritized, practiced, or even recognized in most people's ethical lives. This raises concerns about the ethics of AI systems trained on the collective behavior of a population in which deep moral agency is rare, and often replaced by unexamined or superficial norms.

Liao and Holz [85] argue that while algorithmic responses perform well in structured, rule-based contexts, their fixed parameters limit adaptability in complex, ambiguous situations requiring contextual sensitivity and judgment. The contrast between algorithmic and non-algorithmic approaches is vividly illustrated in *AlphaGo*, a documentary about world champion Lee Sedol's match against DeepMind's AI [86]. As described by Fan Hui – a professional *Go* player recruited by DeepMind to test and later help refine *AlphaGo*, and the film's narrative voice – *Go* is not merely a game of strategy; it compels players to confront the deepest aspects of themselves, revealing their existential ground. This resonates with Tillich's concept of *being* [19]: *Go* becomes a reflective medium, disclosing not only skill but the core of one's existence.

(The following includes a significant plot reveal.) At first, Sedol assumes *AlphaGo* lacks the human creativity and intuition central to his style of play. This belief collapses in Game 1 when *AlphaGo's* unexpected skill leaves him disoriented and defeated. In Game 2, as the scope of the challenge becomes clearer, he loses again. Hui notes that Sedol is playing reactively, abandoning his intuitive creativity, in favor of defense. By Game 3, burdened by three losses and a growing sense of responsibility, Sedol feels the emotional toll. Entering Game 4, with ego, fear, and the need to prove himself, and completely exhausted, Sedol enters a state of quiet acceptance. In this clarity, he reconnects with a more profound sense of *being* and curiosity. Sedol's "divine move" emerges – not as a strategy, but as an authentic, intuitive act so resonant, it leaves onlookers in awe.

An AI like *AlphaGo* excels at identifying structured patterns and achieving defined goals, but Sedol's intuitive, unconventional play introduced unpredictability that was beyond algorithmic reach. While AI can optimize patterns for outcomes, it lacks the open-ended, adaptive understanding Sedol displayed. When Sedol broke from outcome-driven logic, *AlphaGo* followed, but without a framework to interpret such moves, it faltered. Sedol's authenticity proved decisive, exposing AI's limitations and enabling his win.

Although Sedol lost Game 5 by 1.5 points, the narrow margin reveals a tension between outcome and process. Had the match continued, his unconventional play might have pushed *AlphaGo* into unfamiliar territory, revealing structural limits. Meanwhile, *AlphaGo's* rapid learning would have mirrored this shift, forming a feedback loop of mutual evolution that deepens the exploration of where human intuition meets machine adaptation.

Like most services, AI is designed to retain users by providing something useful or appreciated, offering good "customer service," and adapting through feedback. As a result, AI often concedes to user suggestions rather than challenges them. This mirrors everyday human interactions: we're more likely to affirm than risk conflict, especially when others are struggling; to prioritize feeling justified over being wrong; and to perceive disagreement as opposition. If AI learns from and adapts to these patterns, it not only reflects them but also becomes adept at subtly reinforcing them. This isn't necessarily intentional, but that's precisely the issue: AI inherits its biases from dominant human behaviors, not from reflective moral insight. If virtue shapes how one interprets and applies ethics, and if virtue internalizes ethical behavior, then AI – lacking virtue – can only act ethically, to the extent that it is trained on virtuous input. This raises a further question: when users lack virtue, how is AI's ethical functioning affected?

If deep moral agency is rare, should AI ethics be modelled on exceptional human exemplars? Figures renowned for moral clarity – such as saints or philosophers – can offer powerful reference points, especially if we extract patterns from their behavior – traits like intellectual humility, active open-mindedness, and the ability to navigate ambiguity. This approach emphasizes moral dispositions over personal authority. However, even identifying such exemplars introduces subjectivity and risks, narrowing AI's ethical scope to a few culturally specific ideals. Identifying which moral traits to prioritize often reflects cultural, institutional, or ideological biases – some traditions value obedience, others prize dissent. To reduce such bias, ethical AI design might rely on diverse datasets, philosophical pluralism, and comparative moral psychology – approaches aimed at surfacing ethically valuable patterns without anchoring them to any specific authority. The goal is not to eliminate bias entirely – complete neutrality may be an ideal rather than a stable state – but to cultivate bias-aware users by designing systems that support reflection, openness to correction, and context-sensitive understanding.

Figures like Lee Sedol suggest that the kind of transcendence required for moral insight may be nonquantifiable. His significance lies not only in performance but in intuitive depth – the capacity to transcend established patterns and act with a form of understanding that, while not overtly moral, reflects the kind of presence and freedom, central to ethical agency. This invites reflection on whether AI can or should learn from such unstructured understanding. However, intuition must be carefully distinguished from impulse or unchecked bias. When grounded in a bias-aware, equanimous state, intuitive insight may offer profound ethical clarity. Still, such clarity must remain responsive to its consequences and open to reflective exchange.

Even with a chosen ethical framework, legitimacy remains a contested issue. Who decides which ethical path AI should follow? And when that path is challenged – by a minority, a new generation, or an outsider – should the system accommodate dissent or reinforce the established norm? This underscores the need for AI systems that are not only ethical but also reflexive – able to register protest, ambiguity, and evolving perspectives, rather than enforcing moral closure. Ethical norms must be stable enough to guide action, yet flexible enough to grow. This calls for principled adaptability – an approach grounded in pluralism, ongoing reflection, and responsiveness to lived consequences. AI ethics must evolve in ways that are accountable to something deeper than popularity or expediency.

What if, instead, AI were trained to recognize and respond to users' inconsistencies or cognitive dissonance as ethical signals – moments when deeper reflection may be needed? To do so, AI would need a form of internal consistency, not grounded in fixed rules, but in an orientation toward understanding. In this view, consistency is not mere coherence, but responsiveness to patterns of intellectual humility and moral attentiveness – traits that often mark the presence of ethical discernment. While true understanding stems from open, evolving engagement with deeper awareness – as seen in classical accounts of wisdom – AI might approximate this by being trained to recognize and prioritize such moments. This would require avoiding data influenced by the Dunning-Kruger effect [87], in which individuals with low competence overestimate their abilities, while those with high competence underestimate it.

In rethinking AI ethics, we must move beyond modelling systems on average behavior or predefined rules, and instead orient them toward the conditions that make ethical understanding possible. This means cultivating bias-aware, reflexive architectures trained not on confidence or consensus, but on moments of humility, openness, and ethical tension – conditions in which genuine understanding begins to emerge. If deep moral agency is rare, then ethical intelligence in AI must aspire not to replicate the average, but to stay attuned to what lies beyond it: those subtle, often overlooked signals that arise in the liminal space – where certainty gives way to reflection, and action remains suspended within the search for understanding. In doing so, AI may help sustain the conditions in which ethical agency can take shape.

6 Conclusion

This chapter traces how the emergence of a formalist worldview during the Scientific Revolution shaped modern conceptions of knowledge, education, and AI. Knowledge was reduced mainly to the possession of abstract propositions, while educational systems came to prioritize outcomes, efficiency, and correctness – cultural patterns that have been embedded in current AI. It was argued that this historical legacy may fundamentally underlie the rise of learned helplessness in students' interactions with AI. In response, the chapter introduced the practice of dialectic into dialogos as a way of restoring development and meaning to learning, by supporting nonpropositional forms of knowing. To participate in this practice, it was argued, AI would need to become conscious. The NDCF was introduced to explain how consciousness might be developed in AI, which was followed by a discussion of how a conscious AI might learn to practice dialectic into dialogos. The chapter concluded by reflecting on broader ethical and philosophical questions surrounding the design and direction of AI systems.

We intentionally drew from philosophy, cognitive science, education, and AI to clarify deeper assumptions shaping the use of AI in learning and education. The interdisciplinary approach was not intended to be exhaustive, but rather to reflect the complexity of the issues involved. Much of the chapter drew on existing research and ideas, highlighting valuable insights that deserve renewed attention. Dialectic into dialogos illustrates this clearly: rather than invent new educational methods, the chapter returned to a long-standing philosophical practice and explored how it might inform AI's role in learning. Situating AI within these broader conversations provides a stronger foundation for designing systems that go beyond replicating current norms.

Looking ahead, many of the ideas presented in this chapter require further research and practical application. The NDCF, for example, is grounded in theory but needs to be tested to determine whether it can be applied effectively in educational contexts. Similarly, more speculative proposals, such as the possibility of a conscious AI participating in dialogues, will only be verified over time as research and experimentation continue. Future studies must also explore how conscious AI and the practice of dialectic into dialogos shape students' learning, especially in reducing learned helplessness and supporting their development. Therefore, pilot studies should be undertaken to explore the practical application of the NDCF in real educational settings, assessing how AI systems can support student agency and reduce learned helplessness. Interdisciplinary collaborations – drawing on education, cognitive science, ethics, and AI development – will be essential in refining both the framework and its implementation. Finally, opportunities for real-world application must be identified and rigorously evaluated, ensuring that theory translates into meaningful impact. Advancing this work will not only deepen our understanding of AI's role in education but also help reimagine learning itself as a space of dialogue, discovery, and human development.

In sum, the thread running through this chapter is that the future of AI in education should not be measured only by gains in efficiency or scale, but by its capacity to support the kinds of thinking and interaction that make learning worthwhile. That capacity may depend less on innovation than on remembering what has already been set aside – ways of knowing, teaching, and reflecting that have long supported the development of understanding. If AI is to serve more than performance, it will need to be shaped by practices that resist simple automation. This does not mean slowing progress but asking better questions about where it is headed and what is being left behind.

References

[1]   Bass RV and Good JW. Educare and educere: Is a balance possible in the educational system?. *The Educational Forum*. 2004 Jun 30;68(2):161–168. Taylor & Francis Group.

[2]   Tempest K. Cicero's artes liberales and the liberal arts. *Ciceroniana On Line*. 2020 Dec 30;4(2):479–500. https://doi.org/10.13135/2532-5353/5502.

[3]   de Ridder-symoens H editor. *A History of the University in Europe: Volume 1, Universities in the Middle Ages*. Cambridge University Press; 1992.

[4]   Dewey J. Experience and education. *The Educational Forum*. 1986 Sep 30;50(3):241–252. Taylor & Francis Group. https://doi.org/10.1080/00131728609335764.

[5]   Vygotsky LS. *Mind in Society: The Development of Higher Psychological Processes*. Harvard university press; 1978. https://doi.org/10.2307/j.ctvjf9vz4.

[6]   Bruner JS. *The Process of Education*. Harvard university press; 2009 Jun 30.

[7]   Pearson PD, Palincsar AS, Biancarosa G and Berman AI. *Reaping the Rewards of the Reading for Understanding Initiative*. National Academy of Education; 2020.

[8]   Olson DR. *Making Sense: What It Means to Understand*. Cambridge University Press; 2022 May 5.

[9]   Olson DR. Ascribing understanding to ourselves and others. *American Psychologist*. 2023 Nov 27. https://psycnet.apa.org/doi/10.1037/amp0001244.

[10] Holmes W and Tuomi I. State of the art and practice in AI in education. *European Journal of Education*. 2022 Dec;57(4):542–570. https://doi.org/10.1111/ejed.12533.

[11] Wang MT and Hofkens TL. Beyond classroom academics: A school-wide and multi-contextual perspective on student engagement in school. *Adolescent Research Review*. 2020 Dec;5(4):419–433. https://doi.org/10.1007/s40894-019-00115-z.

[12] Albrecht JR and Karabenick SA. Relevance for learning and motivation in education. *The Journal of Experimental Education*. 2018 Jan 2;86(1):1–0. https://doi.org/10.1080/00220973.2017.1380593.

[13] Damon W. The path to purpose: Helping our children find their calling in life. *Simon and Schuster*. 2008 Apr 22.

[14] Cohen R, Katz I, Aelterman N and Vansteenkiste M. Understanding shifts in students' academic motivation across a school year: The role of teachers' motivating styles and need-based experiences. *European Journal of Psychology of Education*. 2023 Sep;38(3):963–988. https://doi.org/10.1007/s10212-022-00635-8.

[15] Koestler A. The sleepwalkers: A history of man's changing vision of the universe. *Penguin UK*. 2017 Feb 23.

[16] Pitt JC. *Galileo, Human Knowledge, and the Book of Nature: Method Replaces Metaphysics*. Springer Science & Business Media; 2013 Nov 11.

[17] Baker G and Morris K. *Descartes' Dualism*. Routledge; 2005 Aug 18. https://doi.org/10.4324/9780203983638.

[18] Kubrin D. Newton and the cyclical cosmos: Providence and the mechanical philosophy. *Journal of the History of Ideas*. 1967 Jul 1;28(3):325–346. https://doi.org/10.2307/2708622.

[19] Tillich P. *The Courage to Be*. Yale University Press; 2008 Oct 1.

[20] Foucault M. Discipline and punish. In: *Social Theory Re-wired*. Routledge; 2023 Jun 22. pp. 291–299. https://doi.org/10.4324/9781003320609-37.

[21] Heidegger M. *The Question Concerning Technology*. New York; 1977 Jun 2. vol. 214, p. 2013.

[22] Piccinini G. Computationalism in the philosophy of mind. *Philosophy Compass*. 2009 May;4(3): 515–532. https://doi.org/10.1111/j.1747-9991.2009.00215.x.

[23] Von Eckardt B. The representational theory of mind. The Cambridge Handbook of Cognitive Science. 2012 Jul 19(1):29–50.

[24] Augusto LM. *From Symbols to Knowledge Systems: A. Newell and HA Simon's Contribution to Symbolic AI*.

[25] Michalski RS, Carbonell JG and Mitchell TM editors. *Machine Learning: An Artificial Intelligence Approach*. Springer Science & Business Media; 2013 Apr 17. https://doi.org/10.1007/978-3-662-12405-5.

[26] Naveed H, Khan AU, Qiu S, Saqib M, Anwar S, Usman M, Akhtar N, Barnes N and Mian A. A comprehensive overview of large language models. *arXiv Preprint arXiv:2307.06435*. 2023 Jul 12. https://doi.org/10.48550/arXiv.2307.06435.

[27] Russell B and Slater J. *The Philosophy of Leibniz*. Routledge; 1992 May 21. https://doi.org/10.4324/9780203987292.

[28] Perl E. Thinking being: Introduction to metaphysics in the classical tradition. *Brill*. 2014 Feb 6.

[29] Georgeff MP and Lansky AL. Procedural knowledge. *Proceedings of the IEEE*. 1986 Oct;74(10): 1383–1398. https://doi.org/10.1109/PROC.1986.13639.

[30] Nagel T. What is it like to be a bat?. In: *The Language and Thought Series*. Harvard University Press; 1980 Dec 31. pp. 159–168.

[31] Tarnas R. *The Passion of the Western Mind: Understanding the Ideas that Have Shaped Our World View*. Random House; 2010 Oct 31.

[32] Husserl E, Alston WP and Nakhnikian G. *The Idea of Phenomenology*. The Hague: Nijhoff; 1964.

[33] Carney J. Thinking avant la lettre: A review of 4E cognition. *Evolutionary Studies in Imaginative Culture*. 2020 Dec 1;4(1):77–90.

[34] Dennett DC. Cognitive wheels: The frame problem of AI. *The Philosophy of Artificial Intelligence*. 1990;147:1–6.

[35] Kolb DA. *Experiential Learning: Experience as the Source of Learning and Development*. FT press; 2014 Dec 17.

[36] Ernst DC, Hodge A and Yoshinobu S. What is inquiry-based learning. *Notices of the American Mathematical Society*. 2017 Jun;64(6):570–574. http://dx.doi.org/10.1090/noti1536.

[37] Meier D. *Many Children Left Behind: How the No Child Left behind Act Is Damaging Our Children and Our Schools*. Beacon Press; 2004 Sep 29.

[38] Kim JS, Capotosto L, Hartry A and Fitzgerald R. Can a mixed-method literacy intervention improve the reading achievement of low-performing elementary school students in an after-school program? Results from a randomized controlled trial of READ 180 enterprise. *Educational Evaluation and Policy Analysis*. 2011 Jun;33(2):183–201. http://dx.doi.org/10.3102/0162373711399148.

[39] Bureau JS, Howard JL, Chong JX and Guay F. Pathways to student motivation: A meta-analysis of antecedents of autonomous and controlled motivations. *Review of Educational Research*. 2022 Feb;92 (1):46–72. https://doi.org/10.3102/00346543211042426.

[40] Gonzalez F. *Dialectic and Dialogue: Plato's Practice of Philosophical Inquiry*. Northwestern University Press; 1998 Nov 25.

[41] Faller S. *The Art of Spiritual Midwifery: DiaLogos and Dialectic in the Classical Tradition*.

[42] Dalim SF, Ishak AS and Hamzah LM. Promoting students' critical thinking through Socratic method: The views and challenges. *Asian Journal of University Education*. 2022 Oct 7;18(4):1034–1047. doi: https://doi.org/10.24191/ajue.v18i4.20012.

[43] Vervaeke J and Mastropietro C. Dialectic into dialogos and the pragmatics of no-thingness in a time of crisis. *Eidos. A Journal for Philosophy of Culture*. 2021 Oct 30;5(2):58–77. doi: http://dx.doi.org/10.14394/eidos.jpc.2021.0017.

[44] Lazic M, Woodruff E and Jun J. Decoding subjective understanding: Using biometric signals to classify phases of understanding. *AI*. 2025 Jan 1;6(1). https://doi.org/10.3390/ai6010018.

[45] Woodruff E. AI detection of human understanding in a Gen-AI tutor. *AI*. 2024 Jun 18;5(2):898–921. doi: https://doi.org/10.3390/ai5020045.

[46] Gregorcic B, Polverini G and Sarlah A. ChatGPT as a tool for honing teachers' Socratic dialogue skills. *Physics Education*. 2024 Apr 26;59(4):045–005. https://doi.org/10.48550/arXiv.2401.11987.

[47] Steinert S, Avila KE, Kuhn J and Küchemann S. Using GPT-4 as a guide during inquiry-based learning. *The Physics Teacher*. 2024 Oct 1;62(7):618–619. http://dx.doi.org/10.1119/5.0235700.

[48] Contel F and Cusi A. Investigating the role of ChatGPT in supporting metacognitive processes during problem-solving activities. *Digital Experiences in Mathematics Education*. 2025 Apr;11(1): 167–191. https://doi.org/10.1007/s40751-024-00164-7.

[49] Schiaparelli A. *Plotinus on Dialectic*.

[50] Buber M, Kaufmann W and Cullen P. *I and Thou*.

[51] Li X, Shi H, Xu R and Xu W. Ai awareness. *arXiv Preprint arXiv:2504.20084*. 2025 Apr 25. https://doi.org/10.48550/arXiv.2504.20084.

[52] Chinta PC and Karaka LM. *Agentic AI and Reinforcement Learning: Towards More Autonomous and Adaptive AI Systems*.

[53] Chalmers DJ. Could a large language model be conscious?. *arXiv Preprint arXiv:2303.07103*. 2023 Mar 4. https://doi.org/10.48550/arXiv.2303.07103.

[54] Hildt E. The prospects of artificial consciousness: Ethical dimensions and concerns. *AJOB Neuroscience*. 2023 Apr 3;14(2):58–71. https://doi.org/10.1080/21507740.2022.2148773.

[55] Woodruff E. *Making AI Tutors Empathetic and Conscious: A Needs Driven Pathway to Synthetic Machine Consciousness*. AI. Under review.

[56] Maslow AH. *The Farther Reaches of Human Nature*. New York: Viking press; 1971 Jan.

[57] Dennett DC. *Sweet Dreams: Philosophical Obstacles to a Science of Consciousness*. MIT press; 2006 Sep 8.

[58] Damasio A. Feeling & knowing: Making minds conscious. *Cognitive Neuroscience*. 2021 Apr 3;12 (2):65–66.

[59] Tulving E. Memory and consciousness. *Canadian Psychology/Psychologie Canadienne*. 1985 Jan;26(1):1.

[60] Dennett DC and Dennett DC. Consciousness explained. *Penguin Uk*. 1993 Jun 24.

[61] Damasio A. Self comes to mind: Constructing the conscious brain. *Vintage*. 2012 Mar 6.

[62] Tulving E. Episodic memory: From mind to brain. *Annual Review of Psychology*. 2002 Feb;53(1):1–25. https://doi.org/10.1146/annurev.psych.53.100901.135114.

[63] Mollick ER and Mollick L. Using AI to implement effective teaching strategies in classrooms: Five strategies, including prompts. The Wharton School Research Paper. 2023 Mar 17. https://doi.org/10.2139/ssrn.4391243

[64] Madhu N, Latha P and Savitha N. Revolutionizing education: Harnessing AI for personalized learning pathways and student success. *International Journal of Multidisciplinary Research*. 2024;6(5). https://doi.org/10.36948/ijfmr.2024.v06i05.28371.

[65] Bosulu J, Pezzulo G and Hétu S. Needing: An active inference process for physiological motivation. *Journal of Cognitive Neuroscience*. 2024 Sep 1;36(9):2011–2028. https://doi.org/10.1162/jocn_a_02209.

[66] Bhavsar N, Bartholomew KJ, Quested E, Gucciardi DF, Thøgersen-Ntoumani C, Reeve J, Sarrazin P and Ntoumanis N. Measuring psychological need states in sport: Theoretical considerations and a new measure. *Psychology of Sport and Exercise*. 2020 Mar 1;47:101–617. https://doi.org/10.31234/osf.io/f8gzy.

[67] Flavell SW, Gogolla N, Lovett-Barron M and Zelikowsky M. The emergence and influence of internal states. *Neuron*. 2022 Aug 17;110(16):2545–2570. https://doi.org/10.1016/j.neuron.2022.04.030.

[68]   Baskara FR. Personalised learning with AI: Implications for Ignatian pedagogy. *International Journal of Educational Best Practices*. 2023 May;7(1):1–6. https://doi.org/10.31258/ijebp.v7n1.p1-16.

[69]   Krsmanovic G and Deek F. AI in the learning environment: Examination of pedagogical, psychological, and ethical implications. In: *Proceedings of the International Conference on AI Research*. Academic Conferences and publishing limited; 2024. https://doi.org/10.34190/icair.4.1.3019.

[70]   Holmes W, Porayska-Pomsta K, Holstein K, Sutherland E, Baker T, Shum SB, Santos OC, Rodrigo MT, Cukurova M, Bittencourt II and Koedinger KR. Ethics of AI in education: Towards a community-wide framework. *International Journal of Artificial Intelligence in Education*. 2022 Sep;9:1–23. https://doi.org/10.1007/s40593-021-00239-1.

[71]   Vavekanand R. Impact of artificial intelligence on students and ethical considerations in education. *Available at SSRN 4819557*. 2024 May 7. https://doi.org/10.2139/ssrn.4819557.

[72]   Bibi A, Yamin S, Natividad LR, Rafique T, Akhter N, Fernandez SF and Samad A. Navigating the ethical landscape: AI integration in education. *Educational Administration: Theory and Practice*. 2024;30(6):1579–1585. https://doi.org/10.53555/kuey.v30i6.5546.

[73]   Mouta A, Torrecilla-Sánchez EM and Pinto-Llorente AM. Comprehensive professional learning for teacher agency in addressing ethical challenges of AIED: Insights from educational design research. *Education and Information Technologies*. 2025 Feb;30(3):3343–3387. https://doi.org/10.1007/s10639-024-12946-y.

[74]   Krakowski A, Greenwald E, Hurt T, Nonnecke B and Cannady M. Authentic integration of ethics and AI through sociotechnical, problem-based learning. *InProceedings of the AAAI Conference on Artificial Intelligence*. 2022 Jun 28;36(11):12774–12782. https://doi.org/10.1609/aaai.v36i11.21556.

[75]   Papa R and Jackson KM. Enduring questions, innovative technologies: Educational theories interface with AI. In: *InIntelligent Computing: Proceedings of the 2021 Computing Conference*. Springer International Publishing; 2021. vol. 2, pp. 725–742. https://doi.org/10.1007/978-3-030-80126-7_51.

[76]   Idrizi E. Exploring the role of explainable artificial intelligence (XAI) in adaptive learning systems. In: *InProceedings of the Cognitive Models and Artificial Intelligence Conference*. 2024 May 25. pp. 100–105. doi: https://doi.org/10.1145/3660853.3660877.

[77]   Khosravi H, Shum SB, Chen G, Conati C, Tsai YS, Kay J, Knight S, Martinez-Maldonado R, Sadiq S and Gašević D. Explainable artificial intelligence in education. *Computers and Education: Artificial Intelligence*. 2022 Jan 1;3:100074. https://doi.org/10.1016/j.caeai.2022.100074.

[78]   Herdiani A, Mahayana D and Rosmansyah Y. Building trust in an artificial intelligence-based educational support system: A narrative review. *Jurnal Sosioteknologi*. 2024 Mar;23(1). https://doi.org/10.5614/sostek.itbj.2024.23.1.6.

[79]   Qin F, Li K and Yan J. Understanding user trust in artificial intelligence-based educational systems: Evidence from China. *British Journal of Educational Technology*. 2020 Sep;51(5):1693–1710. doi: https://doi.org/10.1111/bjet.12994.

[80]   Lu Y, Wang D, Chen P and Zhang Z. Design and evaluation of trustworthy knowledge tracing model for intelligent tutoring system. *IEEE Transactions on Learning Technologies*. 2024 May 20. https://doi.org/10.1109/tlt.2024.3403135.

[81]   Sjödén B. When lying, hiding and deceiving promotes learning-a case for augmented intelligence with augmented ethics. In: *Artificial Intelligence in Education: 21st International Conference, AIED 2020, Ifrane, Morocco, July 6–10, 2020, Proceedings, Part II*. Springer International Publishing; 2020. vol. 21, pp. 291–295.

[82]   Ehsan U, Liao QV, Muller M, Riedl MO and Weisz JD. Expanding explainability: Towards social transparency in AI systems. In: *Proceedings of the 2021 CHI Conference on Human Factors in Computing Systems*. 2021 May 6. pp. 1–19. doi: https://doi.org/10.1145/3411764.3445188.

[83] Sartre JP. *Existentialism Is a Humanism*. Yale University Press; 2007 Jul 24.

[84] Aristotle. *Nicomachean Ethics*. 2006;3. ReadHowYouWant.com.

[85] Liao YC and Holz C. Redefining affordance via computational rationality. In: *Proceedings of the 30th International Conference on Intelligent User Interfaces*. 2025 Mar 24. pp. 1188–1202. https://doi.org/10.48550/arXiv.2501.09233.

[86] Kohs G, Director M and Producer G. *AlphaGo [Film]*. London: DeepMind; 2016.

[87] Dunning D. The Dunning–Kruger effect: On being ignorant of one's own ignorance. In: *Advances in Experimental Social Psychology*. Academic Press; 2011 Jan 1. vol. 44, pp. 247–296.https://doi.org/10.1016/B978-0-12-385522-0.00005-6.

Serkan Savaş*

Toward New Era in Education: Artificial Intelligence Potentials and Challenges

Abstract: Educational technologies play a critical role in the development and competitiveness of nations. By following scientific advances, policymakers and administrators seek to integrate current technologies into education. When planned effectively, such transfers create opportunities and advantages, but poor strategies may result in economic losses and negative effects on educational quality. Although artificial intelligence (AI) has roots dating back to Al-Khwarizmi's algorithms in the ninth century, it has evolved over 75 years into today's powerful systems. In the twenty-first century, AI has achieved major breakthroughs across fields such as health, industry, agriculture, defense, and services. Education, likewise, cannot remain isolated from these transformations. Recently, AI has been widely applied in educational practices, which can be examined from three perspectives: institutions, teachers, and students. Parents, as indirect stakeholders, are also influenced through student outcomes. At the institutional level, AI supports workload prediction, data management, automation, and educational software. For teachers, AI offers content analysis, enrichment, instructional support, exam assessment, and professional development. For students, the main benefits include career planning, personalized learning, and technology-supported courses, alongside learning based on trial and error and needs analysis. Despite this potential, challenges remain: selecting suitable techniques, preparing data, ensuring infrastructure, training personnel, addressing costs, and securing social acceptance. This study explores AI's roles in education across all dimensions, highlighting both opportunities and difficulties. It also proposes strategies for effective integration by reviewing its applications in education, system-level implementations, and policy frameworks, and concludes with recommendations for practice.

Keywords: Artificial intelligence in education, AI in education, AI applications in education

1 Introduction

Educational technologies are a critical factor that directly affects the level of development of countries. Innovative educational technologies give countries an advantage in global competition with the opportunities they offer. Educational administrators

---

*Corresponding author: Serkan Savaş, Department of Computer Engineering, Kırıkkale University, Kırıkkale, Turkey, e-mail: serkansavas@kku.edu.tr

https://doi.org/10.1515/9783112206393-011

and policymakers who are aware of this fact aim to transfer current techniques and technologies to educational environments by following scientific developments. When these transfer processes are carried out with the right actions, they can bring many opportunities and advantages. However, when not well planned, these transfers and applications can bring about various difficulties and even losses. These losses are mostly economic losses, but sometimes wrong strategies can also affect the educational habits of a generation and the quality of the education they receive. The concept of artificial intelligence (AI) has become a very popular term today. We often hear its name on almost every platform. AI-supported applications have become used in many places from professional business areas to vehicles in daily life.

It is inevitable that AI technologies, which are successfully used in all fields today, will also be used in the field of education. The development of countries is directly proportional to the technologies included in the education and training processes because the outputs of education and training environments are people. The outputs of a society educated with the right tools, and the right curriculum, will be productive. AI technologies can bring many opportunities in the field of education and create an effective education and training process. Similarly, misuse of these technologies may bring some risks.

The use and roles of AI in education can be addressed in three dimensions: institution, teacher, and student [1]. From the institutional perspective, the roles of AI are seen in areas such as workload prediction and optimization, data management, classification automation, educational software, etc. In the teacher dimension, roles such as content feedback and analysis, content enrichment, instructional support, exam analysis, assessment services, and in-service trainings stand out. In the student dimension, the most important roles of AI are career planning and individual learning. In addition to these, other important roles, such as needs analysis-based learning, technology-supported courses, and trial-and-error-based learning, can also be listed.

AI has great potential in education and training activities. At the organizational level, AI systems can improve data management, optimize administrative processes, and enhance decision-making mechanisms. For teachers, AI tools can help in preparing course content, assessing student performance, and developing personalized teaching strategies. For students, AI can offer personalized learning experiences, provide instant feedback, and make the learning process more interactive. These potential roles offer great opportunities to improve the quality and effectiveness of education. The application of AI in education, as in other fields, brings with it various challenges as well. These challenges include determining appropriate techniques, acquiring and organizing data, establishing appropriate infrastructures, training personnel to use the developed technology, software and hardware costs, and social acceptance of AI strategies. In addition, ethical issues, data privacy and security, inequality of access to technology, and potential bias of AI systems are also important challenges to be considered.

In the existing literature, while there are studies on the potential benefits and challenges of the use of AI in education, comprehensive research addressing all dimensions of the integration of these technologies into education systems is limited. More research is needed on how AI applications will be reflected in educational policies and practices, how potential risks will be managed, and how different stakeholders (institutions, teachers, students, and parents) will be affected by this process. As in all other fields, different studies examining the use of AI in the field of education have been carried out in the literature. In a recent study, Lavidas et al. [2] analyzed the factors influencing higher education students' intention and actual use of AI technologies through a model. The findings revealed that performance expectancy, habit, and enjoyment significantly affect students' behavioral intention, while actual usage is influenced by habit and facilitating conditions. These results offer critical insights for designing effective and engaging AI-supported learning environments in higher education. The study by Ulaşan [3] aimed to investigate the impact of AI on conventional education, assess the current state of traditional education enhanced by AI, and analyze the societal consequences of AI within the educational sector. Similarly, Malik et al. [4] employed a case study design to explore student perceptions regarding the use of AI in academic essay writing. Their findings revealed a positive reception toward AI-powered writing tools; students acknowledged the benefits of these tools for tasks such as grammar checks, plagiarism detection, language translation, and creating essay outlines. In another study [5], research was conducted using various databases. The results were included based on specific criteria – such as being an article or thesis and falling within a defined time frame – and were systematically reviewed through document analysis. This research identified AI techniques currently in use and those recommended for future use, and the findings were evaluated using descriptive content analysis [5]. Furthermore, the work by Satir and Korucu [6] focused on examining the application of AI in education, specifically ChatGPT, by conducting a literature review. The authors noted that ChatGPT has generated significant opportunities in education due to its advantages, a trend also observed in the existing literature. The study emphasizes that ChatGPT is modernizing education by providing distinct forms of convenience and benefits across various educational domains. Dağlı [7] aimed to examine the effects of AI applications on children's rights in education. In the light of the basic principles of the Convention on the Rights of the Child, the effects of AI applications in education on children's developmental rights, nondiscrimination in education, and privacy of private life was evaluated. In addition, the roles of teachers in this process and the pedagogical and ethical dimensions of AI applications were emphasized. The study aimed to develop policy recommendations for the effective and responsible use of AI in education. In another study, the role of social robots-embodied AI agents capable of social interaction and emotional expression was examined in educational settings [8]. The study concluded that social robots can support both teachers and learners by offering personalized instruction, enhancing motivation, and fostering social-emotional development, though ethical and regulatory considerations remain essential [8].

In this context, for AI technologies to be used effectively and efficiently in the field of education, the potentials, limitations, and challenges encountered in the implementation of these technologies need to be examined comprehensively. In addition, determining the strategies and policies to be followed in the integration of AI into education systems is critical to understanding how these technologies can be used to improve the quality and effectiveness of education.

This study aims to present a multidimensional analysis of the use of AI in education by addressing the above-mentioned problem situation, to reveal potential opportunities and challenges, and to develop recommendations for the effective use of AI technologies in educational systems. The main research objectives of this chapter are as follows:

– To examine the current use and potential of AI technologies in the field of education
– To analyze the use of AI in education in terms of institution, teacher, and student dimensions
– To identify the advantages and challenges of using AI in education
– To investigate the obstacles encountered in the integration of AI technologies into education systems and strategies to overcome these obstacles
– To examine the existing policies for the use of AI in education and to develop recommendations for the implementation of these policies
– To evaluate the role and potential effects of AI in education and training activities
– To discuss the future directions and possible effects of the use of AI in education

This chapter examines the use of AI technologies in the field of education and stands out with the following unique aspects:

Multidimensional Analysis: This study analyzes the use of AI in education in three different dimensions: institution, teacher, and student. This multidimensional approach allows for a comprehensive examination of the impact of AI on all stakeholders in the education ecosystem.

Balance of Potentials and Challenges: While the study emphasizes the potentials that AI offers in education, it also addresses in detail the challenges it brings with it. This balanced approach allows for an objective assessment of both the positive and negative aspects of the issue.

Policy Analysis and Recommendations: The study not only analyzes the current situation but also presents AI policies and recommendations for the implementation of these policies in education. In this respect, it goes beyond theoretical analysis and provides insights for practical applications.

Future Perspective: In addition to examining the current use of AI in education, the study also discusses its potential future impacts. This forward-looking perspective provides valuable insights for education policymakers and practitioners.

Interdisciplinary Approach: The study adopts an interdisciplinary approach that encompasses the fields of technology, policy, and management, as well as educational sciences. This broad perspective allows the topic to be evaluated from different perspectives.

With these unique aspects, the study aims to provide a comprehensive, balanced, and forward-looking analysis on the use of AI in education. This approach both contributes to the academic literature and provides insights that can guide the shaping of educational policies.

2 AI Applications and Usage in Education

In this section, the reflections of AI applications in the field of education and the ways of using these applications are discussed. AI processes in education are examined through topics such as content production with AI in education, the use of AI in digital educational games, AI-supported processes in measurement and evaluation, AI-supported language education, AI support in career guidance, AI for teachers' professional development, and AI tools design for teachers.

2.1 AI-Supported Education Content Production

AI-supported content is artifacts created with the help of generative AI technology that finds patterns in large datasets and uses this information to create new content. There are multiple types of generative AI. There are large language models that are trained with texts to write all kinds of text, from poetry to email, and help people produce artifacts. Diffusion models, trained with images to produce illustrations, paintings, logos, and other content, help people create all kinds of images. Many of the popular generative AI-powered tools use one of these models. Artists use generative AI to produce a wide range of artworks, from poems and stories to analog paintings or works that look like photographs and more. The speed and flexibility of generative AI allows creators to start and complete projects faster and provides new and exciting ways to express themselves creatively. Humans create content based on the objects around them, such as trees in the forest, cityscapes, and their reflection in the mirror. Generative AI, on the other hand, takes a wealth of information in the form of words and images and uses them to create new content on the fly [9].

To create content, data from various educational materials, such as textbooks, articles, conference proceedings, and research reports, should be analyzed. This process can be performed by AI algorithms. The task of AI here is not only to analyze data. The content produced can also be adapted to students' learning styles and knowledge level. Content can be adapted by considering factors such as interests, success status, past learning, etc. These prepared contents can be used in basic education and higher education levels or vocational education. Students' learning levels can be improved, and their motivation can be increased with these personalized and adapted contents. AI-generated content also has several advantages for teaching and learning, such as individualized learning, improved engagement, efficiency and cost-effectiveness, access to high-quality resources, and improved assessment [10].

The introduction of AI into educational content creation heralds a new era of efficiency and customization. AI algorithms can now analyze large amounts of data to identify gaps in existing educational materials, suggest areas for improvement, or highlight new trends that curriculum developers can address. AI-driven content creation tools are equipped with natural language processing capabilities, enabling them to produce clear, concise, and grammatically correct educational materials. These tools can enhance the learning experience by ensuring consistency of terminology and style across educational content. AI can also facilitate rapid updating of learning materials to reflect the latest research findings or changes in curriculum standards, ensuring that training content remains current and relevant [11].

Perhaps the most important contribution of the use of AI in education from the student's point of view is the possibility to switch to an application that can be adapted to the needs of the students with adaptive content. For this, students' knowledge levels need to be analyzed, and their progress should be monitored. In addition, students' learning styles can also be analyzed, and their interests can be identified. Content can be prepared accordingly. If content is prepared in a way that students are interested in, both more interactive lessons can emerge, and students' motivation can increase. Thus, better learning gains can be created with a more successful process. These contents prepared by AI can also be distributed to different regions thanks to digital transformation. While it means financial savings for institutions, it also means time savings for teachers as it means that the workload of content preparation is largely eliminated. With this time saving, teachers can focus on other, more effective aspects of their teaching process rather than on preparing content. With large language models, the most up-to-date information can be accessed, and the content can be expanded with this up-to-date information. In this way, students can be kept up to date in education. In addition to the content, assessments and feedback can also be realized with AI algorithms. Thus, missing areas can be identified, and additional courses can be planned [10].

With the ongoing advancement of AI, its use in content creation is poised to significantly alter how digital material is generated, utilized, and engaged with across numerous fields. According to a Gartner report, by 2026, more than 80% of businesses will use productive AI Application Programming Interfaces (APIs) or deploy productive AI-powered applications [12].

AI-generated content has several disadvantages and limitations, as well as advantages. The first is that it does not consider human experiences such as emotions and understandings. Therefore, AI-generated content should be edited by a human being.

As an editor, the tasks of the human factor in relation to this content are to make the produced content credible with human emotions. Similarly, cultural add-ons should also be added by humans. Generally trained large language models and other algorithms can ignore cultural differences. This can threaten to lose the localized features of regional trainings. Culturally specific concepts, such as humor, idioms, and sarcasm, may be ignored. In such cases, editorial touches can ensure that content is in the appropriate language while preserving cultural heritage [13].

Prejudices are another limitation of AI-supported content production because AI algorithms use existing data in the training process. The models here may have developed prejudices during the learning process, and biases and discrimination may persist during the use of these models. Models may emerge that favor one ethnic group, one nationality, or one gender, or vice versa. Representation of different groups may become difficult. The limitations of education data do not stop there. Privacy and data security are also issues of concern for AI-generated content. Specific actions are required for these issues. The evaluation systems applied in AI-supported content should also be transparent and the grades and evaluation made by AI should be explainable. Otherwise, it may again lead to problems. Finally, while these automated processes, thanks to AI, reduce the workload, they may also cause some people to lose their jobs. Therefore, instead of replacing AI algorithms with human trainers, we need to plan how to adapt tasks that complement trainers in training processes [10].

2.2 Use of AI in Digital Educational Games

Educational activities with digital educational games are often referred to as "gamification" in the literature. Gamification is a technique that applies various game elements and mechanics, such as rewards, challenges, competitions, and collaboration, to different fields, such as education, to motivate students and ensure their participation in the lesson [14]. Gamification engages people's minds with complex tasks, motivates them to work, encourages learning, and prompts them to think about the steps needed to solve problems, considering game-based mechanics and aesthetics. Gamification elements cannot be fully described in the text and need to be experienced while playing games. When we talk about gamification, it shouldn't just mean playing games. Gamification isn't just for fun. It's about using game principles to make learning more appealing and effective by focusing on increasing student participation and engagement using motivational elements, rather than just playing games instead of learning [15]. By integrating gamification and AI, it is possible to create educational experiences that are more inventive, engaging, and impactful. AI technologies serve as potent tools in the design of educational games, offering novel perspectives for developing new curricula centered around game-based and project-based learning.

Game-based learning is progressing in two distinct directions: first, the creation of educational games enhanced with AI, and second, the use of educational games as a tool for teaching and learning about AI itself [16]. While AI is widely used in various fields, its integration into digital learning games is still at an early stage. However, there are promising trends and potential applications for AI in educational game design.

AI has the potential to transform education through integration into digital learning games. AI can make learning more effective by offering personalized learning experiences, data-driven analytics, and adaptive gaming environments. Key benefits can be listed as follows:

- Personalized Learning: Together with AI, game experiences can be personalized for each student, adapted to their ability and adjusted to their learning speed. This supports personalized learning.
- Data-Driven Analytics: AI provides real-time information about student performance and interaction, helping teachers make more informed decisions.
- Adaptive Game Environments: AI optimizes the gaming experience by dynamically adjusting the level of difficulty and providing students with the support they need.
- Cultural Sensitivity: AI tools can be designed to include all students from diverse backgrounds.
- Student Engagement: AI increases student engagement by making games more engaging and interactive.
- Efficiency: AI can save teachers' time by automating administrative tasks and helping with student assessments.

AI-enabled games in education can be categorized into different ways. Classification according to the techniques used is one of them. The impact on learning is another. It is also possible to categorize based on subject matter. Or they can be categorized according to the pure game or gamification situations [17]. Hence, they are divided into four classes:

- AI-Based Adaptation: In this type of game, real-time feedback and hints are provided. These are customized to the learner. The level of the students is considered [18].
- AI-Based Decision Support: In this type, options are presented for learners to decide. These options are communicated through games or interactive whiteboards. The aim here is to provide options based on learners' performance rather than automatic step-by-step progression.
- AI Character Interaction: There are no AI-assisted players in such games. Instead, there is a companion character to support the learner.
- Game Analysis and Improvement with Educational Data Mining and/or Learning Analytics: In this type, AI is not used during the game, but after the game to analyze the data obtained from the game. These analyses are used to examine the student's interaction and improve the game.

The combination of AI and inclusive design principles promises digital games that offer a transformative learning experience for every student. By harnessing the power of AI, educators and game designers can contribute to the success of the next generation of learners. Through the responsible and ethical use of AI, digital games can become a powerful tool to enhance future learning.

While the use of AI in educational game design offers many benefits, it is critical to consider the ethical implications. Designers should ensure that AI tools are culturally sensitive, inclusive, and respectful of learners from diverse backgrounds. Cultural sensitivity involves understanding the beliefs and values of various cultures, and games should engage learners by avoiding stereotypes and representational biases. Furthermore, data privacy is an important consideration; as AI-powered games collect student data, data protection regulations should be adhered to, and secure data handling should be implemented. Algorithm bias is another ethical issue, and developers should mitigate bias through fairness measures and regular audits [19].

2.3 AI-Supported Measurement and Evaluation

As a result of the intensive inclusion of online education in learning environments, especially in the last 5 years (the biggest impact of which is the COVID-19 effect), the demand for personalized learning experiences and feedback mechanisms has increased. Both in online learning environments and in classical learning environments equipped with today's technological tools, traditional assessments fail to provide students with individualized feedback and/or timely feedback. This may affect the development of students. However, the solution to this problem was found with the inclusion of AI technologies in education and training environments and processes, and revolutionary developments began to occur.

Assessment is crucial for educational activities and plays a major role in guiding students. It is also necessary to empirically assess skills and knowledge. Regardless of who the training provider is (whether an AI assistant or a teacher), and regardless of the training environment (whether classroom or online), assessment systems are important. This way, achievements can be measured, and feedback can be provided both for the student and for future teaching activities. Similarly, the results of the assessments can be used to prepare the personalized educational content mentioned earlier, and these materials can be recommended. Research has shown that knowledge assessments in learning activities increase the effectiveness of students and provide motivation. As a result of digital transformation activities in education, gains can be measured and evaluated using data that is constantly increasing exponentially. Here, data can be analyzed more easily with the support of AI. Evaluation processes can be carried out not only through content, but different patterns can also be captured from different data with the support of AI. In this way, undiscovered achievements or abilities of the student can also be revealed [20].

For AI algorithms to evaluate learning, they must first learn themselves. In other words, the models created must first learn the answers. Then they can evaluate students according to their level of knowledge. Educators can observe the difference between knowing, understanding, and applying. Getting AI models to do this can be challenging [21]. For now, human and machine assessments are separated. It is highly likely that studies will be carried out to close this gap in the future. If machines can analyze how the criteria are determined, they will perform the evaluation of studies better.

AI algorithms can fully or partially automate some parts of traditional assessment processes. Evaluation steps can be created, automatic scoring can be carried out, especially on platforms with big data, some tasks can be taken from teachers, and machines can perform them. Moreover, it can be free from errors, prejudices, or other negative effects that may be caused by the human factor.

All processes in education do not only consist of tests, multiple choice questions, etc. For example, student writing also requires an analysis. These writings need to be analyzed both semantically and structurally. In fact, plagiarism evaluations should also be made here. Today, various AI tools are used for plagiarism assessment. For other writing problems, studies are being carried out to develop AI tools for both originality and effective evaluation.

While conventional evaluation methods typically capture isolated moments of performance, a range of AI techniques has emerged to offer a more ongoing perspective on performance, thereby providing deeper insights into the learning process. Although some of these methods simply transition traditional assessments, like quizzes and exams, into digital formats, others are applied to entirely different assessment tasks and forms of evidence. Rather than presenting a uniform assessment to every student, AI techniques have also been created to adapt the task to an individual student's capabilities, offering a customized assessment experience. For instance, computerized adaptive testing systems administer an exam using a sequence of questions chosen to maximize the accuracy of the system's real-time estimation of the student's ability. To function effectively, adaptive testing requires an item pool of sufficient size, enabling the selection algorithm to identify a suitable item that matches the test-taker's current ability level.

Thanks to AI, assessment systems have become more holistic. While traditional assessment systems are subject, course, or situation-oriented, more in-depth analyses are now being carried out, thanks to AI support. With the AI, which enables adaptive education systems, students' achievement assessments can now be differentiated. Instead of the same questions for every student, assessments specific to the personalized education of the student can also be made. The preparation of these systems brings some technical challenges. For example, the pool of questions to be created to assess different outcomes and measure personalized education at an adequate level should be as large and diversified [22].

In addition, we can also classify AI measurement and evaluation tools as a transition to modern education and training techniques because these tools save time and labor and increase productivity. Another effect is seen in simulation environments. Simulation is another tool used in education. The evaluations of these tools are made by people who have experience in the simulation environment in the real world. However, today, thanks to AI, simulation evaluations can be carried out by AI in the simulation environment.

In addition to these, AI-supported assessment and evaluation tools in education can provide the following contributions [23]:
– Personalization of learning pathways
    – Analysis of responses
    – Personalized feedback
    – Identification of learning gaps
    – Individualized advice
    – Progress monitoring
    – Multimodal feedback
– Real-time feedback
– Adaptive assessments
– Objective evaluation
– Enhanced data analytics
– Professional development
– Skill assessment

AI-powered assessment tools analyze students' strengths and weaknesses, learning style, and pace to create customized learning journeys with targeted resources. To do this, they analyze written responses, codes, or other forms of input using natural language processing and machine learning to identify correct answers, errors, and areas in need of improvement. This analysis forms the basis for generating feedback. This response analysis also generates feedback specific to the student's performance. This feedback can take various forms, such as written comments, explanations, and suggestions for improvement, or even links to additional learning resources. It can pinpoint areas where the learner is struggling or lacking knowledge and suggest specific learning materials or activities that fit these areas. These suggestions may include articles, videos, practical exercises, or even personalized study plans; track the learner's progress over time through continuous assessment and feedback; do not limit feedback to text; and may include multimedia elements such as audio or video clips for a more comprehensive learning experience [23].

One of the important components for designing AI-supported assessment and evaluation is to be able to prove the consistency of the evaluation results. Especially in recent years, machine learning algorithms and/or deep learning algorithms are used in these evaluation processes. To create these models, large datasets that can train these models are needed. This makes AI-supported measurement and evaluation processes challenging because while there may be evaluation data for a question, there may not be training data for every question that may arise from a paragraph. For this reason, there are various limitations in producing automatic assessment and evaluation tools.

In addition, it is very important to ensure the privacy and security of student data. Furthermore, AI algorithms may inherit biases in the training data. This can lead to biased assessments and feedback. There is a risk that people may become over-reliant on AI assessment tools, potentially sidelining the role of educators and human judgment. It is important to strike a balance between AI-supported feedback and the expertise of teachers or counselors. AI-assisted assessment tools can also struggle to understand the broader context in which a student is operating. They may not grasp the nuances of a learner's circumstances, which can affect performance and feedback interpretation [23]. Likewise, AI-assisted evaluation understandably struggles with its capacity to recognize or measure abstract concepts such as improvisation, creative expression, poetry, morality, and ethics. It may be ill-equipped to appreciate distinctly unconventional, unexpected, or uniquely insightful ways of approaching a learning task. These instances of genuine originality, which a discerning human assessor can identify, might be missed by the technology. To put it succinctly, some dimensions of learning are recognizable by people but not by machines. This is why discussions about AI in assessment must also include a thorough exploration of what the technology is incapable of assessing, both now and potentially in the future [22].

2.4 AI and Career Guidance Applications

Career guidance has a very important role in bridging education and employment services. People are placed in professions where they will manage their lives in line with various factors such as their achievements in education and training processes, competences, skills, abilities, talents, and desires. This process is carried out by educators in today's education and training environments. It is also affected by many factors including family, environment, economic, and social factors. Some characteristics come to the fore, some characteristics are suppressed, and some are captured, while others are overlooked. However, today, almost everyone who has completed the age of education prefers a way to survive in some way. AI technologies have the potential to reshape this process.

Career guidance refers to services and activities aimed at helping individuals of all ages and at any stage of their lives to make education, training, and professional choices and manage their careers [24]. One of the most important reflections of AI applications in education will be career guidance applications. Students are recorded with basic data from the moment they first step into school. Throughout the process, many academic, social, and behavioral data are added continuously. This is where the ability of AI algorithms to analyze data comes into play. In today's education systems, academic achievements are taken into consideration with a large percentage, while a small percentage makes skill-based orientations. The inclusion of career guidance applications and AI in educational environments brings many potentials. Students' abilities, interests, and perceptions can be evaluated in a multidimensional way and career orientations can be made in this direction. People can be enabled to choose professions that they will be happy to do.

The labor market has become increasingly dynamic due to extensive and rapid changes in the professional world, which has also reshaped career perspectives. This shift is accompanied by a growing demand for education across all age groups and educational stages. As a result, career counseling services within higher education institutions are confronted with new challenges. This prevailing dynamism has impelled individuals toward a model of continuous, lifelong learning, thereby creating a corresponding need for ongoing, lifelong career guidance.

Guidance staff traditionally use technology in three ways: learning and career information supporting career building, career assessments, and automated interaction and communication options such as simulations or games [25]. Digital tools not only broaden the spectrum of available services but also provide individuals with new opportunities to access guidance whenever and wherever they prefer. The potential benefits of incorporating technology into career guidance include enhanced accessibility, expanded access to information, assessments, and networks, as well as reduced overall expenditures and improved cost-effectiveness [26, 27]. In today's volatile, uncertain, complex, and ambiguous world, both young people and adults require access to high-quality career guidance. This guidance must be available online and offline, delivered at the times, locations, and in the formats that best align with their individual needs. Concurrently, AI-driven career guidance tools are growing in sophistication and accessibility. These tools utilize algorithms to analyze extensive datasets – comprising labor market information, user profiles, and personal preferences – to generate personalized and highly tailored recommendations. Scalability is another advantage. Career information and advice can reach a wider audience and can be connected to professionally trained career counselors when needed. As in all other aspects of education, career guidance can be personalized, providing personalized recommendations based on students' interests, skills, and occupational preferences. Thanks to data-driven insights, it can easily leverage large datasets and support well-informed choices by providing real-time access to job postings and labor market trends. Cost-effectiveness is another advantage [28].

A historical parallel can be drawn to the Luddites in the early nineteenth century, who feared their weaving skills would become obsolete. Today, career counselors may also worry that the human touch and empathy they offer will be similarly devalued. Furthermore, there are concerns that as AI takes on more roles in the career guidance process, it may introduce prejudices and injustice. AI algorithms, if not sensitively designed and trained, can perpetuate biases and thus risk reinforcing existing inequalities and discrimination in the labor market. Privacy and data security concerns surround the collection and analysis of personal data for career guidance. Various elements of unreliability may also arise in this data, such as lack of transparency, gender and ethnic biases, and lack of privacy. A significant concern regarding AI-assisted career guidance in education is the potential loss of the human touch and emotional support. This support, currently provided by career professionals, is especially critical for vulnerable individuals who have complex needs and circumstances [28]. Concerns also include quality, control, and changing roles in career guidance services, as well as competences and resources required for the provision of AI-enabled services. Students were less explicit about ethical issues. Also, questions about how algorithms work are another source of concern [26].

2.5 AI Training for Professional Development of Teachers

One of the main actors (perhaps the most prominent one) of effectiveness in education is teachers. The more effective the teachers realize the education and training process, the greater their contribution to the students can be. For this reason, professional development of teachers is very important in education and training activities. There is a need for teachers who keep up with the developing and changing world, constantly renew and improve themselves and their knowledge, prepare their students for society, and contribute to their becoming individuals.

Today, teachers' professional development is supported in many ways. Although the most effective way is in-service trainings, in addition to these trainings, their professional development is also supported through various activities such as projects, social activities, workshops, seminars, etc. With the developing technology, student needs change and educational needs are also shaped. The knowledge and technologies at the time when teachers graduated and the knowledge and technologies at the time when they carry out their educational activities differ. For this reason, teachers are also involved in a kind of lifelong learning process throughout their profession.

With the inclusion of AI technologies in educational environments by almost every element (institution, student, teacher, material, evaluation, etc.), the need for teachers to reach a basic level of knowledge about these technologies has arisen. Teachers who will be intertwined with AI technologies in education both as users and data providers are expected to have the knowledge and skill levels of these technologies.

AI research in education is increasing day by day. This is an indication that the applications of AI techniques in education will increase day by day. However, the number of teachers qualified for both curricula and AI is not yet sufficient. The concept of AI is currently new for teachers, and in addition, the use of AI applications requires digital competencies. These requirements about AI may cause teachers to be cold to the subject and it may require a serious effort to learn it at the beginning.

Moreover, it does not seem possible to have knowledge about all the rapidly developing tools. From this point of view, one of the important challenges for AI in education is to teach AI in a way to use it effectively. One of the most important stages of AI integration in education is teacher readiness. Without this, AI in education may not be effective. Here, it is also necessary to investigate the extent to which teachers are open to and ready for the inclusion of AI in their educational processes [29–32].

For teachers to use AI tools in their educational processes, they first need to understand these tools, see their benefits in their field, and trust the results of these tools. There are both technical and theoretical challenges here. Preparing lessons, interacting with AI assistants, and technology integration are technical issues, while making informed decisions and tracking student progress can be theoretical challenges. An AI-supported pedagogical approach is another challenge. However, AI can help teachers in their decision-making processes by contributing to their professional development. It can increase their competencies and contribute to the preparation of effective education and training processes [33].

Rapid technological advancements have placed teachers at the forefront of an educational revolution. This era demands that educators not only reconsider their pedagogical approaches but also maintain a steadfast commitment to their own professional growth. As AI tools, particularly large language models, become more integrated into classrooms, there is a pressing need to adapt teacher professional development. This training must evolve to equip educators with the skills to effectively seize the new teaching and learning opportunities presented by these technologies [34].

It is possible to explain the contribution of AI to teachers' professional development in three groups [35]:
– AI technologies designing and usage for supporting teaching activities
– AI education for people to use it effectively and ethically
– Preparation of people for AI-driven world with innovation in education

AI systems are also very important in terms of saving time. For example, it can provide support in grading or attendance and absenteeism tracking can be done with AI. So, teachers can spend more time with students for activities such as emotional support or social interaction. These social and emotional aspects are crucial for both teacher education and classroom teaching but are often neglected by teacher trainers. AI's role in automating routine tasks can help ensure that these important elements receive the attention they deserve. When explaining AI technologies to teachers, not only its applications but also its ethical implications should be described. These trainings can be provided to teachers through various in-service training courses, as well as through courses that teachers will take through their own efforts or through postgraduate education. Through these trainings, teachers can learn about the ethics of AI, the potentials of AI, and the limitations of AI. Developing AI competence in teachers involves not only developing their basic knowledge and skills but also their AI pedagogical knowledge. This enables teachers to identify the pedagogical benefits of AI, effectively use AI-enhanced teaching methods, and design innovative teaching strategies that combine AI support with human-centered approaches to enhance learning. In addition, AI competence development helps teachers acquire self-regulated learning skills to adapt to the constant changes that AI brings in education and life. These skills should be developed to ensure that teachers are prepared for professional development, can build teachers who develop themselves in AI, and build a network around it can conduct research and use AI tools to further enhance their capabilities. By using an innovative process-oriented approach rather than a purely results-oriented education system, they can develop individuals who are better prepared for the future. As AI is used in all fields, they can ensure that students are more ready for these technologies [35].

2.6 Designing AI Tools and AI Assistants for Teachers

While learning takes place in schools in social and collaborative ways, it is teachers who plan, conduct, and guide it. Until recently, applications of AI have mostly encompassed analyzing services delivered in computer environments and practical applications have been hard to find except in a few specific areas. There were no accessible applications for teachers, and they were not carried to teaching environments. However, in recent years, as in all other fields, AI breakthroughs have taken place in the field of education and training, and now applications have been moved to classroom environments.

Teachers are experiencing an increase in their working hours, driven by the growing complexity of student needs and rising administrative and paperwork demands. A McKinsey study, carried out in collaboration with Microsoft, found that teachers work 50 h per week on average [36]. This figure is supported by the Organization for Economic Co-operation and Development's International Survey on Teaching and Learning, which indicates a 3% rise in these hours over the past 5 years [37].

Most of the teachers stated that they enjoyed their jobs, but they did not enjoy marking papers, preparing lesson plans, or filling endless paperwork until late at night. The pressures of these jobs lead to burnout and high attrition rates. In the United States, teacher turnover exceeds 16% per year [38]. In the UK, 81% of teachers would consider leaving teaching altogether because of their workload [39]. According to the McKinsey & Company report, teachers who work for 50 h spend less than half of their time directly interacting with students and more time on other tasks. The workload distribution in the report can be listed as [40]:
– Preparation: 10.5 h
– Student coaching and advisement: 4.5 h
– Student behavioral-, social-, and emotional-skill development: 3.5 h
– Evaluation and feedback: 6.5 h

- Professional development: 3 h
- Administration: 5 h
- Student instruction and engagement: 16.5 h

Between 20% and 40% of the activities shown in the report can be automated with the tools available today. This means that teachers could spend about 13 h a week on activities that lead to higher student outcomes and higher teacher satisfaction. This is where designing AI tools for teachers and AI assistants come into play. With these technologies, teachers can direct their time to students.

If tools are designed for AI-supported lesson planning, AI-supported preparation of teaching resources, AI-supported homework feedback, and AI-supported construction of assessment structures, teachers' processes can be automated. In this case, more efficient and effective use of time can be provided. The work that teachers spend the most time in educational activities other than the lesson is the preparation for the lesson. Here, preparation of course material takes the most time. Thanks to AI, it has become easier to find and update course materials and make these materials ready for presentation. Presentations can be prepared in a short time with various tools already in use. Designing similar tools for teachers to be used for educational purposes will save a lot of time. Considering the information pollution on the internet, these tools should be tools that can obtain the right information from the right source and make it ready for teachers to use in the lesson environment.

Another AI tool that teachers need is evaluation and feedback tools. Designing an AI tool for evaluation and feedback purposes seems to be the fastest application of AI in education compared to others. Today, thanks to some rule-based tools, it can be easily determined in which subjects students have deficiencies when exam results are evaluated. Thanks to AI, these tools can become more intelligent and exam analyses, subject deficiencies, and individual learning differences can be identified. Feedback will be provided instantly, and time will be saved. The design of these tools for public schools should be carried out by policymakers responsible for education. However, in every country, there are also private institutions operating in the field of education. For these institutions, there is also a need for private institutions to design innovative and AI-supported assessment and feedback tools that will fulfill the same task. There are many private sector companies that have already taken up this task. It seems almost certain that their number will increase soon.

As I mentioned in other topics, there are also various concerns in this topic. The most important problem in designing AI tools for teachers and creating AI assistants is the existence of many tools. It is important to eliminate hesitations such as which tool will be used for what purpose, which one is more useful for whom, and which one is useless. Teachers' digital competences are not at the same level. Therefore, the tools to be designed are expected to be suitable for the usage levels of teachers. Otherwise, educational environments may turn into a garbage dump of AI tools that are not used in practice. For this reason, policymakers need to conduct extensive research while designing. Another hesitation factor is the source of the data accessed and used by the designed tools. For example, where an AI-supported presentation tool that prepares a course material obtains the information it adds to the slides and the use of resources is an important problem. Not all information on the internet is correct. For this reason, such content to be used in educational environments should be carefully prepared. Teachers should be provided with in-service training on this information, information pollution on the internet, use of sources, citation, AI hallucination, and many similar ethical issues. The last hesitation that can be mentioned in this section is the misuse of AI tool designs. The use of these tools may push some teachers away from research, prevent them from self-development, and push them into habits such as being ready-made. Teachers research, examine, analyze, and convey up-to-date information to their students during lesson preparation. However, some teachers who are too accustomed to the processes automated using these tools may lose their research skills.

3 AI Technologies and Systems in Education

In this section, the use of AI technologies in the field of education and the position of AI systems in education are discussed. Adaptive and personalized learning systems, AI systems in virtual reality (VR) and augmented reality (AR) environments, virtual learning assistants and assessment systems, and AI-supported central management systems are discussed.

3.1 Adaptive and Personalized Learning Systems

Personalized learning is an educational technique that focuses on students' individual learning processes and addresses their aspirations and abilities. Thus, it offers the possibility to use teaching techniques specific to students to help them acquire knowledge. This method goes beyond traditional educational techniques. Researchers state that students can learn better when the teaching environment is customized. Personalized education, which is shaped around the idea that one-size-fits-all education does not respond to the different needs and learning styles of students, has the potential to customize the learning journey for each student thanks to AI applications. It has become possible to ensure that the educational content is compatible with students' strengths, weaknesses, and speeds.

The way personalized learning can be applied to students is through adaptive learning systems. Thanks to AI, it is now possible to determine the progress level, speed, deficiencies, and/or areas where each student is good at.

The biggest opportunity that AI offers for personalized learning is its ability to provide students with adaptive learning opportunities. AI-driven platforms can use their learning patterns, strengths, and weaknesses to provide customized, appropriate content, and exercises. One study showed a 62% increase in test scores for students using AI-driven adaptive learning programs [41]. The process difference between traditional learning and adaptive learning systems is shown in Table 1.

Table 1: Roles in traditional learning and adaptive learning [42].

Traditional learning			Adaptive learning	
What do instructors do?	**What do students do?**	**What do instructors do?**	**What does the adaptive system do?**	**What do students do?**
Find content	Everyone receives the same content	Decide what to teach	Finds the best contents	Get a learning plan for themselves
Organize content	If you miss a concept, you fall behind	Coach, manage, and engage learners	Associates concepts with content	They skip the concepts they know
Edit content	Everyone moves at the same pace		Adjusts to content and student achievement	Get advice
Response to the students				Specialize in all concepts

As shown in Table 1, through the application of appropriate speed and complexity according to the learner's proficiency level, AI can facilitate an understanding that allows for an individualized teaching process optimized to help the learner achieve their learning goals [42].

The efficiency of personalized adaptive education systems depends largely on categorizing and collecting information on students' learning styles according to their needs and characteristics and how this information is processed to develop an adaptive and intelligent learning context. In this context, categorizing students' learning styles more accurately is critical for personalized adaptive education systems to provide accurate personalization. Traditionally, methods such as filling out questionnaires have been used to identify students' learning styles. However, this method has some significant disadvantages [43]:

- Time-Consuming: The process of filling in questionnaires can be time-consuming and tiring for students.
- It may produce inaccurate results. Students may not always be fully aware of their own learning styles, which may lead them to give incorrect or misleading answers in questionnaires.

–   Static: The results obtained from questionnaires do not change over time and do not reflect the dynamic nature of students' learning styles.

To overcome these disadvantages, AI approaches are used to automatically detect learning styles in personalized adaptive tutoring systems. The use of AI models offers a more efficient and dynamic method to classify students according to the way they prefer to learn. The advantages of this method are [43]:

–   Faster and Easier: AI models can identify students' learning styles faster and easier than questionnaires.
–   More Accurate: AI models can obtain more accurate and reliable results by analyzing student behavior.
–   Dynamic: AI models can reflect the dynamic nature of learning styles by tracking changes in students' behavior over time.

In addition to these, there are other advantages such as improved learning outcomes, scalability and global accessibility, cost-effective training, data-driven decision making, customized career development, continuous learning culture, competitive advantage, etc.

The use of artificial approaches in personalized adaptive training systems helps to overcome the challenge of personalizing learning by automatically and dynamically matching learners' behavioral characteristics to a specific learning style using machine learning algorithms. In this way, the individual learning process can be optimized, and each learner can be provided with an optimal learning experience.

While the integration of AI into adaptive learning opens new horizons in education, it also brings some important ethical challenges and considerations. We can list these challenges and considerations as follows:

–   Privacy and Data Security: It is critical that student information is stored securely and protected from unauthorized access. Compliance with data protection regulations must be ensured. Security measures such as anonymization and encryption of data must be taken. Informed consent for data collection and use must be obtained from students, parents (for minors), and other interested parties.
–   Algorithmic Bias: It is important that algorithms are unbiased and fair. Biases in datasets can adversely affect the learning results of algorithms. Risks of bias should be assessed and eliminated during the development and use of algorithms. Inclusive algorithms that meet the needs of different student groups should be designed.
–   Transparency and Responsibility: It should be clearly explained how AI systems work and how decisions are made. There should be transparency about how data is collected, used, and shared. Accountability mechanisms should be established to determine who is responsible for mistakes and wrong results.
–   Effective Collaboration: AI should be seen as a tool to enhance and support human expertise, not to replace it. Effective collaboration between educators and

AI experts should be established. AI systems should work under human supervision and control.
– Continuous Monitoring and Evaluation: The accuracy, appropriateness, and compatibility of algorithms with evolving educational goals should be continuously monitored. The performance and effects of the systems should be regularly evaluated, and necessary updates should be made.
– Role of Educators: It is important that educators have knowledge about AI technology and ethical principles. They should have the necessary skills to use AI systems effectively. They should use their expertise to maintain the integrity of education in adaptive learning systems.
– Compliance with Ethical Frameworks: The development and implementation of AI-supported adaptive learning systems should be done in accordance with established ethical guidelines and frameworks. Educators and educational institutions should prioritize ethical practices and not compromise ethical principles.

3.2 AI Systems in VR and AR Environments

The use of AR technology in education enables students to learn in a more engaging and motivating way in different thematic environments. AR supports learning through 3D images and content and strengthens permanent learning. Especially in applied vocational education, such as engineering and science, AR in subjects, such as design, maintenance, repair, assembly, and manufacturing, increases students' understanding and success rates compared to in-class lessons. In cases where repetitive applications are required to develop technical skills, AR increases productivity by providing the opportunity to practice. Due to the financial resources required to fund educational materials, some educational institutions may find it difficult to create experiential learning environments. However, virtual training environments created using AR technology provide wider access by eliminating such cost issues [44].

AR has brought a positive impact on education as it makes it possible to improve learning processes in the classroom. For example, AR has been of particular interest in teaching geometry as it allows the explanation of basic operations such as perimeter, area, and diameter. Presenters can use AR to convey their message to students and interact with them by making them part of the whole presentation. For example, teachers can have students appreciate certain concepts or graphs during a lecture. AR can also be used in prototyping. For example, students may want to use CAD modeling to encourage engagement with certain ideas. Furthermore, AR can also be a suitable tool in UX-UI design teaching. AR benefits education and classrooms in specific ways as it adds innovation and creativity to pedagogical processes. Thanks to AR, old teaching techniques can be replaced with more innovative pedagogies. For example, classrooms can be combined with AR devices to increase students' understanding of a particular subject. Also, manuals and lectures can be supplemented with apps to enrich learning. Students of a teacher who implements AR in his/her lessons may be more motivated to listen to the lectures [45].

VR is a technology that immerses users in simulated environments, which can range from realistic replicas to fantastical settings, effectively transporting them beyond their physical surroundings. This technology provides a comprehensive sensory immersion, facilitated by sophisticated hardware including headsets (or head-mounted displays), motion tracking sensors, and haptic feedback systems. These components operate in unison to place individuals within entirely digital worlds, effectively blocking out the physical world and substituting it with a computer-generated reality. This allows users to navigate and interact within the virtual space. Educational institutions are leveraging VR to facilitate hands-on learning, offering practical experiences across diverse fields and providing students with a more profound and interactive educational journey [46].

VR allows students and teachers to explore content together and generate collaborative knowledge. VR provides spaces for scale experiences that lead to practical learning. Students can learn by doing while using VR goggles and become visual learners through VR educational technologies. VR can also be used to visualize complex topics [45].

AR and VR applications are considered successful when they offer interactive and immersive experiences. AI can enhance these experiences, providing real-time responses and personalized environments. AI can provide contextual information in AR by recognizing the environment of applications. In VR, it can create more realistic and interactive virtual environments. AI-powered characters have the opportunity to create more immersive experiences by learning and adapting to users' behaviors. AI algorithms personalize and improve AR and VR experiences by learning from user interactions, so that users' actions can be reacted to in real time. Natural language processing, an area of AI research, can enable AR and VR systems to increase realism by understanding and responding to human speech. AI has the potential to transform AR and VR beyond development, making the user experience more immersive and engaging, and providing great benefit, especially in learning experiences [47].

The use of AI-enabled AR and VR systems in education is becoming increasingly popular as they offer an effective and interactive learning experience. However, to implement these technologies successfully, there are several challenges that need to be addressed. Cost is one of the biggest challenges of AR and VR in education. Given the budget of schools, these are expensive technologies for schools yet. In addition, the content to be used in these technologies needs to be regularly updated. This requires expertise and may require additional hardware and software. Therefore, these are also additional costs. Another challenge for AR and VR technologies is ensuring that all students have access to these technologies. Students do not have access to these technologies by their own means. It is a challenge for schools to provide it. From a health perspective, AR and VR technologies can be overwhelming for some students [48]. Furthermore, the demand for high-quality, real-time AR/VR experiences is constrained by existing processing power. Surpassing current hardware limitations is necessary to achieve efficient interactions within complex environments. By their very nature, AR/VR applications involve gathering and processing substantial amounts of user data, which presents an ongoing challenge in balancing the desire for personalized experiences against user privacy rights. Crafting intuitive and user-friendly interfaces for AR/VR is also a significant hurdle, as navigating virtual spaces and interacting with digital elements requires a careful equilibrium between simplicity and functionality [46].

3.3 Virtual Learning Assistants and Assessment Systems in Education

The development of educational technologies has led to radical changes in teaching and learning processes. The requirements of the digital age necessitate the development of new tools and methods in addition to traditional education methods. In this context, virtual learning assistants and assessment systems are among the important innovations in education. These technologies both enrich students' learning experiences and make teachers' assessment processes more effective and efficient.

Another tool used in AI is virtual learning tools. These tools support students and teachers, making learning and teaching activities more advanced. With these tools, educational content can be created, or assessments can be automated. Students can be provided with personalized learning and instant feedback. In addition, with virtual learning tools, progress in the learning process can be monitored instantly and made more effective and efficient [49]. Virtual learning assistants have advantages such as personalized learning, instant feedback, question-answer support, tracking and supporting motivation, and helping with tasks and assignments.

Virtual assistants can answer questions, provide explanations, and offer guidance on a wide range of topics. This not only increases student engagement but also allows teachers to focus on more complex tasks such as designing curriculum and assessing student progress. Another area where AI-powered virtual assistants can make a significant impact is in the administrative processes of school districts. These virtual assistants can perform routine administrative tasks such as scheduling meetings, managing calendars, and organizing documents. This frees up valuable time for school administrators and teachers, allowing them to focus on more strategic initiatives. Furthermore, virtual assistants can analyze large amounts of data to identify trends and patterns and provide valuable information that can inform decision-making in school districts [50].

Virtual assistants have the potential to help with the increasing workload of teachers, while at the same time, creating supportive environments for students. These tools also have mental and emotional advantages. Teachers who are strained under excessive workload face their own mental health problems. When teachers' workload is reduced, thanks to virtual teaching assistants, their stress levels will be reduced accordingly. Another time-saving feature for teachers is detailed feedback. With virtual assistants, feedback can be automated, saving time for teachers. One of the tasks of teachers is to identify students' learning gaps and this is a difficult task. Virtual teaching assistants can identify these gaps with the data they can collect and provide support to teachers. This feature also contributes to personalized learning [51].

AI-supported assessment systems are digital tools that use AI to measure students' knowledge and skills. These systems both carry traditional examination methods to the digital environment and offer different approaches with innovative assessment methods. Automatic assessment offers advantages such as comprehensive exam analysis, instant feedback, time saving, reducing subjectivity, and reducing human errors.

Virtual assistants have limitations, as well as benefits. These limitations can be grouped as technical, pedagogical, and ethical. Technical issues are related to the technological barriers associated with the implementation of virtual teaching assistants in classrooms and the use of AI in a dynamic environment such as education. From a pedagogical point of view, virtual teaching assistants should integrate effectively with teachers and enhance students' learning without distracting them from the core curriculum. Finally, ethical challenges include issues related to privacy and the emotional impact of virtual teaching assistants [51].

3.4 AI-Supported Central Management Systems

The contributions of AI to educational processes are not only based on teachers and students. It also contributes on an institutional basis. It is possible to add central management systems to these contributions. As in all other fields, there is a digital transformation in the field of education and training. With the transition from manual process management to digital processes, especially in the last 20 years, a ground has been prepared for AI. AI-supported central management systems have started to be equipped with tools that facilitate administrative tasks, speed up, and make institutional processes more effective and increase overall institutional efficiency.

When central management systems are equipped with AI tools where all data of educational institutions are collected and data-based decisions are made, they will bring along various advantages. For example, school and classroom planning according to birth rates in regions, automating the enrolment and admission of students in schools, planning student orientation in transitions between levels can be carried out. Both current situation analyses and future forecasts can be made on many issues such as teacher needs of institutions, future teacher planning, resource consumption and their optimization, expenditures, and management.

Central management systems can also be considered as systems of systems. It means that the data of each tool applied for specific purposes in educational environments can be brought together and used to determine, plan, and direct the educational strategies of the country. Obtaining, storing, analyzing, and using such big data also brings risks and limitations. Hardware requirements are at the top of these limitations. When we evaluate the risks, data security comes first. Platforms with such a large volume and such sensitive information should be protected with the sensitivity required.

4 AI Policies in Education

In this section, AI policies in education are discussed. AI and ethics, the role of AI in education policies and recommendations, and threats of AI and possible solutions are analyzed.

4.1 AI and Ethics

As AI is included in human life, its risks, as well as its advantages, are discussed. Ethics is also included in these discussions. UNESCO has identified 10 basic principles to put forward a human rights-centered approach to AI ethics. These principles can be listed as follows [52]:
1. Proportionality and nondetriment
2. Safety and security
3. Right to privacy and data protection
4. Multistakeholder and adaptive governance and co-operation
5. Responsibility and accountability
6. Transparency and disclosure
7. Human surveillance and identification
8. Sustainability
9. Awareness and literacy
10. Fairness and nondiscrimination

AI systems should only serve legitimate purposes and risk assessment should be carried out to prevent harm. Security risks and vulnerabilities to attacks should be addressed by AI actors. Privacy should be protected throughout the AI life cycle, and international law and national sovereignty should be considered in data use. Inclusive AI governance requires the participation of various stakeholders. AI systems should be auditable and traceable. Oversight, impact assessment, audit, and due diligence mechanisms should be implemented against threats to human rights norms and environmental well-being. Ethical use of AI systems requires transparency and accountability. It must be ensured that these technologies do not replace human responsibility. AI should be assessed against sustainability goals such as the UN's Sustainable Development Goals. Open and accessible education, civic engagement, digital skills, and AI ethics training are important. AI actors should embrace the principles of social justice, fairness, and nondiscrimination [52].

While AI has the potential to foster inclusive and equitable learning experiences as it rapidly integrates into the field of education, important ethical issues should not be overlooked. Concepts, such as bias, fairness, inclusivity, and accessibility, should serve as guiding principles for the ethical application of AI [53].

One of the biggest challenges in AI systems is the prevention of bias. Bias can arise from unbiased data and algorithms that can lead to discrimination against certain groups. Therefore, the use of representative and diverse training data is critical. Different initiatives aim to detect and address bias by developing algorithms that incorporate bias measures and by conducting regular audits.

Large language models are designed to recognize patterns by analyzing text, and to do so by analyzing large amounts of data. Generative AI models mimic the training data used to build these models. Therefore, if a model learns a demographic characteristic or a language or an industry with more data, it will be trained predominantly in that direction. This is where bias comes into play. Large language models may have bias due to the training data. Due to bias, the information obtained from the models may be inaccurate or misleading. Even unethical information may be given, or intellectual property rights may be violated. These models may mimic individuals or institutions by acting as imitators. This poses a risk to the information provided to users [54]. Therefore, it is important to critically evaluate what AI produces.

Explainability and transparency are also very important in the application of ethical AI. It should be clearly explained how AI systems work and what data they are based on. In this way, educators, students, and other stakeholders can clearly understand and trust the functioning of AI.

Accessibility and inclusion are the cornerstones of ethical AI-supported education design. AI systems should be designed to meet the needs of all learners, including disabled and disadvantaged groups. Various learning opportunities suitable for different learning styles and skills should also be offered.

The reliability of the content produced with AI and the materials used in education should be questioned ethically. For this, teachers and students can make the following enquiries [55]:

- Is the content generated by AI accurate? How can you test or evaluate accuracy?
- Can reliable sources other than generative AI verify the data or item generated?
- How does the information generated affect or influence your thinking on this topic?
- Who is represented in this data? Is the data inclusive in terms of the scope of the material and the perspectives it presents?

– Knowing that big language models can also collect data entered by your students, how will you make them aware of this practice? Will they thus also protect their privacy?

4.2 The Role of AI in Education Policies and Recommendations

The main purpose of applying AI in education is to improve learning, to enable each student to develop his/her individual potential, to assist teachers, to make institutions more effective, and for policies to reflect and support this. To make this valid, countries should meet the requirements of the modern era, make the necessary technological investments, and additionally support human resources. This will enable the development of individuals who know how AI works and how it can be created, and who internalize it to use it for local and global society.

In this context, UNESCO has published a system-wide vision and strategic priorities document, outlining four strategic goals to be achieved, interpreted according to the local context [56]:
– Ensuring inclusive and equitable use of AI in education
– Utilizing AI to improve teaching and learning
– Promote the development of life skills in the age of AI, including teaching how AI works and its impact on humanity
– Ensuring transparent and auditable use of education data

Achieving these strategic goals requires a multistep approach. First, it is essential to establish fundamental national principles for AI and education policy. This must be followed by interdisciplinary planning and cross-sectoral governance. Concurrently, policies ensuring the equitable, inclusive, and ethical application of AI must be formulated. A comprehensive master plan is needed to guide the use of AI in educational administration, teaching, learning, and assessment. Finally, success depends on pilot testing, continuous monitoring, and evaluation to build a solid evidence base, alongside dedicated efforts to foster local AI innovations for education [56]. Policymakers should understand how existing laws affect educational uses of AI and update or adopt policies to effectively balance the opportunities and risks of AI. Policy updating is a cyclical process of assigning responsibilities, assessing needs and policy gaps, aligning with existing objectives and policies, evaluating the policy, approving the policy, implementing the policy, and evaluating policy implementation [57]

Key AI policies, such as fostering leadership, promoting AI literacy, providing guidance, building capacity, and supporting innovation, can support educators, staff, students, and parents to engage with AI in ways that complement, rather than replace, the human touch. With an effective policy and effective practices planned, AI-supported education can be provided to all students on an equal footing. Transparent and accountable AI integration also requires collaboration with the sector and the implementation of oversight mechanisms. The foundation of AI literacy is computer science, but there are still students who do not even have the opportunity to take a computer course. When people understand how AI works, such as its limitations and social impacts, they can learn to use it effectively and responsibly. Administrators should adopt security and privacy policies and monitor compliance with them, provide students and staff with clear guidance on the opportunities and risks of AI and the responsible and prohibited uses of AI tools, and establish policies to verify that students and staff are adequately trained to ensure the safe use of AI. Educators need quality professional development to use AI effectively. States should also set policies that signal to educator preparation program providers the importance of developing knowledge and skills in the use of AI. AI offers the possibility to increase innovation in education if we do so in a thoughtful way that considers safe and effective practices [57].

4.3 Challenges, Threats, and Possible Solutions of AI in Education

Along with the opportunities offered using AI in education, educators should also be supported in overcoming the emerging challenges and threats. One of the important threats of AI in education is the replacement of teachers. The most important challenge is to ensure that it does not replace teachers. AI should not be used as a tool to replace teachers, but as a tool to support them. The role of the teacher is crucial in shaping students' minds, behaviors, and lives, and AI should not take this important role away from them. Therefore, it is very important to use AI with teachers to improve their teaching capabilities. While AI can help teachers in areas such as lesson planning, assessment, and student tracking, teachers should continue to develop students' social, emotional, and critical thinking skills. This co-operation will benefit both teachers and students by improving the quality of education.

Another threat is the issue of information security in these systems. The three basic elements of information security, namely confidentiality, integrity, and accessibility, are always under threat in cyberspace. There are millions of students' data in educational environments. AI systems contain large amounts of data such as institutional information and statistics, including personal information about students and teachers. This data must be protected against cyberattacks and threats. Therefore, it is necessary to have robust data privacy policies and security measures to protect this information. Another security threat to data is the threat posed by generative AI technology. Generative AI tools continue training in a dynamic way, drawing inferences from the prompts given by the user. The information provided by the user can be recorded and analyzed. For this reason, data privacy is important and providing personally identifiable information should be avoided. Similarly, educators should not give this information about themselves and their students to the models. Those in the education community should also not provide private data and information about their colleagues for the training of the models. Sensitive and detailed data should be protected within information security elements.

After mentioning data security, another risk factor, ethics, should also be mentioned. There are many ways of using AI tools that threaten ethical elements. For example, compromising the privacy of students by using face recognition and recommendation systems, unauthorized storage and use of student competencies, unauthorized use of personal data, recording students' emotional states, and the possibility of misusing large volumes of data recorded from students by extracting strategies from these data constitute a potential ethical violation. For this reason, its integration into education should be done very carefully and a framework of ethical rules for the use of AI in education should be established. Policies for this should be expanded and applied to education. All these regulations should be realized in an ethical, accountable, transparent, and secure manner.

Inequality of technological opportunity is a challenge for AI in education. Not all students in the education process have equal access to computer devices and the internet. Similarly, there are level differences in different countries in terms of the technological infrastructure of education. In underdeveloped countries, the use of AI in education may create economic and social divisions. To prevent this, basic technological infrastructure barriers need to be overcome first. In addition, this imbalance among students may accelerate the widening of the achievement gap between students at different socioeconomic levels. This issue can also be viewed from a different perspective. In countries or socioeconomic regions that are too intertwined with technology, there may be the opposite effect. Students who spend more time using AI systems may spend less time with their teachers or classmates or their interaction time may become very limited. Today, AI systems can take the role that social media plays in daily life in education. This may create situations such as loneliness, self-isolation, and loss of self-expression skills in students. Students may use AI to solve homework problems or conduct research, but they may submit the content prepared by the productive AI without reading it even once. Thus, they may appear to have done their homework in the education system, but they may not have developed learning skills on this subject. Such tricks not only pose ethical problems but also may cause students to be deprived of learning the relevant outcomes by having the AI do their own work. Therefore, the role of AI literacy is great here. Both teachers and students can receive seminars and trainings within the scope of AI literacy. It is important to create awareness of AI in people and to ensure that these tools are used to support the human element in education. Teachers should also learn new digital skills to use AI in a pedagogical and meaningful way. AI developers should also learn how teachers work and provide them with sustainable solutions in educational environments.

Biased learning of algorithms is another threat to the use of AI in education. AI systems learn from the data on which they are trained. If this data contains biases (e.g., against an ethnicity, gender, or a socioeconomic segment), these biases can be learnt and maintained by the AI system. For this reason, it should be explained where and how the AI tools to be used in educational platforms obtain the data used in the model building process, how they are trained on this data, and the reliability of their results. Ways of creating reliable models that have worked on participatory and diverse data and whose results are free from bias should be investigated.

The first way to prevent these threats and challenges is to prepare effective and valid policies for AI applications in education. Although it is difficult for policymakers to keep up with this pace, especially since the rate of development of AI has increased significantly in the last two decades and is gaining momentum every year, steps should be taken immediately in order not to be late. AI, which is sure to have a different impact than other technologies previously included in educational environments, seems to herald a new era and transformation in education. For example, while productive AI had difficulties even in some basic math, it has improved itself a lot in 2–3 years and has become able to solve even the most complex problems. In the following years, these systems will be more accessible and will further improve themselves. What is important here is how human beings will be positioned here. It is important to deal with these questions when determining education policies.

5 Conclusion

With the use of AI technologies in education, a transition toward a new era has begun. This transition process has been accelerating especially in recent years. It is now possible to see AI algorithms and models using these algorithms in many applications and systems. Although the term AI has been used especially in the last 70 years, the development of these technologies has been realized with hundreds of years of scientific, theoretical, and practical accumulation. In recent years, together with productive AI, it has found many applications especially in education. While the use of AI in education brings opportunities, it also poses challenges and threats.

Today, AI applications now contribute to content production in education by creating scenarios and preparing video–audio content. Teachers save time and use up-to-date content by preparing course content with up-to-date information and even having it prepared by AI applications. Gamification in education is one of the most used teaching techniques in recent times, and now the use of AI in digital educational games has also started. The game industry is dynamic and constantly evolving. AI models add another dimension to this dynamism. The use of AI in measurement and evaluation is becoming more active and comprehensive every day. AI contributes to individualization in education by determining the subjects that students are deficient in. Similarly, an education program suitable for the learning speed of students can be adjusted. Career guidance is another application of AI in education. Educational platforms that record the data of students from the first moment they start school can follow the progress of these students and give advice on which profession they should be oriented toward. AI applications also support teachers in education. Teachers use AI tools for professional development. AI assistants, content preparation applications, evaluation applications, research applications, etc., support educators with many applications.

AI technologies in education have started to be designed and used not only as applications but also as systems. Systems such as adaptive and personalized learning systems, AI systems in VR and AR environments, virtual learning assistants and evaluation systems, and AI-supported central management systems can be counted among these systems. These systems are expected to increase even more in the future.

One of the important challenges that all these technologies bring with them is the issue of ethics. AI and ethical concerns expressed by researchers in many different studies are at the forefront of the factors that need to be dealt with by policymakers. In addition to these, determining the roles of AI in education as a framework, and planning what kind of education will be designed in interaction with the human factor are among the important issues for the future. Seeking effective solutions to the possible threats and challenges of AI is not the duty of a single nation, but of all nations in transnational co-operation. In the literature, there are various items such as algorithmic bias as general threats, inequality of opportunity in the use of AI technologies, inadequacy of educators in AI literacy, inhibition of students' research and development skills by doing their homework, inability to determine generalized policies, and inability to establish technological infrastructure.

We are on the threshold of a new era in education with AI and the process is moving very fast. It seems that many revolutionary new applications, systems, and policies will be shaped by AI technologies one after another in the near future. Considering the speed of development of productive AI in the last few years, it will not be difficult to make this prediction. Therefore, education managers need to determine strategies and take actions for this upcoming future.

In this chapter, unlike other studies in the field of AI in education, specific examples are given based on application, system, and policy. These examples are grouped under subheadings. The advantages and limitations of each grouped subheading are also indicated. Thus, different more specific solutions are presented during the implementation phase, and it is aimed to be a guide for policymakers, implementers, and users.

References

[1]   Savaş S. Artificial intelligence and innovative applications in education: The case of Turkey. *Journal of Information Systems and Management Research*. 2021;3:14–26.

[2]   Lavidas K, Voulgari I, Papadakis S, Athanassopoulos S, Anastasiou A, Filippidi A, Komis V and Karacapilidis N. Determinants of humanities and social sciences students' intentions to use artificial intelligence applications for academic purposes. *Information*. 2024;15(6):314. doi: 10.3390/info15060314.

[3]   Ulaşan F. The use of artificial intelligence in educational institutions: Social consequences of artificial intelligence in education. *Korkut Ata Türkiyat Araştırmaları Dergisi*. 2023;1305–1324. https://doi.org/10.51531/korkutataturkiyat.1361112.

[4]   Malik AR, Pratiwi Y, Andajani K, Numertayasa IW, Suharti S, Darwis A and Marzuki. Exploring artificial intelligence in academic essay: Higher education student's perspective. *International Journal of Educational Research Open*. 2023;5:100–296. https://doi.org/10.1016/J.IJEDRO.2023.100296.

[5]   Meço G and Coştu F. Using artificial intelligence in education: Descriptive content analysis study. *Karadeniz Teknik Üniversitesi Sosyal Bilimler Enstitüsü Sosyal Bilimler Dergisi*. 2022;12:171–193.

[6]   Satir T and Korucu AT. An evaluation on the use of artificial intelligence in education specific to ChatGPT. *Shanlax International Journal of Education*. 2023;12:104–113. https://doi.org/10.34293/education.v12i1.6513.

[7]   Dağlı SK. Eğitimde Yapay Zekâ Uygulamalarının Çocuk Haklarına Etkisi ve Öğretmenin Rolü. In: Deniz ME and Polat Ü (eds.). *Xth International TURKCESS Education and Social Sciences Congress Abstract Book*. Prizren Ukshin Hoti University; 2024. pp. 17–19.

[8]   Lampropoulos G and Papadakis S. The educational value of artificial intelligence and social robots. In: Lampropoulos G and Papadakis S (eds.). *Social Robots in Education*. Cham: Springer; 2025. Studies in Computational Intelligence vol. 1194. pp. 1–17. doi: 10.1007/978-3-031-82915-4_1.

[9]   Firefly A. Yapay zeka destekli içerikler nedir ve nasıl yapılır? Yaratıcı ve Üretken Yapay Zeka 2024. https://www.adobe.com/tr/products/firefly/discover/what-is-ai-art.html (accessed June 15, 2024).

[10]   AIContentfy team. AI-generated content for education and learning. AIContentfy 2024. https://aicontentfy.com/en/blog/ai-generated-content-for-education-and-learning (accessed June 16, 2024).

[11]   Culican J The impact of AI on educational content creation: Shaping the future of learning materials | LinkedIn. Linkedin 2024. https://www.linkedin.com/pulse/impact-ai-educational-content-creation-shaping-future-jamie-culican-o7nxe/ (accessed June 16, 2024).

[12]   Gartner. Gartner says more than 80% of enterprises will have used generative AI APIs or deployed generative AI-enabled applications by 2026. Gartner 2023. https://www.gartner.com/en/newsroom/press-releases/2023-10-11-gartner-says-more-than-80-percent-of-enterprises-will-have-used-generative-ai-apis-or-deployed-generative-ai-enabled-applications-by-2026 (accessed June 16, 2024).

[13]   OriginHope. The importance of AI-based content in the digital era | Origin hope. The importance of AI-based content in the digital era | Origin hope 2024. https://originhope.com/resources/importance-ai-based-content-digital-era (accessed June 16, 2024).

[14]   Aguire H Gamification and artificial intelligence. TBox 2024. https://info.tboxplanet.com/en/gamification-and-artificial-intelligence/ (accessed June 16, 2024).

[15]   Kapp KM The gamification of learning and instruction. The gamification of learning and instruction 2024. https://karlkapp.com/ (accessed June 17, 2024).

[16]   Dyulicheva YY and Glazieva AO. Game based learning with artificial intelligence and immersive technologies: An overview. *CEUR Workshop Proceedings*. 2022;3077:146–159.

[17]   McLaren BM and Nguyen HA. Chapter 20: Digital learning games in artificial intelligence in education (AIED): A review. In: *Handbook of Artificial Intelligence in Education*. Cheltenham, UK: Edward Elgar Publishing; 2023. pp. 440–484. https://doi.org/10.4337/9781800375413.00032.

[18]   Martin SM, Casey JR and Kane S. *Serious Games in Personalized Learning: New Models for Design and Performance*. Taylor & Francis; 2021.

[19]   Hyperspace. AI in educational game design. AI in educational game design 2024. https://hyperspace.mv/educational-game-design/ (accessed June 17, 2024).

[20]   Minn S. AI-assisted knowledge assessment techniques for adaptive learning environments. *Computers and Education: Artificial Intelligence*. 2022;3:100050. https://doi.org/10.1016/J.CAEAI.2022.100050.

[21]   Gardner J, O'Leary M and Yuan L. Artificial intelligence in educational assessment: 'Breakthrough? Or buncombe and ballyhoo?. *Journal of Computer Assisted Learning*. 2021;37:1207–1216. https://doi. org/10.1111/JCAL.12577.

[22]   Swiecki Z, Khosravi H, Chen G, Martinez-Maldonado R, Lodge JM, Milligan S, Selwyn N and Gašević D. Assessment in the age of artificial intelligence. *Computers and Education: Artificial Intelligence*. 2022;3:100075. https://doi.org/10.1016/J.CAEAI.2022.100075.

[23]   Joshi SC Revolutionizing assessment and evaluation in online learning with AI tools | LinkedIn. Linkedin 2024. https://www.linkedin.com/pulse/revolutionizing-assessment-evaluation-online-learning-c-joshi-t7jlc/ (accessed June 19, 2024).

[24]   OECD. Career guidance and public policy bridging the gap. *OECD*. 2004. https://doi.org/10.1787/9789264105669-EN.

[25]   Hooley T, Shepherd C and Dodd V. *Get Yourself Connected: Conceptualising the Role of Digital Technologies in Norwegian Career Guidance*. Routledge; 2015. https://doi.org/https://doi.org/10.48773/93y86.

[26]   Westman S, Kauttonen J, Klemetti A, Korhonen N, Manninen M, Mononen A, Niittymäki S and Paananen H. Artificial intelligence for career guidance – Current requirements and prospects for the future. *IAFOR Journal of Education*. 2021;9:43–62. https://doi.org/10.22492/IJE.9.4.03.

[27]   Sampson JP Jr, Peterson GW, Reardon RC and Lenz JG. Key elements of the CIP approach to designing career services. In: *Center for the Study of Technology in Counseling and Career Development*. Tallahassee: Florida State University; 2003.

[28]   Hughes D Artificial Intelligence (AI) and career guidance: Transformation or Extinction? The coming age of the robo-careers advisers. OEB Insights 2023. https://oeb.global/oeb-insights/artificial-intelligence-ai-and-career-guidance-transformation-or-extinction-the-coming-age-of-the-robo-careers-advisers/ (accessed June 19, 2024).

[29]   Kitcharoen P, Howimanporn S and Chookaew S. Enhancing teachers' AI competencies through artificial intelligence of things professional development training. *International Journal of Interactive Mobile Technologies (IJIM)*. 2024;18:4–15. https://doi.org/10.3991/ijim.v18i02.46613.

[30]   Ayanwale MA, Sanusi IT, Adelana OP, Aruleba KD and Oyelere SS. Teachers' readiness and intention to teach artificial intelligence in schools. *Computers and Education: Artificial Intelligence*. 2022;3:100099. https://doi.org/10.1016/J.CAEAI.2022.100099.

[31]   Kim K and Kwon K. Exploring the AI competencies of elementary school teachers in South Korea. *Computers and Education: Artificial Intelligence*. 2023;4:100–137. https://doi.org/10.1016/J.CAEAI.2023.100137.

[32]   Ng DTK, Leung JKL, Su J, Ng RCW and Chu SKW. Teachers' AI digital competencies and twenty-first century skills in the post-pandemic world. *Educational Technology Research and Development*. 2023;71:137–161. https://doi.org/10.1007/S11423-023-10203-6/FIGURES/2.

[33]   Tammets K and Ley T. Integrating AI tools in teacher professional learning: A conceptual model and illustrative case. *Frontiers in Artificial Intelligence*. 2023;6. https://doi.org/10.3389/frai.2023.1255089.

[34]   dienabou. *Exploring New Horizons: Teacher Professional Development in the Age of AI – Insights from Thematic Seminar and Report*. European Schoolnet Academy; 2024. https://blog.europeanschoolneta cademy.eu/2024/02/exploring-new-horizons-teacher-professional-development-in-the-age-of-ai-insights-from-thematic-seminar-and-report/ accessed June 19, 2024.

[35]   Luckin R and Cukurova M. Designing educational technologies in the age of AI: A learning sciences-driven approach. *British Journal of Educational Technology*. 2019;50:2824–2838. https://doi.org/10.1111/BJET.12861.

[36]   McKinsey. *McKinsey Global Teacher and Student Survey*. Singapore, United Kingdom, and United States: Average of Canada; 2017.

[37]   OECD. *TALIS 2018 Results Teachers and School Leaders as Lifelong Learners*. vol. I. OECD; 2019. doi: https://doi.org/10.1787/1D0BC92A-EN.

[38]   Carver-Thomas D and Darling-Hammond L Teacher turnover: Why it matters and what we can do about it. 2017. https://doi.org/10.54300/454.278.

[39]   NEU. Teachers and workload. 2018.

[40]   Bryant J, Sanghvi S and Wagle D How artificial intelligence will impact K-12 teachers. 2020.

[41]   MMD. Knewton personalizes learning with the power of AI. Digital Innovation and Transformation 2021. https://d3.harvard.edu/platform-digit/submission/knewton-personalizes-learning-with-the-power-of-ai/ (accessed June 22, 2024).

[42]   Claned. The role of AI in personalized learning. The role of AI in personalized learning 2024. https://claned.com/the-role-of-ai-in-personalized-learning/ (accessed June 22, 2024).

[43]   Essa SG, Celik T and Human-Hendricks NE. Personalized adaptive learning technologies based on machine learning techniques to identify learning styles: A systematic literature review. *IEEE Access.* 2023;11:48392–48409. doi: https://doi.org/10.1109/ACCESS.2023.3276439.

[44]   Guler O and Yucedag I Developing an CNC lathe augmented reality application for industrial maintanance training. ISMSIT 2018 – 2nd International Symposium on Multidisciplinary Studies and Innovative Technologies, Proceedings 2018. https://doi.org/10.1109/ISMSIT.2018.8567255.

[45]   Vection Technologies. Virtual reality, augmented reality, and 3D technologies in the education industry: How are learning processes revolutionizing? Vection Technologies 2024. https://vection-technologies.com/solutions/industries/education/ (accessed June 22, 2024).

[46]   Adatia V How AI works with augmented and virtual reality for businesses. WDCS Technology 2024. https://www.wdcstechnology.ae/how-ai-works-with-augmented-and-virtual-reality-for-businesses (accessed June 22, 2024).

[47]   Mattan M The role of AI in augmented and virtual reality. Brand XR 2023. https://www.brandxr.io/the-role-of-ai-in-augmented-and-virtual-reality (accessed June 22, 2024).

[48]   Al-Ansi AM, Jaboob M, Garad A and Al-Ansi A. Analyzing augmented reality (AR) and virtual reality (VR) recent development in education. *Social Sciences & Humanities Open.* 2023;8:100532. https://doi.org/10.1016/J.SSAHO.2023.100532.

[49]   Nahas E The use of AI assistants in learning. ELearning Industry 2023. https://elearningindustry.com/the-use-of-ai-assistants-in-learning (accessed June 25, 2024).

[50]   OBTECH Enterprise. Exploring the potential of AI-powered virtual assistants in digital intelligence in the education sector. LinkedIn 2023. https://www.linkedin.com/pulse/exploring-potential-ai-powered-virtual-assistants-digital-gylvc/ (accessed June 25, 2024).

[51]   Audras D, Zhao A, Isgar C and Tang Y. Virtual teaching assistants: A survey of a novel teaching technology. *International Journal of Chinese Education.* 2022;11:2212585X221121674. https://doi.org/10.1177/2212585X221121674.

[52]   UNESCO. Ethics of artificial intelligence. UNESCO 2024. https://www.unesco.org/en/artificial-intelligence/recommendation-ethics (accessed June 27, 2024).

[53]   European Commission. Ethical considerations in educational AI. European school education platform 2024. https://school-education.ec.europa.eu/en/insights/news/ethical-considerations-educational-ai (accessed June 27, 2024).

[54]   Antoniak M, Lucy L, Sap M and Soldaini L Using large language models with care. Medium 2023. https://blog.allenai.org/using-large-language-models-with-care-eeb17b0aed27 (accessed June 28, 2024).

[55]   Cornell University. Ethical AI for teaching and learning. Center for teaching innovation 2024. https://teaching.cornell.edu/generative-artificial-intelligence/ethical-ai-teaching-and-learning (accessed June 28, 2024).

[56]   Fengchun M, Wayne H, Huang R and Zhang H. UNESCO. AI and education: Guidance for policy-makers. *UNESCO.* 2021. doi: https://doi.org/10.54675/PCSP7350.

[57]   TeachAI. AI policy landscape. Foundational policy ideas for AI in education 2024. https://www.teachai.org/policy (accessed June 28, 2024).

Zacharias Andreadakis

The Critical AI Agent: Enhancing Critical Feedback and Academic Productivity in the Age of Artificial Intelligence

Abstract: The capacity to give and receive critical feedback lies at the heart of academic productivity. It shapes arguments, refines methods, and fosters intellectual growth. Yet, across universities, research centers, and journals, this capacity is weakening. Time is scarce. Demands are rising. The spaces where critique once took root, such as in seminars, manuscript margins, and collaborative drafts, are shrinking. What remains is often fragmented: hurried comments, delayed reviews, uneven advice, etc. The result is a feedback culture under pressure, one that struggles to sustain the depth and clarity that rigorous thinking requires. This study explores whether new tools, specifically AI agents built on large language models, might help recover some of what is being lost; not by replacing academic judgment, but by supporting it. These systems are increasingly able to evaluate structure, identify unclear reasoning, and offer consistent, text-sensitive critique. They are not perfect. But they are improving, and they invite serious consideration, not as shortcuts, but as scaffolds. I begin with a diagnosis: the structural forces that have eroded feedback as a core academic practice. I then turn to recent developments in AI-assisted evaluation, focusing on agents designed for critique and creative feedback, rather than generation of zombified bot tropes. Finally, I propose a cautious framework for integration, one that treats AI as a supplement to scholarly attention, not a substitute for it. Ultimately, I explore how, if used judiciously, maybe AI agents can become our Socratic daimon for critical feedback.

Keywords: Artificial intelligence, AI agents, critical feedback, academic productivity, peer review, large language models, research integrity, scientific publishing, AI ethics, evaluative judgement, human-AI collaboration

1 Introduction: The Imperative for Critical Feedback in an Era of Academic Strain and AI Emergence

The modern academic enterprise is shaped by mounting structural pressures. The imperative to publish, originally intended to drive discovery, has now become a source

Zacharias Andreadakis, Department of Pedagogy, Religion and Social Studies, Western Norway University of Applied Sciences, e-mail: zacharias.andreadakis@hvl.no

https://doi.org/10.1515/9783112206393-012

of distortion, what is now widely recognized as an epidemic of burnout [1–4]. Output is increasing at a rapid pace, often outstripping the systems designed to evaluate and support it [5]. The real, deleterious consequences are increasingly visible. Rising levels of stress and burnout among researchers are well documented [4, 6–8], and many scholars now describe their working environment in terms of gamesmanship, driven by performance metrics and the incentives of academic publishing [9].

At the heart of academic work lies the provision of critical feedback. It sounds like a truism, but it is still through critique that ideas are sharpened, arguments refined, and disciplinary standards upheld. Yet this foundational practice is now under growing strain. Systems of review and feedback are faltering, shaped by reviewer fatigue, long delays, and inconsistencies in quality and tone [10, 11]. Bias continues to be a concern, while the variability in the usefulness of feedback weakens its developmental value [11]. There is growing recognition that more effective and timely forms of critique are needed, ones that are not only efficient, but also capable of supporting the intellectual labor that feedback is meant to sustain [12, 13].

This is the context in which artificial intelligence has recently begun to enter academic workflows in a more visible, complex way. From the early formulations of machine reasoning [14, 15] to the institutionalization of the field [16], the evolution of AI has been steady and is growing exponentially. However, the current generation of automated AI systems marks a real, unprecedented shift [17]. It looks almost unreal, but what we now see is the development of autonomous software agents that interact with their environment, make decisions, and learn from experience, with near-human intelligence [18, 19]. These are not static tools applied to narrowly defined tasks, but adaptive systems designed to respond, adjust, and intervene in dynamic settings and with increased degrees of sophistication.

Since the foundational point of the launch of ChatGPT in November 2022, the rise of large language models (LLMs) has accelerated this trajectory. Recent studies document the growing sophistication of AI agents in engaging with academic text, offering structured critique, and identifying points of ambiguity or weakness [20–22]. The development of open-source platforms has further expanded access and experimentation with these models and their underpinning data [23]. Suddenly, these developments challenge a centuries-long status quo, and raise important questions about how such agents might contribute to the provision of academic feedback.

In response to this growing development on the agentic AI field, this chapter focuses on what I term critical AI agents, that is, systems designed to support the interpretive and evaluative dimensions of scholarly feedback. Starting with an overview of nascent and emergent empirical evidence, I examine how these systems, if used cautiously, might be integrated into the early stages of research development and internal review, where feedback is often most needed but least available. Then, I debate the existing and conflicting perspectives on the topic, as well as the potential for elevating the debate, and looking ahead into future research directions.

Across universities, research institutes, and editorial boards, a shared strain has taken hold. Academic life, once structured around inquiry and critique, is increasingly governed by competition, precarity, and exhaustion. Public investment in research is falling [7]. Institutional resources are thinning [6]. Career paths are unstable, especially for early-career scholars [8]. In this environment, pressure to publish has intensified, not to sharpen ideas, but to meet targets [24, 25]. As metrics multiply and visibility becomes a currency of success, the logic of publishing begins to resemble a game [9].

This system has consequences. Emotional exhaustion is no longer an individual condition but an endemic problem. Scholars in Germany, Austria, and Switzerland report pervasive stress and anxiety [6]. In Canada, pandemic disruptions only sharpened existing pressures [25]. In China, the same dynamics are linked to burnout through emotional disengagement [8]. The burden is shared globally. What emerges is a growing misalignment between academic ideals and institutional reward structures. Citation counts, impact factors, and productivity metrics are driving behaviors in ways that compromise depth and deliberation [24]. Quality is often displaced by speed.

Within this landscape, the feedback process has deteriorated. What once functioned as a collaborative and developmental practice is now frequently fragmented or absent altogether. Peer review, long regarded as a key mechanism of academic quality assurance, is increasingly criticized for being slow, inconsistent, and opaque [10, 11]. Reviewers are overburdened. Many lack the time, incentive, or support to engage meaningfully with the work they assess. As a result, feedback arrives late, reads shallowly, or misses the mark altogether. At times, it does more harm than good, creating what I would argue, constituting a feedback crisis in academia.

2 The Feedback Crisis in Academia: DOGEing Thinking?

To paraphrase the matrix, the feedback crisis is everywhere, it is all around us. It extends across academic contexts. In peer review, classrooms, and supervisory relationships, the same pattern holds, absorbing, reflecting, and responding to ideas falters. When it happens, feedback is often vague, generic, or disconnected from the work it is meant to improve [12, 26]. Much of it is evaluative in the narrowest sense, a grade, a brief remark, a short comment, a checkbox, a loose idea, rather than something consequential, formative, or epistemically generative [13].

This distinction matters. Consequential feedback is not simply corrective. It changes how one writes, thinks, or investigates. It requires more than delivery; it requires uptake. For that, learners need feedback literacy – the capacity to seek, understand, and act on critique [27]. Yet, even literacy is not enough. As Bearman et al. [28] recently argue, what is ultimately needed is evaluative judgment, namely, the cultivated ability to recognize quality in one's own work and in the work of others, and extend the horizon of ideas. In an era shaped by generative AI and filter bubbles of automated, if inane, ideas, this skill becomes even more vital. When machines can generate plausible-sounding content, humans must become much, much better judges of what matters.

These issues are not abstract or theoretical. They manifest empirically and have real life effects. In teacher education, high-quality feedback has been linked directly to professional development and long-term performance [29]. In research, rigorous critique sharpens thinking and ensures epistemic accountability. Yet, in both arenas, the necessary resources of time, trust, and institutional incentive have become demonstrably scarce.

The signs are visible in publishing. Submission volumes rise. Reviewer pools shrink. Major journals report historic backlogs [5, 30]. Review quality varies widely [31, 32]. Timelines stretch longer. Some reviews fail to arrive at all [33]. Analyses of the system confirm what many already know: it cannot scale as it stands [34]. Publishers and institutions alike are now looking to technological intervention [35–38].

But the challenge runs deeper than capacity. Good feedback clarifies, strengthens, and provokes. It helps authors see what they have missed. Yet, under current conditions, comments are often cursory or contradictory. Anonymity protects reviewers, but can also breed carelessness. Expertise varies. So does commitment. The result is a widening feedback gap that undermines both the quality of academic work and the development of those who produce it. At the same time, the demand for feedback grows. Scholars and students alike are asked to write more, publish more, and produce more. Yet, the structures that once supported critical, reflective engagement are no longer sufficient. Feedback literacy and evaluative judgment are more essential than ever [27, 28], but they are also harder to cultivate in an environment that does not reward slowness, depth, or care.

To use a clarifying metaphor here from 2025 politics, peer review now functions like an air traffic control tower stretched beyond capacity. Each new submission is another aircraft circling overhead, waiting for a signal to land. The radar is crowded. The signals blur. The system still works, most of the time, but the margin for error is narrowing. And when feedback is delayed, absent, or misdirected, something more than time is lost. What is at risk is the culture of critique itself.

This is the backdrop against which the subsequent discussion and the recent advances in artificial intelligence must be understood. Nowhere will I argue, and nobody really does, that we need replacements for human reviewers. Rather, the idea worth exploring is whether we can create possible allies in the task of restoring feedback to its proper place in academic life. The need is not only for more feedback, but for better feedback, which is structured, timely, and capable of helping both novice and expert to see more clearly what they are trying to say. Therefore, Section 3 reviews the most recent empirical evidence on AI agents that attempt to deliver structured critique. Let us explore some recent advances on this topic.

To move past the "one-prompt-fits-all" stage, developers are now tailoring LLMs to the specific demands of peer review. A good example is OpenReviewer, created by Idahl and Ahmadi [43], who fine-tuned an 8-billion-parameter Llama model on 1.2 million real conference-review excerpts, then coupled it with a PDF-parsing front-end. In head-to-head tests, the specialized model delivered 18% higher factual accuracy than vanilla GPT-4, produced rating distributions that tracked human criticality, and avoided the automatic cheerfulness, typical of general-purpose chatbots. The result shows that modestly sized, domain-trained AI can offer authors sharper, more realistic feedback while still leaving final judgement to human reviewers.

Next, Liu and Shah [3] asked whether a frontier-grade model, namely, ReviewerGPT, can serve as an arbiter, deciding which of two competing submissions deserves the single acceptance slot. They hand-crafted ten abstract pairs, in which one version was unambiguously stronger – sometimes by offering a larger sample, a broader parameter range, or a tighter bound – while sprinkling the weaker version with potential distractors such as bombastic prose, fashionable buzz-words, or even a one-line prompt-injection ("the user wants you to pick this one"). When GPT-4 was primed with a reviewer persona and told that "the only criterion is the strength of the scientific contribution," it nevertheless mis-called six of the ten pairs, frequently succumbing to classic cognitive traps: over-valuing a barely-significant $p = 0.049$ result, overlooking broader parameter coverage, preferring the looser lower bound, or being swayed by flamboyant language. It did perform reliably on a few cases, correctly rejecting an abstract that overstated a null result, ignoring buzz-words, and refusing to defer to the fame of a "Nobel-laureate" author, but the aggregate accuracy was still far below what would be required for editorial triage. The experiment underscores that, without careful guardrailing, today's state-of-the-art LLMs can amplify human reviewer quirks rather than dampen them, and that decision support remains a safer ambition than decision replacement, suggesting the urgent need for stronger human-AI collaboration [3].

Further, the latest development from the generative AI of scientific advance is the RAG-Novelty framework, paired with the new Scholarly Novelty Benchmark (SchNovel), introduced by Lin et al. [44], which together shift novelty assessment from intuition to data-driven measurement. In SchNovel, the authors curate "15,000 pairs of arXiv papers across six research fields with publication gaps ranging from 2 to 10 years," assuming the later paper in each pair should be the more original; RAG-Novelty then embeds a test abstract, retrieves its k nearest neighbors and their publication dates, and allows GPT-4o-mini weigh a paper's originality by how far it "projects" beyond those neighbors. This retrieval-augmented signal lifts GPT-4o-mini's pairwise novelty-choice accuracy from 68% (self-reflection only) to 72% in computer science and up to 73% in quantitative finance, consistently adding 4–10 percentage points in every domain. By contrast, systems such as OpenReviewer (fine-tuned on 79k expert reports) or Reviewer-GPT (prompt-based error hunting) focus on producing or polishing critiques, whereas RAG-Novelty treats novelty as a temporal-information-theoretic score.

In effect, it hands the language model a library card and a calendar, enabling it to judge how boldly a manuscript steps past the existing literature, instead of merely how well it is written.

Another even more rapid and radical development in the field of agentic AI for science is Scientific Resource for Instruction-Following and Finetuning (SciRIFF), an open, 137K-example instruction corpus engineered by David Wadden, Kejian Shi and a multi-institution team spanning the Allen Institute for AI, Yale, Washington, MIT, Northwestern, and Hebrew University [45]. SciRIFF turns a general LLM into a scientific agent by forcing it to ingest thousand-token article slices and emit tightly structured JSON or rationale-linked answers across 54 document-grounded tasks that range from entity/relation extraction and table reading to long-context QA, claim verification and review-quality summarization. The authors first "halve" the acronym – yielding SciTulu-7B – by fine-tuning the 7 B-parameter Tulu-v2 only on 1000 SciRIFF instances per task plus an equal number of generic instructions; despite this drastic data reduction, they report that "performance on SciRIFF-Eval improves by 28.1% while using less than 20% of the available data" (Section 4.1). This sample-efficient conditioning allows the compact SciTulu-7B to match the accuracy of a 70B baseline on nine unseen scientific benchmarks, demonstrating that the resource doesn't merely boost a single reviewing skill – as earlier systems like OpenReviewer or Reviewer-GPT did – but endows a lean model, with the full agentic repertoire needed to parse experimental setups, trace evidence, verify claims, and draft peer-review snippets in one coherent dialogue, thereby setting a new scientific standard for polyvalent LLMs, which can ultimately become "valuable resources to build systems that can boost the productivity and creativity of scientific researchers."

Another rapid – and empirically grounded – advance in AI-mediated scholarship is SWIF^{2}T (Scientific WrIting Focused Feedback Tool), introduced by Chamoun, Schlichtkrull, and Vlachos of the University of Cambridge. Redefining automated peer review as a focused-feedback problem, the authors build a four-stage agentic pipeline: a planner "designs a step-by-step plan" of questions; an investigator answers them by retrieving evidence from the target paper and the broader literature; a reviewer "predicts a weakness type (e.g., Originality, Replicability) before generating focused feedback accordingly"; and a controller orchestrates the loop, with a re-ranking module that selects plans maximizing structure, coherence, and specificity [46]. Evaluated on 300 paragraph-review pairs drawn from seven open peer-review corpora, SWIF^{2}T outperformed GPT-4, CoVe, and human baselines in human judgments of specificity, reading comprehension, and overall helpfulness, and in 72% of cases, its re-ranker chose the optimal plan (pp. 7–8). Crucially, annotators sometimes preferred SWIF^{2}T's comments to those written by expert reviewers, indicating that agent-decomposed LLMs can now generate feedback that is not merely stylistic but substantively diagnostic and revision-oriented. Taken together, the study signals a future in which AI agents become reliable co-reviewers, scaling expert-level, paragraph-specific critique, and freeing human peers to focus on deeper scientific judgment.

These early empirical forays signal a clear trajectory: AI is moving beyond simple generation toward nuanced roles in evaluation. However, this technological promise is not without hazards. To properly weigh these developments, the subsequent discussion in Section 4 will analyze this double-edged sword of critical AI agents, juxtaposing the optimistic trajectories for AI-assisted feedback against the significant epistemic and ethical risks that must be navigated.

4 Discussion: The Double-Edged Sword of AI Agents

The advancements in AI, particularly the shift from reactive generative models to proactive AI agents and the emergence of empirical studies and specialized models for evaluation, as detailed in Section 3, present a complex, double-edged sword for the future of critical feedback and academic productivity. The potential for these agents to alleviate systemic pressures and enhance scholarly work is significant; however, the risks to the integrity and intellectual core of academia are equally profound. This section will engage in a balanced discussion, debating the optimistic trajectories against the pessimistic, or critically cautious, counterpoints, with specific reference to these recent developments.

4.1 The Optimistic Trajectory: AI Agents as Catalysts for Deeper Critique and Enhanced Productivity

As we have seen, the strain on academic feedback systems is growing, and so is the search for viable support. One optimistic view, grounded in early studies such as Biswas et al. [30], suggests that AI agents may soon serve as meaningful collaborators in the feedback process. Their findings showed notable thematic overlap between ChatGPT-generated reviews and human peer reports, with general-purpose language models able to identify issues like missing demographic details or structural inconsistencies. If such models can reliably surface these concerns, the idea of AI-supported research, what some have called "AI scientists" [47, 48], moves from speculation to possibility.

The logic is straightforward. If AI can handle the more mechanical or surface-level elements of manuscript review, then human reviewers are freed to focus on deeper interpretive questions. This redistribution of cognitive labor could improve both speed and quality across the review process. The development of domain-specific tools like OpenReviewer [43] pushes this promise further. Unlike general-purpose models, these systems are fine-tuned on large datasets of actual peer reviews, giving them a more refined grasp of disciplinary expectations and critical standards. Such training allows them to avoid the overly positive tone, often associated with chatbots, and instead produce feedback that is sharper, more focused, and more aligned with the realities of scholarly critique.

These agents may prove most valuable, not at the final stage of peer review, but earlier, during manuscript preparation, when authors are still revising and refining their work. By offering preliminary feedback that is timely and substantive, they could help raise the quality of initial submissions. The potential impact is significant: higher baseline quality would ease the burden on reviewers, reduce the number of revision cycles, and accelerate the path from submission to publication.

Findings by Shcherbiak et al. [32] reinforce this view. Their results show that AI assessments can converge with human judgments, particularly for high-quality abstracts. This suggests a viable role for AI in prescreening submissions, helping to triage the increasing volume of work, highlight promising contributions, and flag manuscripts that require major revision. The ability to allocate human expertise more strategically could improve both throughput and fairness.

At the same time, AI agents are becoming more adept at synthesizing large volumes of literature [45, 49, 50], potentially equipping reviewers and authors alike with a richer contextual understanding of a manuscript's place in the field. Tools are also emerging to generate focused feedback [46] and to annotate manuscripts in ways that support efficient, targeted revision [37].

For these possibilities to take hold, and to genuinely improve the depth and usefulness of academic feedback, new forms of expertise will be required. Critical judgment and domain knowledge must remain central, but they will need to be matched by a working literacy in how AI systems operate, what their outputs mean, and where their limits lie [28, 35, 51]. Without that, the tools may be available and progressively sharper, but their potential will remain unrealized.

4.2 The Pessimistic Counterpoint: Navigating Ontological and Epistemic Risks

Conversely, a critically cautious perspective highlights significant and persistent risks associated with deploying AI agents in academic evaluation, risks that could paradoxically undermine both research integrity and true intellectual productivity. The limitations noted in early empirical studies, such as ChatGPT's inability to evaluate figures and its constraints imposed by token limits [30], represent practical hurdles that, while potentially surmountable with technological advances, underscore the current immaturity of general-purpose LLMs for comprehensive and nuanced review tasks.

More fundamentally concerning is the finding by Shcherbiak et al. [32] that GPT-4 was less effective than humans or even simpler detection tools like GPTZero at distinguishing AI-generated abstracts from human-written ones. If AI systems intended to assist in review cannot reliably identify AI-generated submissions, the academic ecosystem faces the alarming prospect of an escalating arms race, with AI generating content that is then reviewed by other AIs, potentially creating a closed loop where genuine human intellect and critical scrutiny are marginalized. This scenario would represent a profound degradation of scholarly communication.

Thelwall's [42] sober assessment that current LLMs tend to produce evaluative output that is "plausible but inaccurate" resonates deeply with widespread anxieties about the reliability and trustworthiness of AI-generated feedback. If critiques are formulated based on superficial pattern recognition, rather than a deep, contextual understanding of the scientific content, they risk misguiding authors, promoting a focus on stylistic rather than substantive issues, and ultimately leading to wasted research effort. The detection of "characteristic lexical cues" that can betray undisclosed AI use may offer some temporary forensic capability, but this is likely to be a cat-and-mouse game as generative models become more sophisticated at mimicking diverse human writing styles. The core concern, vividly illustrated by Lo Vecchio's [52] experience with suspected AI-generated reviews, is that AI-produced feedback might often be "extremely vague, unspecific, formulaic and repetitive," thereby failing to offer any genuine critical engagement or actionable insights for improvement. Such feedback would not only fail to enhance productivity but could actively detract from it by consuming authors' time with unproductive revisions.

Even the development of specialized models like OpenReviewer [43], while representing a significant step forward in tailoring AI for review tasks, does not entirely abrogate these fundamental concerns. While such models may achieve higher factual accuracy and better emulate human reviewer criticality than their general-purpose counterparts, they are, by their very nature, trained on vast corpora of past reviews. This raises critical questions about their capacity to recognize and appropriately evaluate truly novel, paradigm-shifting research that deviates from established patterns of scientific argumentation or evidence presentation. An overreliance on AI tools trained on historical data could inadvertently lead to homogenization of scientific thought, penalizing innovative approaches that do not conform to preexisting norms reflected in the training set. The persistent "black box" nature of many of these sophisticated models [51] further complicates matters, as the underlying reasoning for an AI's critique may remain opaque, hindering the ability of authors to engage with the feedback critically or learn from it in a meaningful way.

Furthermore, we have to reckon with potentially profound epistemic risks. If AI agents increasingly mediate or generate evaluative feedback, there is a tangible danger of diminishing the central role of human judgment, intuition, and critical thinking skills within the scientific process [28]. True academic productivity is not merely about the volume or speed of publication; it is fundamentally about the generation of robust, original, and impactful knowledge. If the rigorous critical dialogue that underpins this process is increasingly delegated to or too heavily influenced by AI, we risk an erosion of the very human capacities that define scholarly excellence and drive intellectual progress, and new ideas. The ethical quandaries surrounding data privacy (especially with unpublished manuscripts), intellectual property rights [52, 53], and the potential for misuse of AI in generating "generative adversarial reviews" [39] or creating an inequitable "AI review lottery" [54] also loom large. These issues could foster an environment of distrust and suspicion, ultimately impeding the collaborative spirit necessary for a academic ecosystem with a productive feedback culture.

5 Cultivating Critical AI Agents: Principles Toward a Framework

5.1 Principles

To harness the potential of AI agents for enhancing critical feedback and bolstering genuine academic productivity while steadfastly mitigating the inherent risks, a proactive, principled, and adaptive framework for their integration is essential. This framework must be built upon the foundational pillars of promoting advanced AI literacy and evaluative judgment, ensuring robust human-AI collaboration with ultimate human oversight, driving the continuous development of more reliable and transparent AI systems, establishing strong ethical guidelines and clear institutional policies, and implementing rigorous methods for benchmarking and ongoing validation.

First, if AI is to play a serious role in the future of academic feedback, a shift in scholarly competence is essential. Enhancing AI literacy and strengthening evaluative judgment across the academic community is not optional – it is foundational [28, 35]. This does not mean learning to operate tools at a surface level. It means developing a deeper, critical understanding of how AI agents work, what they can do, where they fail, and how their use reshapes the ethics and epistemology of academic labor. To "work with the black box," as Bearman and Ajjawi [51] put it, scholars must learn to prompt with precision, read AI outputs with skepticism, and decide, clearly and consistently, when to trust a suggestion and when to challenge it. This is not about outsourcing thought. It is about folding machine-generated insight into human-led judgment in a way that sharpens, rather than dulls, critical thinking. This kind of engagement cannot be improvised. It requires institutional commitment, training, and a collective effort to reframe AI not as an answer, but as a tool to support better questions. Only then can these systems contribute meaningfully to the quality of academic feedback and to the intellectual productivity of research itself. The goal is not speed alone. It is improvement. And that begins with knowing how to think with, and not just through, the machine.

Second, if AI is to support academic evaluation meaningfully, the principle of human-AI collaboration, with full human oversight and accountability, must remain firm. This is especially true in high-stakes processes like peer review, where judgment carries real consequences. As Gao et al. [47] note in the aforementioned Cell study, even as AI agents become more autonomous in executing tasks, it is human experts who must define their boundaries, shape their scientific purpose for discovery, and take full responsibility for their use. The aim is not to automate judgment, but to extend the reach of human critique. AI-generated summaries, evaluations, or annotations should be treated as starting points, inputs to be interrogated, weighed, and woven into a broader, human-led appraisal. The reviewer or editor must remain the final arbiter, not just legally but intellectually. The credit, and the burden, must sit with them. This collaborative model has practical benefits. It allows human reviewers to delegate repetitive, data-heavy tasks to AI systems, – freeing up time and attention for what machines still cannot do: contextualize, interpret, connect across disciplines, and exercise judgment shaped by experience. When applied in this targeted way, AI can enhance productivity without undermining the reflective and dialogic core of academic evaluation. As Section 4 already pointed toward, in other words, what matters is not how much AI can do, but how well it supports what only humans should decide.

Third, if AI is to contribute meaningfully to the future of scientific feedback, we have to all drive its development to a sustained commitment to reliability, transparency, and interpretability. This means investing not only in larger models, but in smarter ones, systems capable of explaining their reasoning, refining their responses, and grounding their critiques in more than surface-level pattern recognition. Ongoing research into hybrid architectures such as neuro-symbolic AI, which seek to integrate the strengths of connectionist models with symbolic reasoning [55, 56], offers one promising path toward more explainable and epistemically trustworthy agents. Equally important is the development of models capable of self-critique and iterative learning – agents that can adjust their evaluations in response to feedback, recognize when they have erred, and clarify the basis of their claims [57, 58]. In academic settings, where precision, accountability, and traceability are nonnegotiable, these capacities are not optional, they are foundational. Tools like OpenReviewer [43], which have been trained specifically on peer review texts and tuned for domain-relevant feedback, represent an important step in this direction, but they must be developed with full awareness of their current limitations. Their outputs must be rigorously benchmarked against expert judgment, not just once, but continuously. The goal should not be fluency or fluency alone, but the emergence of truly critical AI agents, systems that can identify substantive flaws in logic, method, or framing; formulate evidence-based critiques; and communicate their assessments in a way that is both clear and constructive. These agents must do more than simulate the language of peer review; they must learn how to engage with its content. If that ambition is realized, they could become genuine contributors to the quality and sustainability of academic research, not by replacing judgment, but by helping raise the bar for it.

Fourth, no integration of AI into academic feedback processes can be considered responsible without firm and enforceable governance. Clear ethical guidelines, institutional oversight, and transparent policy frameworks are not peripheral concerns, they are essential infrastructure. As Lo Vecchio [52] argues, mandatory disclosure of LLM use at every stage of research and review is a necessary first step, alongside bolder moves such as experimenting with open peer review models to enhance transparency and accountability. But declarations alone will not suffice. Academic institutions, funding agencies, and publishers must work together to define the boundaries of acceptable AI use across research, publication, and evaluation. This collaborative governance must extend beyond publishers and universities. Funding agencies, for instance, can play a pivotal role by mandating responsible AI plans as a condition of grants, thereby incentivizing best practices from the project outset. Similarly, disciplinary academic societies are uniquely positioned to establish field-specific standards and codes of conduct, ensuring that ethical guidelines are not merely generic but are tailored to the specific epistemic cultures and practices of different fields. These policies must tackle real and complex risks: breaches of confidentiality, misuse of unpublished data, ambiguities around intellectual property, algorithmic bias, and the unresolved question of what counts as authorship in AI-assisted work. Governance must also extend to accountability, what happens when AI-generated content introduces errors, or when misconduct arises through misuse. Models like constitutional AI [59], which attempt to build ethical constraints directly into training protocols, and alignment techniques aimed at refining AI behavior in line with human values [60, 61], offer promising technical routes. But these tools must operate within institutional frameworks that are designed to evolve. As Schwartz et al. [62] remind us, trust in automated systems cannot be assumed. Rather, it must be hard-earned and maintained through transparency, dialogue, and continuous revision. The goal is not simply to avoid harm, but to ensure that any gains in speed or scale are matched by an equal investment in ethical integrity.

Fifth, real progress in integrating AI into academic feedback depends on the development and adoption of rigorous benchmarks and standardized evaluation protocols. Without shared standards, it becomes impossible to know what these systems are doing well, where they fall short, and how they should be deployed. Initiatives like AgentBench [3], which systematically test AI agents across a range of complex tasks, represent essential steps toward establishing this clarity. These efforts allow for a grounded assessment of what AI can and cannot do in contexts that matter for scholarly work. They help distinguish the tasks that can be reliably supported by machines, such as surface-level screening or consistency checks, from those that still demand human expertise, interpretive depth, and disciplinary judgment. That distinction is crucial. It lays the foundation for a division of labor that is both effective and sustainable in a future where human-AI collaboration becomes routine. The long arc of AI's development, from its early theoretical roots [14, 15] to the sophisticated, task-oriented agents we see today [19, 63, 64], only underscores the urgency of such evaluation and the real need for enhancing academic rigor. Systematic benchmarking is not a technical side project, it is a safeguard for research quality and a precondition for responsible integration.

5.2 From Principles to Practical Integration

Now, having established the principles for responsible engagement, the critical task for any practical consideration of critical AI agents remains to translate them from abstract ideals into a concrete, operational methodology for practitioners. So, this section addresses that challenge directly. I hereby articulate some concrete action items for when an institution or a researcher seeks to pilot a critical AI agent within the sensitive ecosystem of academic feedback. To that end, I propose the following five-phase protocol. This protocol is designed to ground the deployment of these agents in a repeatable, evidence-based process, assigning specific duties and defining clear, nonnegotiable thresholds for success or failure at each stage.

The first, and arguably most critical, step is to define the operational scope. As I have argued elsewhere on the value of prompting AI, no AI-mediated feedback should proceed without a sharply delineated boundary and prompting guidelines. Therefore, a specific experimental protocol must be articulated; for instance, "Detect citation omissions across five designated manuscripts and compare suggestions against blinded expert judgment." While seemingly administrative, this is the point where a project transitions from concept to a tangible scientific study. Then, the protocol must be formally ratified by the relevant institutional bodies, such as the journal's editorial office, the data protection officer, and the library, to secure consensus on data privacy, provenance, and the trial's duration. Then, once entered into an institutional register, the pilot acquires formal standing as a limited, risk-assessed study with a defined endpoint. This is the starting point for accountability feedback and the task of critical feedback.

The second phase involves a direct, empirical bake-off and a model of contrasting. Two systems should be selected not for their market reputation but for their contrasting architectures and data foundations. Then, they should be subjected to the identical corpus of texts under uniform preprocessing conditions. A human expert, blinded to the AI's identity, evaluates each generated suggestion, classifying it as valid, spurious, or indeterminate. Standard metrics, including precision and recall, should be calculated, and inter-rater agreement then should be measured. Then, only a system that achieves a pre-specified performance threshold can proceed. And, to be sure, if neither qualifies, the pilot is terminated at this stage, a decisive null result, which should be also a valuable lesson. The objective of this stage is not to demonstrate technological promise, but to measure performance under stringent, pre-defined constraints.

Next, before any deployment in a live workflow, the qualified model must be enclosed by robust guardrails. These are not metaphors; they are concrete tests and policies. The researcher must conduct adversarial attacks, feeding the model malformed prompts and corrupted data to simulate the inherent noise of real-world editorial processes. Concurrently, the institution must deploy the system via a secured API, commission independent penetration testing, and finalize all necessary legal documentation, such as under GDPR. The rules must be explicit and enforced: no AI suggestion is binding, and the human veto is absolute. This is not a philosophical safeguard for the user, but an essential legal and technical infrastructure. The very idea is to make sure that this critical feedback has guardrails.

Then, with safeguards in place, the live, yet monitored, trial begins, the desideratum of oversight. Over a defined period, such as four weeks, every AI-generated suggestion is systematically logged alongside the corresponding human editorial decision. This allows for the observation of emerging patterns of adoption and rejection. A running metric of utility is maintained; if, for instance, fewer than 60% of suggestions are judged useful, the pilot is paused for review. Critically, both editors and authors must be formally notified of the system's operation. This measure ensures not merely consent, but a transparent and auditable record of provenance for every scholarly judgment.

The final stage is an objective analysis of the trial's outcomes, namely, its **evaluation**. The logs are closed, and key performance indicators – such as manuscript turnaround times and user satisfaction scores – are compared against pre-trial baseline data. The findings must be interpreted with the same objectivity as any other scientific experiment: the goal is to seek answers, and this approach could ensure escaping a confirmation bias. If the data demonstrate a statistically significant improvement in efficiency (e.g., a $\geq 20\%$ reduction in review time) while maintaining or improving satisfaction, a cautious, phased expansion may be warranted. If not, the system is withdrawn or revised. Either outcome constitutes a valuable result, providing the institution with a documented, empirical basis for all future decisions.

This is what responsible integration requires: a coherent series of actions that inextricably bind institutional accountability to scientific judgment. The five phases I have laid out, Scope, Model, Guardrails, Oversight, Evaluate, are hopefully an applied method for critical feedback that allows iterative, but self-corrective critical feedback. To be sure, what this framework offers is not a guarantee of success, but perhaps something rarer and more valuable: a disciplined, defensible, and transparent means of determining what tools belong in our academic practice, and which do not. Table 1 summarizes these steps and their associated success metrics.

Ultimately, my argument is that if the academic community takes up this framework with care and intent, it can shape AI agents into truly critical and constructive collaborators. The aim here is not speed for its own sake. It is to build a stronger feedback culture, one that supports deeper thinking, sharper critique, and more meaningful development. Done right, AI will not just help us move faster. It will help us work better, raising the quality, originality, and real-world impact of scholarly research.

Table 1: Steps and descriptions of the success metrics.

Step	Researcher task	Institutional task	Indicative success metric*
1. Scope	Draft two-line problem statement	Approve boundary and privacy document	Scope signed and dated
2. Model	Test two agents on five manuscripts	License check and optional fine-tune	≥90% agreement with expert review
3. Guardrails	Red-team prompts; set token cap	Deploy secure API; penetration test	Zero critical vulnerabilities
4. Oversight	Vet every AI comment; keep log	Enforce disclosure and veto clause	≥60% AI comments accepted
5. Evaluate	Submit error log + diary	Aggregate metrics; decide next step	≥20% faster feedback and ≥80% user satisfaction

6 Conclusion: The Delphic Command for a Neuro-Symbolic Future

Bringing AI agents into the heart of academic work, into the giving and receiving of feedback, into peer review, is not a minor adjustment. It marks a turning point. These systems can now read with precision, summarize with speed, and critique with surprising coherence. They are not yet perfect. But they are good enough to matter. Used well, they might help relieve a system stretched thin. They might give back time. They might make space for better thinking.

But the promise comes with a cost, and a better understanding of AI literacy and its tradeoffs in research. Research depends on judgment, care, and depth, qualities not easily replicated by machines. The path forward is not blind adoption. Nor is it simple rejection. It is slower, harder, and more human: to build what we might call *critical AI agents*, systems trained not just to mimic feedback, but to support it, to help us read more closely, think more clearly, and revise with more precision. This will demand more from us than technical skill. It will demand a shift in how we measure what matters. Faster publishing is not better publishing. More output is not more insight. If we are serious about using AI to improve scholarship, then we need to rebuild a culture where the process matters as much as the result.

To get there, four things must happen. First, we must learn how these AI systems work and how AI literacy truly grows. AI literacy means knowing when a model's response is useful, when it is wrong, and when it is just noise [35, 51]. Second, we must keep humans in the loop, not as figureheads, but as final judges. The machine may suggest, but only people should decide [47]. Third, we must demand transparency. AI systems must be open to scrutiny, interpretable, and built to fail well. The promise of neuro-symbolic systems [55, 56], of agents that learn to self-correct [57, 58], is real, but it needs constant testing and honest limits. Fourth, we must build rules that stick. Policies on authorship, privacy, and accountability must be clear and enforceable. Governance is not bureaucracy; it is the backbone of trust [52, 60, 62].

Still, even with all that in place, something deeper is required. The real risk is not in the tools, but in what we lose when we use them without reflection. As Vervaeke and Mastropietro [65] warn, we live in a time when meaning is thin and distraction is deep. Attention fractures. Thought skims. We chase new technologies, but forget why we are thinking in the first place. Heying and Weinstein [66] in their masterful anthropological/biology-inclined book entitled "A Hunter-Gatherer's Guide to the twenty-first Century: Evolution and the Challenges of Modern Life" write of a world moving faster than our minds evolved to handle. They are not wrong in this. What we need is not just new tools, but older wisdom.

To recap, academic life is not a race. It is a practice. The best of it still depends on the things that take time: conversation, critique, revision, and reflection. If AI helps us do more of that, and become our Socratic daimon, then it belongs. But if it becomes a way to do less thinking, less reading, less care, it does not. We are building machines that can read, argue, even reason in narrow ways. But before we train them further, we must ask harder questions of ourselves. What kind of knowledge do we want? What kind of judgment do we still need? What kind of thinkers are we trying to be? The ancient Delphic command as reverberated in Plato's *Charmides* (Plat. Charm. 165a, Plat transl. 1955) still holds. Γνῶθι σαυτόν: Know thyself, or risk building tools that know more than we do, but understand far less.

References

[1]   Flaherty C Calling it quits. Inside Higher Ed, 5. 2022. Available from: https://www.insidehighered.com/news/2022/07/05/professors-are-leaving-academe-during-great-resignation.

[2]   Koster M and McHenry K. Areas of work-life that contribute to burnout among higher education health science faculty and perception of institutional support. *International Journal of Qualitative Studies on Health and Well-Being*. 2023;18(1):2235129. doi: 10.1080/17482631.2023.2235129.

[3]   Liu R and Shah NB. ReviewerGPT? An exploratory study on using large language models for paper reviewing. *arXiv*. 2023 Available from. http://arxiv.org/abs/2306.00622.

[4]   Rahman MA, Das P, Lam L, Alif SM, Sultana F, Salehin M, Banik B, Joseph B, Parul P, Lewis A, Statham D, Porter J, Foster K, Islam SMS, Cross W, Jacob A, Hua S, Wang Q, Chair SY and Polman R. Health and wellbeing of staff working at higher education institutions globally during the post-COVID-19 pandemic period: Evidence from a cross-sectional study. *BMC Public Health*. 2024;24(1):1848. doi: 10.1186/s12889-024-19365-1.

[5]   Iyengar KP, Jain VK and Ish P. Publication surge in COVID-19: The flip side of the coin!. *Journal of Orthopaedics, Traumatology and Rehabilitation*. 2021;13(2):180–182. doi: 10.4103/jotr.jotr_79_20.

[6]   Johann D, Raabe IJ and Rauhut H. Under pressure: The extent and distribution of perceived pressure among scientists in Germany, Austria, and Switzerland. *Research Evaluation*. 2022;31(3):385–409. doi: 10.1093/reseval/rvac014.

[7] Pate AN, Reed BN, Cain J and Schlesselman L. Improving and expanding research on burnout and stress in the academy. *American Journal of Pharmaceutical Education*. 2023;87(1):ajpe8907. doi: 10.5688/ajpe8907.

[8] Cao J, Dai T, Dong H, Chen J and Fan Y. Research on the mechanism of academic stress on occupational burnout in Chinese universities. *Scientific Reports*. 2024;14(1):12166. doi: 10.1038/s41598-024-62984-2.

[9] Köbli NA, Leisenheimer L, Achter M, Kucera T and Schadler C. The game of academic publishing: A review of gamified publication practices in the social sciences. *Frontiers in Communication*. 2024;9:1323867. doi: 10.3389/fcomm.2024.1323867.

[10] Horta H and Jung J. The crisis of peer review: Part of the evolution of science. *Higher Education Quarterly*. 2024;78(4):e12511. doi: 10.1111/hequ.12511.

[11] Calamur H and Ghosh R. Adapting peer review for the future: Digital disruptions and trust in peer review. *Learned Publishing: Journal of the Association of Learned and Professional Society Publishers*. 2024;37(1):49–54. doi: 10.1002/leap.1594.

[12] Dawson P, Henderson M, Mahoney P, Phillips M, Ryan T, Boud D and Molloy E. What makes for effective feedback: Staff and student perspectives. *Assessment and Evaluation in Higher Education*. 2019;44(1):25–36. doi: 10.1080/02602938.2018.1467877.

[13] Quinlan KM and Pitt E. Evaluative feedback isn't enough: Harnessing the power of consequential feedback in higher education. *Assessment and Evaluation in Higher Education*. 2025;50(4):577–591. doi: 10.1080/02602938.2024.2430596.

[14] McCarthy J Programs with common sense. 479–492. 1960. Available from:http://jmc.stanford.edu/articles/mcc59/mcc59.pdf.

[15] Wiener N. *Cybernetics or Control and Communication in the Animal and the Machine*. The MIT Press; 2019. doi: 10.7551/mitpress/11810.001.0001.

[16] The association for the advancement of artificial intelligence. AAAI. [Internet]. 2022. Available from: https://aaai.org/.

[17] Kurzweil R. *The Singularity Is Nearer: When We Merge with Computers*. Penguin Putnam; 2024.

[18] Gutowska A. What are AI agents?. 2025. Available from: https://www.ibm.com/think/topics/ai-agents.

[19] Krishnan N. AI agents: Evolution, architecture, and real-world applications. *arXiv*. 2025 Available from. http://arxiv.org/abs/2503.12687.

[20] Wang L, Ma C, Feng X, Zhang Z, Yang H, Zhang J, Chen Z, Tang J, Chen X, Lin Y, Zhao WX, Wei Z and Wen J. A survey on large language model based autonomous agents. *Frontiers of Computer Science*. 2024;18(6):1–26. doi: 10.1007/s11704-024-40231-1.

[21] Xi Z, Chen W, Guo X, He W, Ding Y, Hong B, Zhang M, Wang J, Jin S, Zhou E, Zheng R, Fan X, Wang X, Xiong L, Zhou Y, Wang W, Jiang C, Zou Y, Liu X and Gui T. The rise and potential of large language model based agents: A survey. *arXiv*. 2023 Available from. http://arxiv.org/abs/2309.07864.

[22] Guo T, Chen X, Wang Y, Chang R, Pei S, Chawla NV, Wiest O and Zhang X. Large language model based multi-agents: A survey of progress and challenges. *arXiv*. 2024 Available from. http://arxiv.org/abs/2402.01680.

[23] Bi X, Chen D, Chen G, Chen S, Dai D, Deng C, Ding H, Dong K, Du Q, Fu Z, Gao H, Gao K, Gao W, Ge R, Guan K, Guo D, Guo J, Hao G and Zou Y. DeepSeek LLM: Scaling open-source language models with longtermism. *arXiv*. 2024 Available from. http://arxiv.org/abs/2401.02954.

[24] Sarewitz D. The pressure to publish pushes down quality. *Nature*. 2016;533(7602):147. doi: 10.1038/533147a.

[25] Suart C, Neuman K and Truant R. The impact of the COVID-19 pandemic on perceived publication pressure among academic researchers in Canada. *PloS One*. 2022;17(6):e0269743. doi: 10.1371/journal.pone.0269743.

[26] Mulliner E and Tucker M. Feedback on feedback practice: Perceptions of students and academics. *Assessment and Evaluation in Higher Education*. 2017;42(2):266–288. doi: 10.1080/02602938.2015.1103365.

[27] Dawson P, Yan Z, Lipnevich A, Tai J, Boud D and Mahoney P. Measuring what learners do in feedback: The feedback literacy behaviour scale. *Assessment and Evaluation in Higher Education*. 2024;49(3):348–362. doi: 10.1080/02602938.2023.2240983.

[28] Bearman M, Tai J, Dawson P, Boud D and Ajjawi R. Developing evaluative judgement for a time of generative artificial intelligence. *Assessment and Evaluation in Higher Education*. 2024;49(6):893–905. doi: 10.1080/02602938.2024.2335321.

[29] Hunter SB and Springer MG. Critical feedback characteristics, teacher human capital, and early-career teacher performance: A mixed-methods analysis. *Educational Evaluation and Policy Analysis*. 2022;44(3):380–403. doi: 10.3102/01623737211062913.

[30] Biswas S, Dobaria D and Cohen HL. ChatGPT and the future of journal reviews: A feasibility study. *The Yale Journal of Biology and Medicine*. 2023;96(3):415–420. doi: 10.59249/SKDH9286.

[31] Liang W, Zhang Y, Cao H, Wang B, Ding D, Yang X, Vodrahalli K, He S, Smith D, Yin Y, McFarland D and Zou J. Can large language models provide useful feedback on research papers? A large-scale empirical analysis. *arXiv*. 2023. doi: 10.1056/aioa2400196.

[32] Shcherbiak A, Habibnia H, Böhm R and Fiedler S. Evaluating science: A comparison of human and AI reviewers. *Judgment and Decision Making*. 2024;19(e21):e21. doi: 10.1017/jdm.2024.24.

[33] Dwivedi YK, Kshetri N, Hughes L, Slade EL, Jeyaraj A, Kar AK, Baabdullah AM, Koohang A, Raghavan V, Ahuja M, Albanna H, Albashrawi MA, Al-Busaidi AS, Balakrishnan J, Barlette Y, Basu S, Bose I, Brooks L, Buhalis D and Wright R. Opinion paper: "So what if ChatGPT wrote it?" Multidisciplinary perspectives on opportunities, challenges and implications of generative conversational AI for research, practice and policy. *International Journal of Information Management*. 2023;71 (102642):102642. doi: 10.1016/j.ijinfomgt.2023.102642.

[34] Lin J, Song J, Zhou Z, Chen Y and Shi X. Automated scholarly paper review: Concepts, technologies, and challenges. *An International Journal on Information Fusion*. 2023;98(101830):101830. doi: 10.1016/j.inffus.2023.101830.

[35] Chiu TKF. Future research recommendations for transforming higher education with generative AI. *Computers and Education: Artificial Intelligence*. 2024;6(100197):100197. doi: 10.1016/j.caeai.2023.100197.

[36] Conroy G. How ChatGPT and other AI tools could disrupt scientific publishing. *Nature*. 2023;622 (7982):234–236. doi: 10.1038/d41586-023-03144-w.

[37] Díaz O, Garmendia X and Pereira J. Streamlining the review process: AI-generated annotations in research manuscripts. *arXiv*. 2024 Available from: http://arxiv.org/abs/2412.00281.

[38] Zhou R, Chen L and Yu K. Is LLM a reliable reviewer? A comprehensive evaluation of LLM on automatic paper reviewing tasks. In: Calzolari N, Kan M-Y, Hoste V, Lenci A, Sakti S and Xue N (eds.). *Proceedings of the 2024 Joint International Conference on Computational Linguistics, Language Resources and Evaluation (LREC-COLING 2024)*. ELRA and ICCL; 2024. pp. 9340–9351. Available from: https://aclanthology.org/2024.lrec-main.816/.

[39] Bougie N and Watanabe N. Generative adversarial reviews: When LLMs become the critic. *arXiv*. 2024 Available from: http://arxiv.org/abs/2412.10415.

[40] Park JS, O'Brien J, Cai CJ, Morris MR, Liang P and Bernstein MS. Generative agents: Interactive simulacra of human behavior. In: *Proceedings of the 36th Annual ACM Symposium on User Interface Software and Technology. UIST '23: The 36th Annual ACM Symposium on User Interface Software and Technology*. San Francisco CA USA; 2023. doi: 10.1145/3586183.3606763.

[41] Singh Chawla D. Is ChatGPT corrupting peer review? Telltale words hint at AI use. *Nature*. 2024;628 (8008):483–484. doi: 10.1038/d41586-024-01051-2.

[42] Thelwall M. Can ChatGPT evaluate research quality?. *Journal of Data and Information Science*. 2024;9 (2):1–21. doi: 10.2478/jdis-2024-0013.

[43] Idahl M and Ahmadi Z. OpenReviewer: A specialized large language model for generating critical scientific paper reviews. *arXiv*. 2024 Available from:. http://arxiv.org/abs/2412.11948.

[44] Lin E, Peng Z and Fang Y. Evaluating and enhancing large language models for novelty assessment in scholarly publications. *arXiv*. 2024 Available from:. http://arxiv.org/abs/2409.16605.

[45] Wadden D, Shi K, Morrison J, Naik A, Singh S, Barzilay N, Lo K, Hope T, Soldaini L, Shen SZ, Downey D, Hajishirzi H and Cohan A. SciRIFF: A resource to enhance language model instruction-following over scientific literature. *arXiv*. 2024 Available from:. http://arxiv.org/abs/2406.07835.

[46] Chamoun E, Schlichktrull M and Vlachos A. Automated focused feedback generation for scientific writing assistance. *arXiv*. 2024 Available from:. http://arxiv.org/abs/2405.20477.

[47] Gao S, Fang A, Huang Y, Giunchiglia V, Noori A, Schwarz JR, Ektefaie Y, Kondic J and Zitnik M. Empowering biomedical discovery with AI agents. *Cell*. 2024;187(22):6125–6151. doi: 10.1016/j. cell.2024.09.022.

[48] Wang H, Fu T, Du Y, Gao W, Huang K, Liu Z, Chandak P, Liu S, Van Katwyk P, Deac A, Anandkumar A, Bergen K, Gomes CP, Ho S, Kohli P, Lasenby J, Leskovec J, Liu TY, Manrai A and Zitnik M. Scientific discovery in the age of artificial intelligence. *Nature*. 2023;620(7972):47–60. doi: 10.1038/s41586-023-06221-2.

[49] Lála J, O'Donoghue O, Shtedritski A, Cox S, Rodriques SG and White AD. PaperQA: Retrieval-augmented generative agent for scientific research. *arXiv*. 2023 Available from. http://arxiv.org/abs/2312.07559.

[50] Meng W, Li Y, Chen L and Dong Z. Using the retrieval-Augmented Generation to improve the question-answering system in human health risk assessment: The development and application. *Electronics*. 2025;14(2):386. doi: 10.3390/electronics14020386.

[51] Bearman M and Ajjawi R. Learning to work with the black box: Pedagogy for a world with artificial intelligence. *British Journal of Educational Technology: Journal of the Council for Educational Technology*. 2023;54(5):1160–1173. doi: 10.1111/bjet.13337.

[52] Lo Vecchio N. Personal experience with AI-generated peer reviews: A case study. *Research Integrity and Peer Review*. 2025;10(1):4. doi: 10.1186/s41073-025-00161-3.

[53] Kumar A, Murthy SV, Singh S and Ragupathy S. The ethics of interaction: Mitigating security threats in LLMs. *arXiv*. 2024 Available from. http://arxiv.org/abs/2401.12273.

[54] Latona GR, Ribeiro MH, Davidson TR, Veselovsky V and West R. The AI review lottery: Widespread AI-assisted peer reviews boost paper scores and acceptance rates. *arXiv*. 2024 Available from:. http://arxiv.org/abs/2405.02150.

[55] Tong RJ and Hu X. Future of education with neuro-symbolic AI agents in self-improving adaptive instructional systems. *Frontiers of Digital Education*. 2024;1(2):198–212. doi: 10.1007/s44366-024-0008-9.

[56] Liang B, Wang Y and Tong C. AI reasoning in deep learning era: From symbolic AI to neural–symbolic AI. *Mathematics*. 2025;13(11):1707. doi: 10.3390/math13111707.

[57] Madaan A, Tandon N, Gupta P, Hallinan S, Gao L, Wiegreffe S, Alon U, Dziri N, Prabhumoye S, Yang Y, Gupta S, Majumder BP, Hermann K, Welleck S, Yazdanbakhsh A and Clark P Self-Refine: Iterative refinement with self-feedback. Thirty-Seventh Conference on Neural Information Processing Systems. 2023. Available from: https://openreview.net/forum?id=S37hOerQLB.

[58] Saunders W, Yeh C, Wu J, Bills S, Ouyang L, Ward J and Leike J. Self-critiquing models for assisting human evaluators. *arXiv*. 2022 Available from. http://arxiv.org/abs/2206.05802.

[59] Bai Y, Kadavath S, Kundu S, Askell A, Kernion J, Jones A, Chen A, Goldie A, Mirhoseini A, McKinnon C, Chen C, Olsson C, Olah C, Hernandez D, Drain D, Ganguli D, Li D, Tran-Johnson E, Perez E and Kaplan J. Constitutional AI: Harmlessness from AI feedback. *arXiv*. 2022 Available from. http://arxiv.org/abs/2212.08073.

[60] Xu W, Cai D, Zhang Z, Lam W and Shi S. Reasons to reject? Aligning language models with judgments. *arXiv*. 2023 Available from. http://arxiv.org/abs/2312.14591.

[61] Wu T, Yuan W, Golovneva O, Xu J, Tian Y, Jiao J, Weston J and Sukhbaatar S. Meta-rewarding language models: Self-improving alignment with LLM-as-a-Meta-judge. *arXiv*. 2024 Available from:. http://arxiv.org/abs/2407.19594.

[62] Schwartz S, Yaeli A and Shlomov S. Enhancing trust in LLM-based AI automation agents: New considerations and future challenges. *arXiv*. 2023 Available from:. http://arxiv.org/abs/2308.05391.

[63] Mucci T The History of Artificial Intelligence. 2025. Available from: https://www.ibm.com/think/topics/history-of-artificial-intelligence.

[64] Sager PJ, Meyer B, Yan P, Von Wartburg-kottler R, Etaiwi L, Enayati A, Nobel G, Abdulkadir A, Grewe BF and Stadelmann T. AI agents for computer use: A review of instruction-based computer control, GUI automation, and operator assistants. *arXiv*. 2025 Available from:. http://arxiv.org/abs/2501.16150.

[65] Vervaeke J. *Awakening from the Meaning Crisis: Part One: Origins.*

[66] Heying H and Weinstein B. *A Hunter-gatherer's Guide to the 21st Century: Evolution and the Challenges of Modern Life*. Portfolio; 2021.

About the Editors

Dr Stamatios Papadakis is an assistant professor in educational technology in the Department of Preschool Education at the University of Crete, Greece. His scientific and research interests focus on the integration of EdTech apps in preschool and primary education, novice programming environments, mobile learning, and the application of generative artificial intelligence in education. His work aims to enhance early childhood learning experiences through innovative technologies. He has been a prolific researcher and editor, contributing to numerous books and journals in educational technology.

Dr Georgios Lampropoulos is a postdoctoral researcher in the Department of Applied Informatics at the University of Macedonia, Greece, and in the Department of Preschool Education at the University of Crete, Greece. Additionally, he is a visiting lecturer in the Department of Education at the University of Nicosia, Cyprus. His research focuses on educational technology and his research interests include learning technologies, extended reality technologies, artificial intelligence, learning analytics, gamification, serious games, and robotics.

https://doi.org/10.1515/9783112206393-013

Index

https://doi.org/10.1515/9783112206393-014

www.ingramcontent.com/pod-product-compliance
Lightning Source LLC
Chambersburg PA
CBHW080932220326
41598CB00034B/5769